Advanced Mathematics for Engineering Students
The Essential Toolbox

T0130836

Advanced Mathematics for Engineering Students
The Essential Toolbox

Brent J. Lewis

E. Nihan Onder

Andrew A. Prudil

Butterworth-Heinemann
An imprint of Elsevier

Library of Congress Cataloging-in-Publication Data
A catalog record for this book is available from the Library of Congress

British Library Cataloguing-in-Publication Data
A catalogue record for this book is available from the British Library

ISBN: 978-0-12-823681-9

For information on all Butterworth-Heinemann publications
visit our website at https://www.elsevier.com/books-and-journals

Publisher: Katey Birtcher
Acquisitions Editor: Steve Merken
Editorial Project Manager: Alice Grant
Production Project Manager: Nadhiya Sekar
Designer: Greg Harris

Typeset by VTeX

Preface

This book is derived from class-tested courses presented over many years. It delivers the foundation for an advanced undergraduate course with focused examples from programs of study in computer engineering, materials engineering, chemical engineering, nuclear engineering, and engineering physics. A broad range of topics are covered, including Laplace and Fourier transform theory; analytical methods of solution for ordinary and partial differential equations; numerical solution of ordinary and partial differential equations based on finite difference and finite element techniques; matrices, linear systems, and vector analysis; treatment for experimental results; numerical techniques of finding zeros of functions, interpolation, integration, and differentiation; and complex analysis. It also presents advanced mathematical concepts missing in traditional texts, such as nondimensionalization, nonlinear differential equations, integral equations and the Green's function, and calculus of variations. This book uniquely provides both the underlying theory and its application using state-of-the-art computational tools available to the engineer.

- It demonstrates and summarizes for the engineer new technology for applied mathematical analysis with algebraic, numerical, and statistical commercial software packages for day-to-day analysis.
- Example problems in the chapters are revisited with a demonstration of commercial tools to quickly solve problems introduced throughout the book.

Acknowledgments

This book is derived from a third-year undergraduate course in advanced mathematics at the Royal Military College of Canada in the Department of Chemistry and Chemical Engineering as two individual one-term courses: MAE315: Differential Equations and Fourier Series; and CCE315: Chemical and Materials Engineering Computations. Covered material was also derived from an applied mathematics course PHYS312 at Queen's University in Engineering Physics and from engineering courses ENE6002 (Thermal-Hydraulics of Two-Phase Flow), MTH6201 (Numerical Methods), and MEC6206 (Fluid Mechanics II) at École Polytechnique of Montreal. The authors would also like to acknowledge help from Dr. Kimberley Lewis on ANOVA analysis and to thank Stephen Merken, Senior Acquisitions Editor at Elsevier, for his invaluable assistance, guidance, and advice.

Contents

List of tables

List of figures

Advanced
Mathematics for
Engineering Students
The Essential Toolbox

Prologue

1.1 Introduction

A differential equation is a mathematical construct involving derivatives (i.e., the rates of change of certain quantities) that includes an expression with one or more independent variables. It may represent a quantity embodied in a physical law of nature that may arise in such fields as mathematics, physics, astronomy, biology, economics, or engineering.

Differential equations can be solved in certain cases as self-contained analytical expressions. For more complex and real-life problems that occur in various fields of study, a solution may require numerical methods on a computer.

1.1.1 History of differential equations

Differential equations were constructs that followed with the invention of differential and integral calculus by Isaac Newton and Gottfried Wilhelm Leibniz c.1670. This advancement involved the use of infinitesimal quantities to determine tangents to curves and the calculation of lengths and areas of curved surfaces. The more modern terms of derivatives and integrals, respectively, were described by Newton as "fluxions" and "fluents," while Leibniz referred to these entities as "differences" and "sums." Newton specifically described two ordinary differential equations and one partial differential equation that could be solved employing an infinite series method (Sasser, 2018):

$$\frac{dy}{dx} = f(x),$$
$$\frac{dy}{dx} = f(x, y),$$
$$x\frac{\partial u}{\partial x} + y\frac{\partial u}{\partial y} = u.$$

The "Bernoulli" ordinary differential equation was proposed by Jacob Bernoulli in 1695, which was further simplified and solved by Leibniz in 1696. The vibrating string problem was mathematically studied by Jean le Rond d'Alembert, Leonhard Euler, Daniel Bernoulli, and Joseph-Louis Lagrange. The one-dimensional wave equation studied in 1746 by d'Alembert was generalized to three dimensions by Euler within a decade. The tautochrone problem (that is, an isochrone curve along which a

Advanced Mathematics for Engineering Students. https://doi.org/10.1016/B978-0-12-823681-9.00009-5

body will fall with uniform vertical velocity), as formulated by Euler and Lagrange in the 1750s, was subsequently solved in 1755 by Lagrange. Both Euler and Lagrange further developed this type of methodology for application to mechanics, leading to the important formulation of Lagrangian mechanics. Fourier also published his mathematical work on the conductive flow of heat in 1822.

A history of ordinary differential equations over 100 years is briefly summarized in Table 1.1.

Table 1.1 History of ordinary differential equations.

Date	Problem identity	Problem description	Mathematician
1690	Isochrone problem	Finding a curve along which a body will fall with uniform vertical velocity	James Bernoulli
1691	Quadrature of the hyperbola	Finding a square equal to the area under the curve on a given interval	G. W. Leibniz
1696	Brachistochrone problem	Finding the path from which a particle will fall from one point to another in the shortest time	John Bernoulli
1698	Orthogonal trajectories	Finding a curve which cuts all curves of a family of curves at right angles	John Bernoulli
1701	Isoperimetric problem	Making an integral a maximum or minimum while keeping constant the integral of a second given function	Daniel Bernoulli
1728	Reduction of second-order to first-order equations	Finding an integrating factor	Leonard Euler
1734	Singular solutions	Finding an equation of a family of curves represented by a general solution	Alexis Clairaut
1743	Determining an integrating factor for a general linear equation	Concept of the adjoint of a differential equation	Joseph Lagrange
1762	Linear equation with constant coefficients	Conditions under which the order of a linear differential equation can be lowered	Jean d'Alembert

1.1.2 **Organization of the book**

The book consists of 14 chapters. It provides an important foundation in applied mathematics and demonstrates the use of modern technology for applied mathematical analysis with commercial software packages (e.g., algebraic, numerical, and statistical). The reader is first introduced to fundamental theory for the analytical and numerical solution of ordinary and partial differential equations. Example problems in the chapters are then revisited with a demonstration of the use of the COMSOL tool to quickly solve all differential equations previously introduced with the use of a finite element method. This work therefore demonstrates the power of modern and state-of-the-art computational technology available to the engineer. Hence, this treatment provides the underlying theory in order to understand how the tools can be applied.

The chapters of the book are organized as follows:

- Chapter 1 provides a historical perspective of the development of differential equations and introduces the organization for the book.
- Chapter 2 discusses ordinary differential equations, including first-, second-, and higher-order equations as well as systems of equations. Solution methods are discussed for common engineering problems. This material is also needed for the solution of partial differential equations in Chapter 5.
- Chapter 3 describes Laplace and Fourier transform methods that provide an alternative analytical method to solve partial differential equations in Chapter 5. Also introduced are discrete Fourier transforms and fast Fourier transforms that find application in signal processing and for solution of boundary value problems.
- Chapter 4 summarizes the use of matrices for solving linear systems of equations. For instance, numerical solution of differential equations can be effectively formulated, reduced, and solved as a matrix problem. Vector calculus is further introduced for the derivation of important theorems for vector fields that arise in engineering.
- Chapter 5 includes a treatise of analytical methods for solving partial differential equations. This treatment considers the standard techniques of separation of variables and transform methods.
- Chapter 6 examines difference numerical methods for differential equations as a numerical approach when the analytical methods of solution in Chapter 5 are not feasible. Such techniques are specifically needed for engineering problems that commonly arise in real-world situations.
- Chapter 7 further examines finite element techniques as an alternative numerical approach for the solution of ordinary and partial differential equations. This technique is used, for instance, in various engineering disciplines. This method of solution is used specifically in the COMSOL commercial software package and can be applied for multiphysics applications.
- Chapter 8 describes the treatment of experimental results, which is an essential component for analyzing experimental data and understanding error propagation

in measurements and experimental design. Regression analysis is further detailed and demonstrated with the Excel commercial software package.

- Chapter 9 details general numerical techniques needed to interpolate and smooth measured data, finding the zeros of functions, and perform numerical integration and differentiation.
- Chapter 10 provides an introduction to complex analysis. For example, the residue integration method can be used to find inverse Laplace transforms that may not be found in standard mathematical handbooks for solution of ordinary and partial differential equation problems. The complex potential that arises in conformal mapping can be used to solve analogous problems in fluid flow, electrostatics, and heat transfer.
- Chapter 11 discusses the concept of dimensional analysis for computing sets of dimensionless parameters to provide a functional relationship between dimensionless groups. It also describes the nondimensionalization of equations used in the scaling of laboratory models to full-scale systems or for simplification of differential equations.
- Chapter 12 extends the discussion to nonlinear differential equations which occur in nature.
- Chapter 13 describes integral equations, where problems can be recast as an integral rather than as a differential equation and vice versa. An integro-differential equation for neutron transport, for example, arises in nuclear reactor design. An integral equation provides a solution method where the boundary conditions of a differential equation can be incorporated into the kernel of an integral equation. This approach leads to the development of the Green's function.
- Chapter 14 describes the calculus of variations that complements ordinary differential calculus. It provides a powerful technique to find an optimum quantity to be minimized (or maximized). It leads to the development of the Euler–Lagrange equation, with application to find the motion or optimal shape of an object, or determine the shortest path on a surface (geodesic).

The relationships and interconnections of the subject matter in the various chapters of the book are highlighted in Fig. 1.1, where DE stands for differential equation, ODE stands for ordinary differential equation, and PDE stands for partial differential equation.

1.1.3 Use of the book for course instruction

For a course of 3 contact hours per week lasting two full-term semesters, the following material was covered. For the first semester, as an introductory review, a few lectures were given on ordinary differential equations for the benefit of the students culminating in a summary of solution methods in Section 2.1.1. This material is needed as more advanced techniques are introduced and was included in the course instruction for completeness. Thus, during the first semester, the complete Chapter 2 (Sections 2.1 to 2.5) and parts of Chapter 3 (Section 3.1) and Chapter 4 (Sections 4.1 to 4.3) were presented. During the next semester, the remaining material in Chapter 3

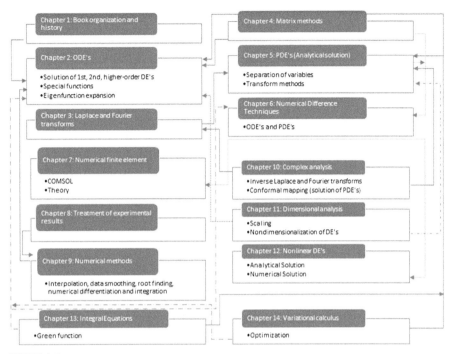

FIGURE 1.1

Interconnection of chapter material.

(Section 3.2) was covered along with Chapter 5 (Sections 5.1 and 5.2), Chapter 6 (Sections 6.1 and 6.2), part of Chapter 7 (Section 7.1), Chapter 8 (Sections 8.1 to 8.4), and Chapter 9 (Sections 9.1 to 9.5). Section 7.1 was essentially a one-period demonstration of COMSOL, where all problems in Chapter 6 were solved again.

The remaining material, not taught at the undergraduate level, includes Chapters 10 to 14, which cover more advanced material that can be used for specialized course instruction in graduate teaching. It can also be used for future reference for practicing engineers.

All problems at the end of the chapters are solved in a separate solution manual that may be made available to the instructor. These problems consist of all assignments, quizzes, and examination questions given for these courses, plus additional problems introduced for advanced material in the book.

Ordinary differential equations

An ordinary differential equation is a differential equation that contains one or more functions of only one independent variable as well as derivatives of these functions. This can be contrasted to partial differential equations in Chapter 5 that contain more than one independent variable. The functions can represent physical laws to model dynamical systems and arise in many problems of biology, economics, physics, chemistry, and engineering.

Techniques for solving first-order, second-order, and higher-order equations are presented. Solution methods for systems of ordinary differential equations are also discussed that result in the formulation of problems involving eigenvectors and eigenvalues. Higher-order ordinary differential equations can also be recast as a system of first-order equations. Using power series methods for the solution of some types of second-order differential equations, including the Frobenius method, this methodology can give rise to important functions as solutions of these equations that occur in engineering. For example, a Bessel function will arise as a solution of a second-order differential equation when the geometry of a body is cylindrical in nature. It is further shown how special functions, such as Bessel and Legendre functions, are special cases of the more general Sturm–Liouville problem. Solutions of Sturm–Liouville problems lead to orthogonal sets of eigenfunctions, which can be used for series expansion of given functions. Such representations include Fourier series, Fourier–Bessel series and Fourier–Legendre series, among others. This type of analysis is important for engineering problems in heat conduction, fluid flow, and mass transfer for instance, where it is used to apply boundary conditions for the solution of partial differential equations.

2.1 First-order equations

Ordinary differential equations are equations involving derivatives and only one independent variable.

For example,

$$y' = \tan(x) + 2, \quad \text{where } y' = \frac{dy}{dx}, \tag{2.1}$$

$$y'' + 2y - 4 = 0, \quad \text{where } y'' = \frac{d^2y}{dx^2}. \tag{2.2}$$

Advanced Mathematics for Engineering Students. https://doi.org/10.1016/B978-0-12-823681-9.00010-1

The <u>order</u> of a differential equation is the order of the highest derivative that appears in the equation (for example, the orders of the examples in Eqs. (2.1) and (2.2) are 1 and 2, respectively).

Hence, <u>first-order differential equations</u> can be written as

$$F(x, y, y') = 0, \tag{2.3}$$

or sometimes [for example, see Eq. (2.1)],

$$y' = f(x, y). \tag{2.4}$$

Solution of first-order differential equations

A solution is a function in the form

$$y = h(x), \text{ (explicit solution)}, \tag{2.5}$$

$$H(x, y), \text{ (implicit solution)}, \tag{2.6}$$

which satisfies the differential equation in Eq. (2.3) on some open interval $a < x < b$.

A <u>general solution</u> of a first-order differential equation involves an (essential) arbitrary constant. A <u>particular solution</u> of a first-order differential equation is obtained by assigning a specific value to the constant.

Example 2.1.1. (i) The formula $y = \cos(x) + c$ is a general solution of $y' + \sin(x) = 0$, and (ii) $y = \cos(x) + 5$ is a particular solution.

A <u>singular solution</u> is a solution which cannot be obtained from the general solution by specifying the value of the constant.

In most applications, one is presented with an <u>initial value problem</u>:

$$y' = f(x, y), \qquad y(x_o) = y_o, \tag{2.7}$$

where x_o and y_o are given numbers in which the <u>initial condition</u> $y(x_o) = y_o$ is used to determine a particular solution.

Separable differential equations

Many first-order differential equations can be reduced to the form

$$g(y)dy = f(x)dx, \tag{2.8}$$

which leads to the following solution by integration on both sides:

$$\int g(y)dy = \int f(x)dx + c. \tag{2.9}$$

Example 2.1.2. Solve $y' + a\,y\,x^2 = 0$ (where a is a constant). This ordinary differential equation arises in the heat transfer problem of Example 5.2.13. By separation of variables, $dy/y = -a\,x^2\,dx$. Using Eq. (2.9), one obtains $y = ce^{-x^3a/3}$. [answer]

Modeling (separable differential equations)

Example 2.1.3. *Initial value problem: radioactive decay.* In radioactive decay, the rate of loss of the number of atoms $(-dN/dt)$ is proportional to N, where the proportionality constant is termed the radioactive decay constant λ such that $dN/dt = -\lambda N$.

Determine (a) the law for radioactive decay subject to the initial condition $N(t = 0) = N_o$ and (b) the relationship between the radioactive decay constant and half-life of the material.

Solution:
(a) Using separation of variables, $dN/N = -\lambda dt$. Analogous to Example 2.1.2, the solution is $N(t) = c\,e^{-\lambda t}$. Hence, the initial condition $N(0) = ce^0 = N_o$ implies $c = N_o$. The radioactive decay law is therefore $N(t) = N_o\,e^{-\lambda t}$. [answer]
(b) The half-life $T_{1/2}$ is defined as the time after which the radioactive substance has decreased to half its original value N_o. To determine the radioactive decay constant, $\cancel{N_o}e^{-\lambda \cdot T_{1/2}} = \frac{1}{2}\cancel{N_o}$, which implies $\lambda = \dfrac{\ln(2)}{T_{1/2}}$. [answer]

Example 2.1.4. *Radiocarbon dating.* In the atmosphere (and living organisms), the ratio of radioactive carbon $^{14}_{6}C$ and $^{12}_{6}C$ is constant. When an organism dies the absorption of carbon ceases. Hence, the age of a fossil can be determined by comparing the carbon ratio in the fossil with that in the atmosphere.

If a fossil contains 25% of the original amount of $^{14}_{6}C$, what is its age?

Solution: For radioactive decay from Example 2.1.3, $N(t) = N_o e^{-\lambda t}$, where N_o is the initial amount of $^{14}_{6}C$ and the half-life of ($^{14}_{6}C$) is 5730 years.

The radioactive decay constant $\lambda = \ln 2/(5730) = 0.000121\ \mathrm{y}^{-1}$. Thus, the time after which 25% of $^{14}_{6}C$ is still present can be calculated by solving $\cancel{N_o}e^{-0.000121 \cdot t} = \frac{1}{4}\cancel{N_o}$.
Hence, the age of the fossil is $t = \dfrac{\ln(1/4)}{-0.000121} = 11,460$ years. [answer]

Example 2.1.5. *Newton's law of cooling.* A container at a temperature of 45°C is placed into a cooler at time $t = 0$, where the cooler is at a temperature $T_o = 5$°C. After 2 minutes, the temperature of the container T is reduced to 25°C. At what time is the temperature of the container equal to 15°C?

Solution.

(i) *Modeling.* Newton's law of cooling is $\dfrac{dT}{dt} = k\,(T - T_o)$, where T_o is the temperature of the surrounding medium. Hence, the differential equation becomes $\dfrac{dT}{dt} = k\,(T - 5)$.

(ii) *General solution.* Separation and integration of the above differential equation yields $\dfrac{dT}{T-5} = kdt$. Thus, $\ln(T - 5) = kt + c'$, which implies $T(t) = ce^{kt} + 5$.

(iii) *Particular solution.* Using the initial condition $T(0) = ce^0 + 5 = 45$ yields $T(t) = 40e^{kt} + 5$.

(iv) *Determine k.* Given $T(2) = 40e^{k\cdot 2} + 5 = 25$, which implies $k = \frac{1}{2}\ln\left(\dfrac{25-5}{40}\right) = -0.3466$ min^{-1}. Therefore, for the answer, $T(t) = 40e^{-0.3466\cdot t} + 5 = 25$, so that $t = 4$ minutes. [answer]

Reduction to separate form

Equations of the form

$$\boxed{y' = g\left(\frac{y}{x}\right)} \tag{2.10}$$

can be made separable by a change of variables. Let

$$\frac{y}{x} = u \quad \text{or} \quad y = xu. \tag{2.11}$$

Product differentiation of Eq. (2.11) gives

$$y' = u + xu', \quad \text{where} \quad u' = \frac{du}{dx}. \tag{2.12}$$

The right-hand side of Eq. (2.10) is

$$g\left(\frac{y}{x}\right) = g(u). \tag{2.13}$$

Therefore, equating Eq. (2.12) and Eq. (2.13) gives $u + xu' = g(u)$ or $\dfrac{du}{g(u) - u} = \dfrac{dx}{x}$. Integrating both sides yields

$$\boxed{\ln x = \int \frac{du}{g(u) - u} + c,} \tag{2.14}$$

where $u = \dfrac{y}{x}$.

Example 2.1.6. Solve $xyy' + y^2 + 4x^2 = 0$. Dividing this equation by x^2 gives $\dfrac{y}{x}y' + \left(\dfrac{y}{x}\right)^2 + 4 = 0$. Setting $u = y/x$ and solving for y' yields $y' = -\dfrac{u^2 + 4}{u} = g(u)$.

Therefore, Eq. (2.14) yields the result $\ln x = -\dfrac{1}{2}\displaystyle\int \dfrac{u\,du}{u^2 + 2} + c'$. Thus, on integrating, the final solution is $y^2 = \dfrac{c}{x^2} - 2x^2$. [answer]

Exact differential equations

A first-order differential equation of the form

$$\boxed{M(x, y)dx + N(x, y)dy = 0} \tag{2.15}$$

is called <u>exact</u> if its left-hand side is the exact differential

$$du = \dfrac{\partial u}{\partial x}dx + \dfrac{\partial u}{\partial y}dy \tag{2.16}$$

of some function $u(x, y)$. Then the differential equation in Eq. (2.15) can be written as $du = 0$, and by integration it follows that

$$\boxed{u(x, y) = c.} \tag{2.17}$$

Comparing Eq. (2.15) and Eq. (2.16), Eq. (2.15) is exact if

$$(a) \quad \dfrac{\partial u}{\partial x} = M \quad \text{and} \quad (b) \quad \dfrac{\partial u}{\partial y} = N. \tag{2.18}$$

If M and N have continuous first partial derivatives, then

$$\dfrac{\partial M}{\partial y} = \dfrac{\partial^2 u}{\partial y \partial x} \quad \text{and} \quad \dfrac{\partial N}{\partial x} = \dfrac{\partial^2 u}{\partial x \partial y}. \tag{2.19}$$

By continuity of the two second derivatives,

$$\boxed{\dfrac{\partial M}{\partial y} = \dfrac{\partial N}{\partial x}.} \tag{2.20}$$

In fact, Eq. (2.20) is not only necessary but also sufficient for Eq. (2.15) to be an exact differential.

On integrating Eq. (2.18)(a) with respect to x,

$$\boxed{u = \int M dx + k(y).} \tag{2.21}$$

To determine $k(y)$, we derive $\partial u/\partial y$ from Eq. (2.21), use Eq. (2.18)(b) to get dk/dy, and then integrate.

Example 2.1.7. Solve

$$(3x^2 y)dx + (x^3 + 3y^2)dy = 0. \tag{2.22}$$

Solution.

(i) *Test for exactness.* We have $M = 3x^2 y$ and $N = x^3 + 3y^2$. Thus, $\dfrac{\partial M}{\partial y} = 3x^2 = \dfrac{\partial N}{\partial x}$. Therefore, Eq. (2.22) is exact.

(ii) *Implicit solution.* From Eq. (2.21) the implicit solution is $u = \int M dx + k(y) = \int \left(3x^2 y\right) dx + k(y) = x^3 y + k(y)$. To find $k(y)$, use Eq. (2.18)(b) such that $\dfrac{\partial u}{\partial y} = x^3 + \dfrac{dk}{dy} = N = x^3 + 3y^2$. Thus, one obtains $\dfrac{dk}{dy} = 3y^2$ with the solution $k = y^3 + c'$. The final solution is

$$u(x, y) = \left(x^3 y + y^3\right) = c. \qquad \text{[answer]} \tag{2.23}$$

(iii) *Check implicit solution* $u(x, y) = c$. Differentiating Eq. (2.23) with respect to x gives $\left(3x^2 y + x^3 y' + 3y^2 y'\right) = 0$. This latter expression simplifies to $3x^2 y + \left(x^3 + 3y^2\right) y' = 0$, which, in turn, yields Eq. (2.22) since $y' = dy/dx$.

Integrating factors

Sometimes an equation

$$P(x, y)dx + Q(x, y)dy = 0 \tag{2.24}$$

is not exact, but it can be made exact by multiplying by an integrating factor $F(x, y)$ such that $F P dx + F Q dy = 0$, where (compare with Eq. (2.20))

$$\frac{\partial (FP)}{\partial y} = \frac{\partial (FQ)}{\partial x}. \tag{2.25}$$

Moreover, if F only depends on one variable, that is, $F = F(x)$, Eq. (2.25) yields

$$F \frac{\partial P}{\partial y} = Q \frac{\partial F}{\partial x} + F \frac{\partial Q}{\partial x}.$$

Dividing through by $F Q$ and rearranging gives

$$\frac{1}{F} \frac{\partial F}{\partial x} = \frac{1}{Q} \left(\frac{\partial P}{\partial y} - \frac{\partial Q}{\partial x} \right). \tag{2.26}$$

Thus, if the right-hand side of Eq. (2.26) only depends on x, that is, $R(x) = \dfrac{1}{Q} \left(\dfrac{\partial P}{\partial y} - \dfrac{\partial Q}{\partial x} \right)$, then the integrating factor can be obtained from

$$F(x) = \exp \int R(x)dx. \tag{2.27}$$

Alternatively, if $F = F(y)$, then instead of Eq. (2.26), one has

$$\frac{1}{F}\frac{\partial F}{\partial y} = \frac{1}{P}\left(\frac{\partial Q}{\partial x} - \frac{\partial P}{\partial y}\right) \tag{2.28}$$

and

$$F(y) = \exp \int \tilde{R}(y)dy. \tag{2.29}$$

Example 2.1.8. Solve the radioactive decay equation of Example 2.1.3 with a constant source term R_c:

$$\frac{dN}{dt} = R_c - \lambda N, \tag{2.30}$$

with the initial condition $N(t = 0) = N_0$.

Solution. Eq. (2.30) can be equivalently written as

$$(R_c - \lambda N)\,dt - dN = 0. \tag{2.31}$$

Thus, one can identify $P = (R_c - \lambda N)$ and $Q = -1$. Hence,

$$R(t) = \frac{1}{Q}\left(\frac{\partial P}{\partial N} - \frac{\partial Q}{\partial t}\right) = -1[-\lambda] = \lambda$$

and

$$F(t) = \exp\left\{\int \lambda dt\right\} = e^{\lambda t}.$$

Therefore, multiplication of Eq. (2.30) by this integrating factor gives

$$\frac{dN}{dt}e^{\lambda t} = R_c e^{\lambda t} - \lambda N e^{\lambda t} \text{ or } \frac{\left(dN e^{\lambda t}\right)}{dt} = R_c e^{\lambda t}.$$

Thus, separation of variables with $N(0) = N_0$ yields the solution:

$$N(t) = \frac{R_c}{\lambda}\left(1 - e^{-\lambda t}\right) + N_0\, e^{-\lambda t}. \quad \text{[answer]}$$

This problem can represent the rate of release of radioactive fission products into a reactor coolant, as shown in Example 9.4.1.

Example 2.1.9. Consider the mixing of salt into a tank containing a volume of water $V(t)$ (m^3) with a mass of salt $m(t)$ (kg). Instantaneous mixing in the tank is assumed with a volumetric flow rate of water into the tank $\dot{v}_i(t)$ (m^3/s) with a salt concentration $c_i(t)$ (kg/m^3) and a volumetric flow rate of water out of the tank $\dot{v}_o(t)$ (m^3/s) with a salt concentration $c_o(t)$ (kg/m^3). The conservation equation for the volume of water in the tank is

$$\frac{dV(t)}{dt} = \dot{v}_i(t) - \dot{v}_o(t).$$

Assuming that the flow rates in and out of the tank are equal and constant, $\dot{v}_i(t) = \dot{v}_o(t) = \dot{v}$, the solution of this differential equation for the volume of water is simply $V(t) = V_o$. Similarly, the conservation equation for the mass of salt in the tank is

$$\frac{dm(t)}{dt} = \dot{v}(c_i(t) - c_o(t)).$$

Moreover, assuming a constant inlet concentration of salt $c_i(t) = c_i$ with an outlet concentration of $c_o(t) = \dfrac{m(t)}{V_o}$, the mass balance for the salt becomes

$$\frac{dm(t)}{dt} = \dot{v}\,c_i - \frac{\dot{v}}{V_o}m(t). \tag{2.32}$$

Solve for the mass of salt in the tank $m(t)$ and the outlet concentration of salt $c_o(t)$.

Solution. From Example 2.1.8, noting that Eq. (2.30) and Eq. (2.32) are of the same form where on comparison $R_c \rightarrow \dot{v}c_i$ and $\lambda \rightarrow \dot{v}/V_o$, the solution follows on applying an integrating factor with the initial condition $m(t=0) = m_o$:

$$m(t) = c_i V_o \left(1 - e^{-(\dot{v}/V_o)t}\right) + m_o\, e^{-(\dot{v}/V_o)t} \qquad \text{[answer]}$$

and

$$c_o(t) = c_i \left(1 - e^{-(\dot{v}/V_o)t}\right) + (m_o/V_o)\, e^{-(\dot{v}/V_o)t}. \qquad \text{[answer]}$$

Linear differential equations

A first-order <u>linear</u> differential equation is of the form

$$\boxed{y' + p(x)y = r(x).} \tag{2.33}$$

If $r(x) = 0$, then the equation is called <u>homogeneous</u>. In this case, by separating variables, the following solution follows: $\dfrac{dy}{y} = -p(x)dx$. Thus, $\ln y = -\int p(x)dx + c'$ or

$$\boxed{y(x) = ce^{-\int p(x)dx}.} \tag{2.34}$$

The <u>nonhomogeneous</u> Eq. (2.33) can be solved with an integrating factor where Eq. (2.33) is written as $(p(x)y - r)dx + dy = 0$, so that $P = py - r$ and $Q = 1$.

Therefore Eq. (2.26) becomes $\dfrac{1}{F}\dfrac{dF}{dx} = p(x)$ and Eq. (2.27) yields the integrating factor

$$F(x) = e^{\int p dx}. \tag{2.35}$$

Thus, multiplying Eq. (2.33) by Eq. (2.35) gives $e^{\int p dx}(y' + py) = \left(e^{\int p dx}y\right)' = e^{\int p dx}r$. On integrating with respect to x, $e^{\int p dx}y = \int e^{\int p dx}r dx + c$ and solving for y,

$$\boxed{y(x) = e^{-h}\left[\int e^{h}r dx + c\right], \quad h = \int p(x)dx.} \tag{2.36}$$

Example 2.1.10. Solve $y' + y = \sinh x$. Here, $p = 1$, $r = \sinh x$, and $h = \int p dx = x$. From Eq. (2.36), $y(x) = e^{-x}\left[\int e^{x}\sinh x\, dx + c\right] = \dfrac{1}{2}\left[\dfrac{e^{x}}{2} - xe^{-x}\right] + ce^{-x}$. [answer]

Reduction to linear form

The Bernoulli equation is given by

$$\boxed{y' + p(x)y = g(x)y^{a}} \quad (a \text{ is a real number}). \tag{2.37}$$

This equation can be reduced to a linear form on letting

$$u(x) = [y(x)]^{1-a}. \tag{2.38}$$

Hence, differentiating Eq. (2.38) and substituting y' from Eq. (2.37) gives

$$u' = (1-a)y^{-a}y' = (1-a)y^{-a}(gy^{a} - py) = (1-a)(g - py^{1-a}) = (1-a)(g - pu).$$

Hence, one obtains the linear equation $u' + (1-a)pu = (1-a)g$ of the form of Eq. (2.33).

2.1.1 Summary of solution methods

Any first-order differential equation can be put in the form

$$\frac{dy}{dx} = f(x,y) \quad \text{or} \quad M(x,y)dx + N(x,y)dy = 0, \tag{2.39}$$

where the general solution contains an arbitrary constant. The following Table 2.1 summarizes various techniques for finding the general solution of Eq. (2.39).

Table 2.1 General solution of various first-order differential equations.

Differential equation	General solution (or method to obtain it)
1. *Separation of variables* $f_1(x)g_1(y)dx + f_2(x)g_2(y)dy$	Divide by $g_1(y)f_2(x) \neq 0$ and integrate to obtain $\int \frac{f_1(x)}{f_2(x)}dx + \int \frac{g_2(x)}{g_1(x)}dy = c$.
2. *Exact equation* $M(x, y)dx + N(x, y)dy = 0$, $\frac{\partial M}{\partial y} = \frac{\partial N}{\partial x}$	The equation can be written as $Mdx + Ndy = du(x, y) = 0$, where du is an exact differential. Thus, the solution is $u(x, y) = c$ or equivalently $\int M\partial x + \int \left[N - \frac{\partial}{\partial y}\int M\partial x\right]dy = c$, where ∂x indicates that the integration is to be performed with respect to x keeping y constant.
3. *Integrating factor* $M(x, y)dx + N(x, y)dy = 0$, $\frac{\partial M}{\partial y} \neq \frac{\partial N}{\partial x}$	The equation can be written as $FMdx + FNdy = 0$, where F is an integrating factor so that $\frac{\partial}{\partial y}(FM) = \frac{\partial}{\partial x}(FN)$ and then method 2 applies.
4. *Linear equation* $\frac{dy}{dx} + p(x)y = r(x)$	An integrating factor is given by $F = e^{\int p(x)dx}$ and the equation can be written $\frac{d}{dx}(Fy) = Fr$ with solution $Fy = \int Frdx + c$ or $y = e^{-h}\left[\int e^h rdx + c\right]$, where $h = \int pdx$.
5. *Reduction to separable form* $\frac{dy}{dx} = g\left(\frac{y}{x}\right)$	Let $y/x = u$, and the equation becomes $u + x\frac{du}{dx} = g(u)$ or $xdu - (g(u) - u)dx = 0$, which is of the form of type 1 and has the solution $\ln x = \int \frac{du}{g(u)-u} + c$, where $u = y/x$. If $g(u) = u$ the solution is $y = cx$.
6. *Bernoulli's equation* $\frac{dy}{dx} + p(x)y = g(x)y^a$, $a \neq 0, 1$	Letting $u = y^{1-a}$ the equation reduces to type 4 with solution $ue^{(1-a)\int pdx} = (1 - a)\int ge^{(1-a)\int pdx} dx + c$. If $a = 0$, the equation is of type 4. If $a = 1$, it is of type 1.

Table 2.1 (*continued*)

Differential equation	General solution (or method to obtain it)
7. Equation solvable for y $y = g(x, p)$, where $p = y'$	Differentiate both sides of the equation with respect to x, $\frac{dy}{dx} = \frac{dg}{dx} = \frac{\partial g}{\partial x} + \frac{\partial g}{\partial p}\frac{dp}{dx}$ or $p = \frac{\partial g}{\partial x} + \frac{\partial g}{\partial p}\frac{dp}{dx}$. Then solve this last equation to obtain $G(x, p, c) = 0$. The required solution is obtained by eliminating p between $G(x, p, c)$ and $y = g(x, p)$. An analogous method exists if the equation is solvable for x.
8. Clairaut's equation $y = px + g(p)$, where $p = y'$	The equation is of type 7 with solution $y = cx + g(c)$. The solution will also have a singular solution in general.

2.1.2 Approximate solution by iteration

Picard's <u>iteration method</u> provides an approximate solution to the initial value problem in Eq. (2.7). By integration, Eq. (2.7) may be written in the form

$$y' = y_o + \int_{x_o}^{x} f(t, y(t))dt. \tag{2.40}$$

Successive approximations to the solution $y(x)$ in Eq. (2.40) are obtained as

$$y = y_o,$$

$$y_1(x) = y_o + \int_{x_o}^{x} f(t, y_o))dt,$$

$$y_2(x) = y_o + \int_{x_o}^{x} f(t, y_1(t))dt,$$

$$\vdots$$

$$\boxed{y_n(x) = y_o + \int_{x_o}^{x} f(t, y_{n-1}(t))dt.} \tag{2.41}$$

Example 2.1.11. Solve $y' = 1 + y$, $y(0) = 0$.

Solution. Here $x_o = 0$, $y_o = 0$, $f(x, y) = 1 + y$, and Eq. (2.41) becomes $y_n(x) = \int_0^x [1 + y_{n-1}(t)]\,dt = x + \int_0^x y_{n-1}(t)dt$. Solving from $y_o = 0$ gives

$$y_1(x) = x + \int_0^x 0 \, dt = x,$$

$$y_2(x) = x + \int_0^x t \, dt = x + \frac{x^2}{2},$$

$$y_3(x) = x + \int_0^x \left(t + \frac{t^2}{2}\right) dt = x + \frac{x^2}{2} + \frac{x^3}{3 \cdot 2} + \frac{x^4}{4 \cdot 3 \cdot 2}. \qquad \text{[answer]} \quad (2.42)$$

Alternatively, the solution can be obtained from separation of variables (see Section 2.1): $\dfrac{dy}{1+y} = dx$. On integrating, $\ln(1 + y) = x + c'$ so that $y = ce^x - 1$. But the condition $y(0) = 0$ yields $y = e^x - 1 = x + \dfrac{x^2}{2!} + \dfrac{x^3}{3!} + \dfrac{x^4}{4!} + \dots$ for the range $(-\infty < x < \infty)$. This latter expression agrees with Eq. (2.42).

2.1.3 Existence and uniqueness of solutions

Given the initial value problem in Eq. (2.7), it is of practical importance that a given model has a unique solution before one tries to compute a solution. In particular, it is important to determine the conditions that:

(i) the initial value problem has at least one solution (existence),
(ii) the initial value problem has at most one solution (uniqueness).

Existence theorem

If $f(x, y)$ is continuous at all points (x, y) in **R**, $|x - x_o| < a$, $|y - y_o| < b$, and bounded in **R**, such that $|f(x, y)| \le K$ for all (x, y) in **R**, then the initial value problem has at least one solution $y(x)$. This solution is defined <u>at least</u> for all x in the interval $|x - x_o| < \alpha$, where α is the smaller of two numbers a and b/K.

Uniqueness theorem

If $f(x, y)$ and $\frac{\partial f}{\partial y}$ are continuous for all (x, y) in **R** and bounded such that (a) $|f| \le K$ and (b) $\frac{\partial f}{\partial y} \le M$ for all (x, y) in **R**, then Eq. (2.7) has at most one solution $y(x)$. Hence, by the *existence theorem* it has precisely one solution which is defined <u>at least</u> in the interval $|x - x_o| < \alpha$. In fact, it can be obtained by Picard's method, where Eq. (2.41) (with $n = 1, 2, \dots$) converges to that solution $y(x)$.

Consider Example 2.1.11 and take **R**: $|x| < 5$ and $|y| < 3$ (so that $a = 5$ and $b = 3$). Thus,

$$|f| = |1 + y| \le K = 4, \quad |\frac{\partial f}{\partial y}| = 1 \le M = 1, \quad \alpha = b/K = 0.75 < a.$$

2.2 Second-order linear differential equations
2.2.1 Homogeneous linear equations

A second-order differential equation is linear if it has the form

$$\boxed{y'' + p(x)y' + q(x)y = r(x).} \tag{2.43}$$

If $r(x) = 0$, the equation is homogeneous:

$$\boxed{y'' + p(x)y' + q(x)y = 0.} \tag{2.44}$$

For example, $(1 - x^2)y'' + 2x^2y' + 5y = 0$.

Eq. (2.44) has the important property that a linear combination of solutions is again a solution (superposition principle or linearity principle). Two linearly independent solutions y_1 and y_2 of Eq. (2.44) form a basis of solutions, where the general solution is given by

$$\boxed{y = c_1 y_1 + c_2 y_2} \tag{2.45}$$

and c_1 and c_2 are arbitrary constants. A particular solution from Eq. (2.45) is obtained if one specifies numerical values for c_1 and c_2 by imposing two initial conditions:

$$\boxed{y(x_o) = K_0, \ y'(x_o) = K_1} \quad (x_o, K_0, \text{ and } K_1 \text{ are given numbers}). \tag{2.46}$$

Together Eq. (2.44) and Eq. (2.46) constitute an initial value problem. If p and q are continuous on some open interval **I** and x_o is in **I**, then Eq. (2.44) has a general solution on **I**, and Eq. (2.44) and Eq. (2.46) has a unique solution on **I** (which is a particular solution; thus, Eq. (2.44) has no singular solutions). Some applications may also lead to the boundary conditions

$$\boxed{y(P_1) = k_1, \ y(P_2) = k_2,} \quad (P_1, P_2, k_1, \text{ and } k_2 \text{ are given numbers}), \tag{2.47}$$

where P_1 and P_2 are endpoints on **I**. Together Eq. (2.44) and Eq. (2.47) constitute a boundary value problem.

Linear independence

Two functions $y_1(x)$ and $y_2(x)$ are linearly independent on an interval **I** if

$$k_1 y_1(x) + k_2 y_2(x) = 0 \text{ on } \mathbf{I} \text{ implies } k_1 = 0 \text{ and } k_2 = 0, \tag{2.48}$$

that is, y_1 and y_2 are not proportional to one another.

Table 2.2 General solution of second-order homogeneous differential equations with constant coefficients.

Case	Roots (Eq. (2.50))	Basis (Eq. (2.49))	General solution
I	Distinct real λ_1, λ_2	$e^{\lambda_1 x}, e^{\lambda_2 x}$	$y = c_1 e^{\lambda_1 x} + c_2 e^{\lambda_2 x}$
II	Real double root $\lambda = -\dfrac{a}{2}$	$e^{-\frac{ax}{2}}, xe^{-\frac{ax}{2}}$	$y = (c_1 + c_2 x)e^{-\frac{ax}{2}}$
III	Complex conjugate $\lambda_1 = -\frac{a}{2} + i\omega$ $\lambda_2 = -\frac{a}{2} - i\omega$	$e^{-\frac{ax}{2}}\cos\omega$ $e^{-\frac{ax}{2}}\sin\omega$	$y = e^{-\frac{ax}{2}}(A\cos\omega x + B\sin\omega x)$

2.2.2 Homogeneous equations with constant coefficients

If $p(x)$ and $q(x)$ are constant (that is, $p(x) = a$ and $q(x) = b$), one obtains the homogeneous linear equation

$$y'' + ay' + by = 0. \tag{2.49}$$

This equation can be solved by substituting $y = e^{\lambda x}$, yielding

$$\left(\lambda^2 + a\lambda + b\right) e^{\lambda x} = 0.$$

Thus, λ is a root of the <u>characteristic equation</u> (or auxiliary equation)

$$\lambda^2 + a\lambda + b = 0. \tag{2.50}$$

The solution of Eq. (2.50) is

$$\lambda_1 = \frac{1}{2}(-a + \sqrt{a^2 - 4b}), \quad \lambda_2 = \frac{1}{2}(-a - \sqrt{a^2 - 4b}). \tag{2.51}$$

Depending on the discriminant $a^2 - 4b$, one obtains three cases:

Case I: two real roots λ_1, λ_2 if $a^2 - 4b > 0$;
Case II: a real double root $\lambda_1 = \lambda_2 = -a/2$ if $a^2 - 4b = 0$;
Case III: complex conjugate roots $\lambda_1 = -\frac{a}{2} + i\omega$ and $\lambda_2 = -\frac{a}{2} - i\omega$, where $\left(\omega = \sqrt{b - \frac{a^2}{4}}\right)$, if $a^2 - 4b < 0$.

The general solution of Eq. (2.49) for the three cases is given in Table 2.2.

Example 2.2.1. (*Case I*) Solve $y'' - 4y = 0$.

Solution. Here $a = 0$ and $b = -4$, which implies $\lambda_1 = 2$ and $\lambda_2 = -2$ from Eq. (2.51). Therefore, the general solution is $y = c_1 e^{2x} + c_2 e^{-2x}$. [answer]

Example 2.2.2. (*Case II*) Solve $y'' + 2y' + y = 0$.

Solution. Here $a = 2$ and $b = 1$, which implies a double root $\lambda = -1$ (since $a^2 - 4b = 0$) from Eq. (2.51). Therefore, the general solution is $y = (c_1 + c_2 x)e^{-x}$. [answer]

Example 2.2.3. (*Case III*) Solve $y'' + 2y' + 5y = 0$.

Solution. Here $a = 2$ and $b = 5$, which implies complex conjugate roots (since $a^2 - 4b < 0$) from Eq. (2.51). Therefore, $\lambda_1 = -1 + i\sqrt{5 - \dfrac{(2)^2}{4}} = -1 + 2i$ and $\lambda_2 = -1 - 2i$. The general solution is $y = e^{-x}(A\cos 2x + B\sin 2x)$. [answer]

Example 2.2.4. (*Kirchhoff's second law*) Consider an electric current $I(t)$ flowing through a circuit containing a resistance R, capacitance C, and inductance L (Fig. 2.1). These quantities are related through Kirchhoff's voltage law as given by the integro-differential equation (see Chapter 13)

$$L I(t)' + R I(t) + \frac{1}{C}\int_{t_0}^{t} I(s)\,ds = 0.$$

Differentiating this equation with respect to time t yields a second-order homoge-

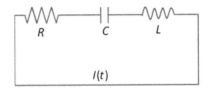

FIGURE 2.1

Diagram of an LRC circuit.

neous ordinary differential equation with constant coefficients:

$$I(t)'' + \left(\frac{R}{L}\right)I'(t) + \left(\frac{1}{LC}\right)I(t) = 0$$

or

$$I'' + 2w_d I' + w_o^2 I = 0.$$

Here $w_d = R/(2L)$ and $w_o^2 = 1/(LC)$. Thus, the characteristic equation is $\lambda^2 + 2w_d + w_o^2 = 0$ with roots $\lambda_{1,2} = -w_d \pm \sqrt{w_d^2 - w_o^2}$. If $R = 0$, then $w_d = 0$ and $\lambda = \pm i\, w_o$. Hence, from Table 2.2, the solution is

$$I(t) = A\cos(w_o t) + B\sin(w_o t).$$

This result implies that with no resistance, the circuit oscillates with no dissipation. On the other hand, if $R < \sqrt{4L/C}$ (such that $w_d^2 < w_o^2$), $\lambda = -w_d \pm i\sqrt{w_o^2 - w_d^2} = -w_d \pm i\beta$, where $\beta = \sqrt{w_o^2 - w_d^2}$. For this case, from Table 2.2,

$$I(t) = Ae^{-w_d t}\cos(\beta t) + Be^{-w_d t}\sin(\beta t).$$

Consequently, the presence of a resistor damps the current oscillations produced by the capacitor and inductor.

Example 2.2.5. *(Damped oscillator/spring)* Consider a mass m at the end of a spring (Fig. 2.2). From Newton's second law of motion, the force on a body is equal to the mass times the acceleration of the body (that is, $my''(t)$). This force is equal to the sum of all forces acting on the body. From Hooke's law for the force exerted by the spring for a displacement y, $F_s = -k \cdot y$, where the spring has a constant $k > 0$. There is a friction force by the air that is proportional to the velocity of the object such that $F_a = -d \cdot y'(t)$, where $d > 0$ is a damping coefficient. Hence, for this given force balance,

$$my'' = -d\,y' - k\,y \text{ or } m\,y'' + d\,y' + k\,y = 0.$$

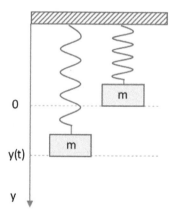

FIGURE 2.2

Diagram of a spring system.

Similar to Example 2.2.4, with $w_d = d/(2m)$ and $w_o^2 = k/m$, the characteristic equation $\lambda^2 + 2w_d + w_o^2 = 0$ has roots $\lambda_{1,2} = -w_d \pm i\beta$, where $\beta = \sqrt{w_o^2 - w_d^2}$. Hence, from Table 2.2, the solution is

$$y(t) = e^{-w_d t}\left(A\cos(\beta t) + B\sin(\beta t)\right).$$

Equivalently, in terms of an amplitude A and phase ϕ for the damped oscillator,

$$y(t) = Ae^{-w_d t}\cos(\beta t - \phi).$$

2.2.3 Euler–Cauchy equation

The Euler–Cauchy equations of the form

$$\boxed{x^2 y'' + axy' + by = 0} \quad (a, b \text{ constants}) \qquad (2.52)$$

can also be solved by an algebraic method by substituting $y = x^m$ into Eq. (2.52), yielding

$$x^2 m (m - 1) x^{m-2} + a x m x^{m-1} + b x^m = 0.$$

Omitting x^m (where $x \neq 0$) yields the auxiliary equation

$$\boxed{m^2 + (a - 1)m + b = 0.} \qquad (2.53)$$

Similarly, three cases arise for the general solution of Eq. (2.52), as shown in Table 2.3.

Table 2.3 General solution of the Euler–Cauchy equation.

Case	Roots (Eq. (2.53))	Basis (Eq. (2.52))	General solution
I	Distinct real m_1, m_2	x^{m_1}, x^{m_2}	$y = c_1 x^{m_1} + c_2 x^{m_2}$
II	Real double root $m = \frac{1}{2}(1 - a)$	$x^{\frac{(1-a)}{2}}$	$y = (c_1 + c_2 \ln x) x^{\frac{(1-a)}{2}}$
III	Complex conjugate $m_1 = \mu + iv$ $m_2 = \mu - iv$	$x^\mu \cos(v \ln x)$ $x^\mu \sin(v \ln x)$	$y = x^\mu [A \cos(v \ln x) + B \sin(v \ln x)]$

Example 2.2.6. (*Case I*) Solve $x^2 y'' - 3xy' + 3y = 0$.

Solution. The auxiliary equation is $m^2 - 4m + 3 = 0$. Hence, $m_1 = 3$ and $m_2 = 1$ from Eq. (2.53). Therefore, the general solution is $y = c_1 x^3 + c_2 x$. [answer]

Example 2.2.7. (*Case II*) Solve $x^2 y'' - xy' + y = 0$.

Solution. The auxiliary equation has a double root $m = 1$. Therefore, the general solution is $y = (c_1 + c_2 \ln x)x$. [answer]

Example 2.2.8. (*Case III*) Solve $x^2 y'' - xy' + 2y = 0$.

Solution. The auxiliary equation is $m^2 - 2m + 2 = 0$, which implies $m_{1,2} = 1 \pm i\sqrt{2 - \dfrac{(-2)^2}{4}} = 1 \pm i$. The general solution is $y = x[(A \cos(\ln x) + B \sin(\ln x)]$. [answer]

2.2.4 Existence and uniqueness of homogeneous equations

Consider the general homogeneous linear equation

$$y'' + p(x)y' + q(x)y = 0, \tag{2.54}$$

subject to the two initial conditions

$$y(x_o) = K_0, \ y'(x_o) = K_1. \tag{2.55}$$

Existence and uniqueness of the initial value problem

If $p(x)$ and $q(x)$ are continuous functions on some open interval **I** and x_o is in **I**, then the initial value problem Eq. (2.54) and Eq. (2.55) has a unique solution on **I**.

Linear independence of solutions

Define the <u>Wronskian</u> of two solutions y_1 and y_2 as

$$W(y_1, y_2) = \begin{vmatrix} y_1 & y_2 \\ y_1' & y_2' \end{vmatrix} = y_1 y_2' - y_2 y_1'. \tag{2.56}$$

If Eq. (2.54) has continuous coefficients $p(x)$ and $q(x)$ on some open interval **I**, then the two solutions y_1 and y_2 are linearly dependent on **I** if and only if $W = 0$ at some x_o on **I**. Furthermore, if $W = 0$ for $x = x_o$, then $W \equiv 0$ on **I**; hence, if there is an x_1 in **I** at which $W \neq 0$, then y_1 and y_2 are linearly independent on **I**.

Example 2.2.9. Show that $y = (c_1 + c_2 x)e^{-x}$ is a general solution of $y'' + 2y' + y = 0$ for Example 2.2.2.

Solution. Substituting in the solutions $y_1 = e^{-x}$ and $y_2 = xe^{-x}$ into the Wronskian in Eq. (2.56), the above theorem implies linear independence since

$$W(e^{-x}, xe^{-x}) = \begin{vmatrix} e^{-x} & xe^{-x} \\ -e^{-x} & (1-x)e^{-x} \end{vmatrix} = (1-x)e^{-2x} + xe^{-2x} = e^{-2x} \neq 0.$$

[answer]

General solution of homogeneous linear differential equations

<u>Existence:</u> If the coefficients $p(x)$ and $q(x)$ are continuous on **I**, then Eq. (2.54) has a general solution on **I**.

<u>General solution:</u> If Eq. (2.54) has continuous coefficients $p(x)$ and $q(x)$ on **I**, then every solution on **I** is of the form

$$y(x) = c_1 y_1(x) + c_2 y_2(x), \tag{2.57}$$

where $y_1(x)$ and $y_2(x)$ form a basis of solutions of Eq. (2.54) on **I** and c_1 and c_2 are suitable constants.

Method to obtain a second basis solution

The method of reduction of order provides a means to obtain a second basis solution of the homogeneous Eq. (2.54). Thus, if y_1 is a solution of Eq. (2.54), let $y_2 = uy_1$. Substituting y_2 and the corresponding derivatives $y_2' = u'y_1 + uy_1'$ and $y_2'' = u''y_1 + 2u'y_1' + uy_1''$ into Eq. (2.54) and collecting terms yields

$$u''y_1 + u'(2y_1' + py_1) + u(y_1'' + py_1' + qy_1) = 0. \qquad (2.58)$$

The last term in brackets $y_1'' + py_1' + qy_1 = 0$ since y_1 is a solution. Thus, dividing Eq. (2.58) by y_1 and letting $u' = U$ (and $u'' = U'$) gives $U' + \left(\dfrac{2y_1'}{y_1} + p\right)U = 0$. By separation of variables, $\ln U = -2\ln y_1 - \int p\,dx$, which implies

$$U = \frac{1}{y_1^2}e^{-\int p\,dx}. \qquad (2.59)$$

Therefore,

$$\boxed{y_2 = uy_1 = y_1 \int U\,dx.} \qquad (2.60)$$

Example 2.2.10. Given that the solution of $y'' + ay' + by = 0$ for the case of a double root is $y_1 = e^{-ax/2}$ (see Table 2.2), find the other solution.

Solution. Here $p(x) = a$. Then in Eq. (2.59) one obtains $-\int p\,dx = -a\int dx = -ax$. Thus, $U = e^{ax} \cdot e^{-ax} = 1$ and $u = \int U\,dx = x$. Therefore, $y_2 = uy_1 = xe^{-ax/2}$, as given in Table 2.2 (Case II). [answer]

2.2.5 Nonhomogeneous equations

Given the nonhomogeneous linear equation in Eq. (2.43) (with $r(x) \neq 0$) and the corresponding homogeneous equation in Eq. (2.44), a general solution of Eq. (2.43) is of the form

$$\boxed{y = y_h + y_p,} \qquad (2.61)$$

where y_h is a general solution of Eq. (2.44) and y_p is any particular solution of Eq. (2.43). A particular solution of Eq. (2.43) is a solution obtained from Eq. (2.61) by assigning specific values of the arbitrary constants c_1 and c_2 in $y_h(x)$.

Example 2.2.11. Solve the initial value problem $y'' - 3y' + 2y = 12e^{-2x}$, $y(0) = 1$, $y'(0) = -1$.

Solution.
(i) *General solution of homogeneous equations.*
The characteristic equation $\lambda^2 - 3\lambda + 2 = 0$ has roots 1 and 2. Therefore, $y_h = c_1e^x + c_2e^{2x}$.

(ii) *Particular solution of nonhomogeneous equations.*
Try $y_p = Ce^{-2x}$ (see the method of undetermined coefficients in the next section).
Then $y_p' = -2Ce^{-2x}$, $y_p'' = 4Ce^{-2x}$, and substitution gives $4Ce^{-2x} - 3(-2Ce^{-2x}) + 2Ce^{-2x} = 12e^{-2x}$. Hence, $4C + 6C + 2C = 12$, so that $C = 1$ and the general solution of the nonhomogeneous equation is $y = y_h + y_p = c_1e^x + c_2e^{2x} + e^{-2x}$.
(iii) *Particular solution satisfying initial conditions.*
We have $y'(x) = c_1e^x + 2c_2e^x - 2e^{-2x}$. The initial conditions yield

$$y(0) = c_1 + c_2 + 1 = 1,$$
$$y'(0) = c_1 + 2c_2 - 2 = -1,$$

so that $c_1 = -1$ and $c_2 = 1$. Therefore, $y(x) = -e^x + e^{2x} + e^{-2x}$. [answer]

In summary, to solve the nonhomogeneous Eq. (2.43), or an initial value problem for Eq. (2.43), one solves the homogeneous Eq. (2.44) (as discussed in Section 2.2.2 and Section 2.2.4) and finds a particular solution y_p of Eq. (2.43). Methods for finding y_p are detailed as follows.

Method of undetermined coefficients

This method applies to the equation with constant coefficients

$$\boxed{y'' + ay' + by = r(x),} \tag{2.62}$$

where $r(x)$ is of a special form as detailed in the following rules:

(a) *Basic rule.* If $r(x)$ in Eq. (2.62) is one of the functions in the first column of Table 2.4, choose the corresponding function for y_p and determine its undetermined coefficients by substitution into Eq. (2.62).

(b) *Modification rule.* If the choice for y_p is also a solution of the homogeneous equation corresponding to Eq. (2.62), then multiply the choice for y_p by x (or x^2 if this solution corresponds to a double root of the characteristic equation of the homogeneous equation).

(c) *Sum rule.* If $r(x)$ is a sum of functions in several lines of Table 2.4, then choose for y_p the corresponding sum of functions.

Example 2.2.12. (*rule a*) Solve $y'' + 5y = 25x^2$.

Solution. Using Table 2.4, $y_p = K_2x^2 + K_1x + K_o$, yielding $y_p'' = 2K_2$. Therefore, $2K_2 + 5(K_2x^2 + K_1x + K_o) = 25x^2$. Equating coefficients on both sides gives $5K_2 = 25$, $5K_1 = 0$, $2K_2 + 5K_o = 0$, so that $K_2 = 5$, $K_1 = 0$, and $K_o = -2$. Therefore, $y_p = 5x^2 - 2$. Thus, the general solution is given by $y = y_h + y_p = A\cos\sqrt{5}x + B\sin\sqrt{5}x + 5x^2 - 2$. [answer]

Example 2.2.13. (*rule b*) Solve $y'' - y = e^{-x}$.

Table 2.4 Choice for the y_p function.

Term in $r(x)$	Choice for y_p
$ke^{\gamma x}$	$Ce^{\gamma x}$
kx^n $(n = 0, 1, ...)$	$K_n x^n + K_{n-1} x^{n-1} + ... + K_1 x + K_o$
$k \cos \omega x$ $k \sin \omega x$	$K \cos \omega x + M \sin \omega x$
$ke^{\alpha x} \cos \omega x$ $ke^{\alpha x} \sin \omega x$	$e^{\alpha x}(K \cos \omega x + M \sin \omega x)$

Solution. The characteristic equation $(\lambda^2 - 1) = (\lambda - 1)(\lambda + 1) = 0$ has the roots 1 and -1, so that $y_h = c_1 e^x + c_2 e^{-x}$. Normally Table 2.4 suggests $y_p = Ce^{-x}$ (which is already a solution of the homogeneous equation). Thus, rule (b) applies, where $y_p = Cxe^{-x}$. As such, $y_p' = C(e^{-x} - xe^{-x})$ and $y_p'' = C(-2e^{-x} + xe^{-x})$. Substituting these expressions into the differential equation gives $C(-2 + x)e^{-x} - Cxe^{-x} = e^{-x}$. Simplifying, one obtains $C = -1/2$. Thus, the general solution is given by $y = c_1 e^x + c_2 e^{-x} - \frac{1}{2}xe^{-x}$. [answer]

Example 2.2.14. (*rules b and c*) Solve $y'' + 2y' + y = e^{-x} - x$, $y(0) = 0$, $y'(0) = 0$.

Solution. The characteristic equation $\lambda^2 + 2\lambda + 1 = (\lambda + 1)^2 = 0$ has a double root $\lambda = -1$. Hence, $y_h = (c_1 + c_2 x)e^{-x}$ (see Section 2.2.2). For the x-term on the right-hand side, Table 2.4 suggests the choice $K_1 x + K_o$. However, since $\lambda = -1$ is a double root, rule b suggests the choice $Cx^2 e^{-x}$ (instead of Ce^{-x}) so that $y_p = K_1 x + K_o + Cx^2 e^{-x}$. As such, $y_p'' + 2y_p' + y_p = 2Ce^{-x} + K_1 x + (2K_1 + K_o) = e^{-x} - x$, which implies $C = \frac{1}{2}$, $K_1 = -1$ and $K_o = 2$. The general solution is $y = y_h + y_p = (c_1 + c_2 x)e^{-x} + \frac{1}{2}x^2 e^{-x} + 2 - x$.

For the initial conditions, $y' = (-c_1 + c_2 - c_2 x)e^{-x} + (x - \frac{1}{2}x^2)e^{-x} - 1$. Therefore, $y(0) = c_1 + 2 = 0$ (so that $c_1 = -2$) and $y'(0) = -c_1 + c_2 - 1 = 0$ (so that $c_2 = -1$). Thus, $y = -(2 + x)e^{-x} + \frac{1}{2}x^2 e^{-x} + (2 - x)$. [answer]

Method of variation of parameters
This method is more general and applies to the equation

$$\boxed{y'' + p(x)y' + q(x)y = r(x).} \tag{2.63}$$

The method gives a particular solution y_p of Eq. (2.63):

$$\boxed{y_p(x) = -y_1 \int \frac{y_2 r}{W} dx + y_2 \int \frac{y_1 r}{W} dx,} \tag{2.64}$$

where y_1 and y_2 form a basis of solutions of the corresponding homogeneous equation,

$$y'' + p(x)y' + q(x)y = 0, \tag{2.65}$$

and W is the Wronskian of y_1, y_2,

$$W = y_1 y_2' - y_1' y_2. \tag{2.66}$$

Example 2.2.15. Solve $y'' - y = \sinh x$.

Solution. The basis solutions of the homogeneous equation are $y_1 = \sinh x$ and $y_2 = \cosh x$. Thus, $W = \sinh x \sinh x - \cosh x \cosh x = -1$. Hence, from Eq. (2.64) (choosing the constants of integration to be zero),

$$y_p = \sinh x \int \cosh x \sinh x \, dx - \cosh x \int \sinh x \sinh x \, dx = \frac{1}{2}(x \cosh x - \sinh x).$$

The general solution is $y = y_h + y_p = [c_1 \sinh x + c_2 \cosh x] + \frac{1}{2}(x \cosh x - \sinh x)$. [answer]

Note: Had the two arbitrary constants of integration c_1 and $-c_2$ been included, Eq. (2.64) would have yielded the additional general solution $c_1 \sinh x + c_2 \cosh x = c_1 y_1 + c_2 y_2$ (which always results).

2.3 Higher-order linear differential equations
Homogeneous linear equations
An ordinary differential equation of nth order is of the form

$$F(x, y, y', ... y^{(n)}) = 0, \tag{2.67}$$

where the nth derivative $y^{(n)} = \dfrac{d^n y}{dx^n}$. The equation is <u>linear</u> if it can be written as

$$\boxed{y^{(n)} + p_{n-1}(x)y^{(n-1)} + ... + p_1(x)y' + p_0(x)y = r(x).} \tag{2.68}$$

The corresponding <u>homogeneous</u> equation written in "standard" form is

$$\boxed{y^{(n)} + p_{n-1}(x)y^{(n-1)} + ... + p_1(x)y' + p_0(x)y = 0.} \tag{2.69}$$

Eq. (2.69) has the important property that a linear combination of solutions is again a solution (<u>superposition principle</u> or <u>linearity principle</u>). A <u>basis</u> of solutions of Eq. (2.69) consists of n linearly independent solutions of $y_1, ..., y_n$. The corresponding general solution is a linear combination:

$$\boxed{y(x) = c_1 y_1(x) + ... + c_n y_n(x)} \quad (c_1, ..., c_n \text{ are arbitrary constants}). \quad (2.70)$$

From Eq. (2.70) one obtains a particular solution by choosing n numbers for $c_1, ...c_n$ by imposing n initial conditions:

$$\boxed{y(x_0) = K_0, \; y'(x_0) = K_1, ..., y^{(n-1)}(x_0) = K_{n-1}} \quad (x_0, K_0, ...K_{n-1} \text{ are given}). \quad (2.71)$$

Together Eq. (2.69) and Eq. (2.71) constitute an initial value problem for Eq. (2.69). If $p_0, ..., p_{n-1}$ are continuous on some open interval **I** and x_0 is in **I**, then Eq. (2.69) has a general solution on **I**, and Eq. (2.69) and Eq. (2.71) have a unique solution on **I** (which is a particular solution).

Similarly, as in Section 2.2.4, the basis functions $y_1, ..., y_n$ are linearly independent on the interval **I** if and only if the Wronskian (extended to nth order), as defined by

$$W(y_1, ..., y_n) = \begin{vmatrix} y_1 & y_2 & \cdots & y_n \\ y_1' & y_2' & \cdots & y_n' \\ \cdot & \cdot & \cdots & \cdot \\ y_1^{(n-1)} & y_2^{(n-1)} & \cdots & y_n^{(n-1)} \end{vmatrix}, \quad (2.72)$$

is different from zero on the interval.

Homogeneous equations with constant coefficients

In the case of constant coefficients, the nth-order homogeneous equation becomes

$$\boxed{y^{(n)} + a_{n-1} y^{(n-1)} + ... + a y' + a_o y = 0.} \quad (2.73)$$

Similarly, substituting $y = e^{\lambda x}$ and its derivatives into Eq. (2.73), one obtains the characteristic equation

$$\boxed{\lambda^n + a_{n-1}\lambda^{n-1} + ... + a_1\lambda + a_o = 0,} \quad (2.74)$$

which has roots $\lambda_1, \lambda_2...\lambda_n$. Four cases must be considered:

Case I: Roots all real and distinct. Then the n solutions are

$$\boxed{y_1 = e^{\lambda_1 x}, ..., y_n = e^{\lambda_n x}.} \quad (2.75)$$

These functions constitute a basis, and the general solution of Eq. (2.73) is

$$\boxed{y = c_1 e^{\lambda_1 x} + ... + c_n e^{\lambda_n x}.} \quad (2.76)$$

Case II: Complex simple roots. Complex roots occur in conjugate pairs since the coefficients of Eq. (2.73) are real, that is, $\lambda = \gamma + i\omega$ and $\bar{\lambda} = \gamma - i\omega$. Then a solution corresponding to the roots λ and $\bar{\lambda}$ is

$$y = e^{\gamma x}(A_1 \cos \omega x + B_1 \sin \omega x). \qquad (2.77)$$

Case III: Real multiple roots. If λ_1 is a root of order (or multiplicity) m, then a solution corresponding to the m linearly independent solutions is

$$y = (c_1 + c_2 x + c_3 x^2 + ... + c_m x^{m-1})e^{\lambda_1 x}. \qquad (2.78)$$

Case IV: Complex multiple roots. Given the complex conjugate pair $\lambda = \gamma + i\omega$ and $\bar{\lambda} = \gamma - i\omega$, a solution corresponding to the roots λ and $\bar{\lambda}$ of order m is

$$\begin{aligned} y =&(A_1 \cos \omega x + B_1 \sin \omega x)e^{\gamma x} + (A_2 \cos \omega x + B_2 \sin \omega x)x e^{\gamma x} + ... \\ &+ (A_m \cos \omega x + B_m \sin \omega x)x^{m-1}e^{\gamma x}. \end{aligned} \qquad (2.79)$$

Example 2.3.1. (*Case I*) Solve $y''' - 6y'' + 11y' - 6y = 0$.

Solution. The roots of the characteristic equation $\lambda^3 - 6\lambda^2 + 11\lambda - 6 = (\lambda - 1)(\lambda - 2)(\lambda - 3) = 0$ are $\lambda_1 = 1$, $\lambda_2 = 2$, and $\lambda_3 = 3$. Using Eq. (2.76), the general solution is $y = c_1 e^x + c_2 e^{2x} + c_3 e^{3x}$. [answer]

Example 2.3.2. (*Case II*) Solve $y''' - 4y'' + 8y' = 0$.

Solution. The characteristic equation is $\lambda^3 - 4\lambda^2 + 8\lambda = 0$ or $\lambda(\lambda^2 - 4\lambda + 8) = 0$, with roots $\lambda_1 = 0$, $\lambda_2 = 2 + 2i$, and $\bar{\lambda}_2 = 2 - 2i$. The general solution given by Eq. (2.76) and Eq. (2.77) is $y = c_1 + e^{2x}(A_1 \cos 2x + B_1 \sin 2x)$. [answer]

Example 2.3.3. (*Case III*) Solve $y^V + 6y^{IV} + 12y''' + 8y'' = 0$.

Solution. The characteristic equation is $\lambda^5 + 6\lambda^4 + 12\lambda^3 + 8\lambda^2 = 0$ or $\lambda^2(\lambda + 2)^3 = 0$. This equation has roots $\lambda_1 = \lambda_2 = 0$ and $\lambda_3 = \lambda_4 = \lambda_5 = -2$. Therefore by Eq. (2.78), the general solution is $y = c_1 + c_2 x + (c_3 + c_4 x + c_5 x^2)e^{-2x}$. [answer]

Example 2.3.4. (*Case IV*) Solve $y^{(7)} - 7y^{(6)} + 29y^{(5)} - 75y^{(4)} + 131y''' - 149y'' + 95y' - 25y = 0$.

Solution. The characteristic equation is $\lambda^{(7)} - 7\lambda^{(6)} + 29\lambda^{(5)} - 75\lambda^{(4)} + 131\lambda^3 - 149\lambda^2 + 95\lambda - 25 = 0$ or $(\lambda - 1)^3(\lambda^2 - 2\lambda + 5)^2 = (\lambda - 1)^3[(\lambda - 1 + 2i)(\lambda - 1 - 2i)]^2 = 0$. This equation has a triple root 1 and double roots $1 + 2i$ and $1 - 2i$. Hence by Eq. (2.78) and Eq. (2.79), the general solution is $y = (c_1 + c_2 x + c_3 x^2)e^x + (A_1 \cos 2x + B_1 \sin 2x)e^x + x(A_2 \cos 2x + B_2 \sin 2x)e^x$. [answer]

Nonhomogeneous equations

For the nth-order nonhomogeneous differential equation given by Eq. (2.68), the nth derivative is $y^{(n)} = \dfrac{d^n y}{dx^n}$ and $r(x)$ is continuous on **I**. Eq. (2.68) has no singular solutions and a general solution is of the form

$$y(x) = y_h(x) + y_p(x), \tag{2.80}$$

where y_h is a general solution of the nth-order homogeneous linear differential equation in Eq. (2.69) and y_p is any particular solution of Eq. (2.68), which can be determined as follows.

Method of undetermined coefficients

Similar to Section 2.2.5 for second-order differential equations, this method gives y_p for the constant coefficient equation

$$y^{(n)} + a_{n-1}y^{(n-1)} + ... + ay' + a_o y = r(x). \tag{2.81}$$

In fact, the only difference from Section 2.2.5 concerns the modification rule since now the characteristic equation of the present homogeneous equation

$$y^{(n)} + a_{n-1}y^{(n-1)} + ... + ay' + a_o y = 0 \tag{2.82}$$

can have multiple roots (that is, greater than simple or double roots). The basic rules are summarized as follows:

(a) Basic rule as in Section 2.2.5 with Table 2.4.
(b) Modification rule. If the choice for y_p is also a solution of the homogeneous Eq. (2.82), then multiply $y_p(x)$ by x^k, where k is the smallest positive integer such that no term of $x^k y_p(x)$ is a solution of Eq. (2.82).
(c) Sum rule as in Section 2.2.5 with Table 2.4.

Example 2.3.5. (*rule b*) Solve $y''' + 6y'' + 12y' + 8y = 6e^{-2x}$.

Solution.
(i) The characteristic equation $\lambda^3 + 6\lambda^2 + 12\lambda + 8 = (\lambda + 2)^3 = 0$ has a triple root $\lambda = -2$. Hence by Section 2.3 (Case III), $y_h = c_1 e^{-2x} + c_2 x e^{-2x} + c_3 x^2 e^{-2x}$.
(ii) With $y_p = Ce^{-2x}$, one obtains $-8C + 24C - 24C + 8C = 6$ (that is, there is no solution). Therefore, by rule (b) choose $y_p = Cx^3 e^{-2x}$. As such, $y_p' = C(-2x^3 + 3x^2)e^{-2x}$, $y_p'' = C(4x^3 - 12x^2 + 6x)e^{-2x}$, and $y_p''' = C(-8x^3 + 36x^2 - 36x + 6)e^{-2x}$. Substituting these expressions into the differential equation gives $(-8x^3 + 36x^2 - 36x + 6)C + 6(4x^3 - 12x^2 + 6x)C + 12(-2x^3 + 3x^2)C + 8x^3 C = 6$. Simplifying, one obtains $C = 1$. Thus, the general solution is given by $y = y_h + y_p = (c_1 + c_2 x + c_3 x^2)e^{-2x} + x^3 e^{-2x}$. [answer]

Method of variation of parameters

This method can be applied to find a particular solution y_p of the nth-order nonhomogeneous linear differential equation in Eq. (2.68). The particular solution on some interval I is determined from

$$y_p(x) = y_1 \int \frac{W_1(x)}{W(x)} r(x) dx + y_2 \int \frac{W_2(x)}{W(x)} r(x) dx + \ldots + y_n \int \frac{W_n(x)}{W(x)} r(x) dx.$$

(2.83)

Here $y_1, \ldots y_n$ is a basis of solutions of the homogeneous equation in Eq. (2.69) on I, with the Wronskian W, and $W_j (j = 1, \ldots, n)$ is obtained from W by replacing the jth column of W by the column $[0\ 0 \ldots 0\ 1]$. For example, when $n = 2$, $W = \begin{vmatrix} y_1 & y_2 \\ y_1' & y_2' \end{vmatrix}$,

$$W_1 = \begin{vmatrix} 0 & y_2 \\ 1 & y_2' \end{vmatrix} = -y_2, \quad W_2 = \begin{vmatrix} y_1 & 0 \\ y_1' & 1 \end{vmatrix} = y_1.$$

Eq. (2.83) becomes identical with Eq. (2.64).

2.4 Systems of differential equations

Introduction of vectors and matrices

In the analysis of linear systems of differential equations, that is, of the form

$$\begin{aligned} y_1' &= a_{11} y_1 + a_{12} y_2 + \ldots + a_{1n} y_n, \\ y_2' &= a_{21} y_1 + a_{22} y_2 + \ldots + a_{2n} y_n, \\ &\qquad \ldots \\ y_n' &= a_{n1} y_1 + a_{n2} y_2 + \ldots + a_{nn} y_n, \end{aligned}$$

(2.84)

one typically appeals to the use of matrices and vectors. A brief review of matrices is given in this section. The coefficients of Eq. (2.84) form an $n \times n$ matrix:

$$A = [a_{jk}] = \begin{bmatrix} a_{11} & a_{12} & \ldots & a_{1n} \\ a_{21} & a_{22} & \ldots & a_{2n} \\ \vdots & & \ldots & \\ a_{n1} & a_{n2} & \ldots & a_{nn} \end{bmatrix}.$$

(2.85)

A column vector x with n components $x_1, \ldots, x_n$ is

$$\begin{bmatrix} x_1 \\ x_2 \\ \vdots \\ x_n \end{bmatrix}.$$

Similarly a <u>row vector</u> v is of the form

$$v = [v_1 \; v_2 \; ... \; v_n].$$

Calculation with matrices and vectors

Let $A = \begin{bmatrix} a_{11} & a_{12} \\ a_{21} & a_{22} \end{bmatrix}$ and $B = \begin{bmatrix} b_{11} & b_{12} \\ b_{21} & b_{22} \end{bmatrix}$, as well as $v = \begin{bmatrix} v_1 \\ v_2 \end{bmatrix}$ and $x = \begin{bmatrix} x_1 \\ x_2 \end{bmatrix}$.

Addition: $A + B = \begin{bmatrix} a_{11}+b_{11} & a_{12}+b_{12} \\ a_{21}+b_{21} & a_{22}+b_{22} \end{bmatrix}$ and $v + x = \begin{bmatrix} v_1 + x_1 \\ v_2 + x_2 \end{bmatrix}$.

Scalar multiplication: $kA = \begin{bmatrix} ka_{11} & ka_{12} \\ ka_{21} & ka_{22} \end{bmatrix}$ and $kv = \begin{bmatrix} kv_1 \\ kv_2 \end{bmatrix}$.

Matrix multiplication: The product $C = AB$ of two $n \times n$ matrices $A = [a_{jk}]$ and $B = [b_{jk}]$ is the $n \times n$ matrix $C = [c_{jk}]$ with entries $c_{jk} = \sum\limits_{m=1}^{n} a_{jm}x_{mk}$ with $j = 1, ...n$ and $k = 1, ...n$. For example, $AB = \begin{bmatrix} 2 & 1 \\ -3 & 0 \end{bmatrix}\begin{bmatrix} 5 & 3 \\ -2 & 1 \end{bmatrix} = \begin{bmatrix} 2 \cdot 5 + 1 \cdot (-2) & 2 \cdot 3 + 1 \cdot 1 \\ (-3) \cdot 5 + 0 \cdot (-2) & (-3) \cdot 3 + 0 \cdot 1 \end{bmatrix} = \begin{bmatrix} 8 & 7 \\ -15 & -9 \end{bmatrix}$. Note $AB \neq BA$ (matrix multiplication is noncommutative). Similarly, $v = Ax$, where A is an $n \times n$ matrix and x is a vector with n components. Therefore, v is a vector with n components: $v_j = \sum\limits_{m=1}^{n} a_{jm}x_m$, where $j = 1, ...n$. For example, $\begin{bmatrix} 2 & 1 \\ -3 & 0 \end{bmatrix}\begin{bmatrix} x_1 \\ x_2 \end{bmatrix} = \begin{bmatrix} 2x_1 + x_2 \\ -3x_1 \end{bmatrix}$.

Differentiation: The derivative of a matrix (vector) with variable components is obtained by differentiating each component. For example, if $y(t) = \begin{bmatrix} y_1(t) \\ y_2(t) \end{bmatrix} = \begin{bmatrix} \sinh 4t \\ \cos t \end{bmatrix}$, then $y'(t) = \begin{bmatrix} y_1'(t) \\ y_2'(t) \end{bmatrix} = \begin{bmatrix} 4\cosh 4t \\ -\sin t \end{bmatrix}$.

Transposition: If $A = \begin{bmatrix} a_{11} & a_{12} & a_{13} \\ a_{21} & a_{22} & a_{23} \\ a_{31} & a_{32} & a_{33} \end{bmatrix}$, then $A^T = \begin{bmatrix} a_{11} & a_{21} & a_{31} \\ a_{12} & a_{22} & a_{32} \\ a_{13} & a_{23} & a_{33} \end{bmatrix}$, and if $v = \begin{bmatrix} v_1 \\ v_2 \end{bmatrix}$, then $v^T = [v_1 \; v_2]$.

Using matrix multiplication and differentiation, Eq. (2.84) can be written as

$$y' = Ay, \text{ where } y' = \begin{bmatrix} y_1' \\ y_2' \\ \vdots \\ y_n' \end{bmatrix}, \quad A = \begin{bmatrix} a_{11} & a_{12} & ... & a_{1n} \\ a_{21} & a_{22} & ... & a_{2n} \\ & & ... & \\ a_{n1} & a_{n2} & ... & a_{nn} \end{bmatrix}, \text{ and } y = \begin{bmatrix} y_1 \\ y_2 \\ \vdots \\ y_n \end{bmatrix}.$$

Linear independence: We say that r vectors $v^{(1)}, ..., v^{(r)}$ with n components are linearly independent if $c_1 v^{(1)} + ... + c_r v^{(r)} = 0$ implies that all scalars $c_1, ..., c_r$ are zero (note that 0 denotes the zero vector).

Inverse of a matrix: $AA^{-1} = A^{-1}A = I$, where I is the unit matrix (that is, the main diagonals with unit entries and all other entries are zero). If A has an inverse, then it is called <u>nonsingular</u>; otherwise it is <u>singular</u>. For example (for $n = 2$),

$$A^{-1} = \frac{1}{\det A} \begin{bmatrix} a_{22} & -a_{12} \\ -a_{21} & a_{11} \end{bmatrix}, \text{ where the } \underline{\text{determinant}} \text{ of } A \text{ is } \det A = \begin{vmatrix} a_{11} & a_{12} \\ a_{21} & a_{22} \end{vmatrix} =$$

$a_{11}a_{22} - a_{12}a_{21}$.

Eigenvalues and eigenvectors: Consider the equation

$$\boxed{Ax = \lambda x,} \tag{2.86}$$

where A is a given matrix, λ is a scalar (real or complex) to be determined, and x is a vector to be determined. A scalar λ where $x \neq 0$ is called an <u>eigenvalue</u> of A, and this vector x is called an <u>eigenvector</u> of A. Eq. (2.86) can be equivalently written as

$$\boxed{(A - \lambda I)x = 0,} \tag{2.87}$$

which corresponds to n linear equations in the n unknowns $x_1, ..., x_n$. For example, for $n = 2$,

$$\begin{bmatrix} a_{11} - \lambda & a_{12} \\ a_{21} & a_{22} - \lambda \end{bmatrix} \begin{bmatrix} x_1 \\ x_2 \end{bmatrix} = \begin{bmatrix} 0 \\ 0 \end{bmatrix}, \tag{2.88}$$

or in components,

$$\begin{aligned} (a_{11} - \lambda)x_1 + a_{12}x_2 &= 0, \\ a_{21}x_1 + (a_{22} - \lambda)x_2 &= 0. \end{aligned} \tag{2.89}$$

For a solution to Eq. (2.87), $A - \lambda I$ must be singular, which results if and only if $\det(A - \lambda I)$ (that is, the <u>characteristic determinant</u> of A) is zero:

$$\begin{aligned} \det(A - \lambda I) = \begin{vmatrix} a_{11} - \lambda & a_{12} \\ a_{21} & a_{22} - \lambda \end{vmatrix} &= (a_{11} - \lambda)(a_{22} - \lambda) - a_{12}a_{21} \\ &= \lambda^2 - (a_{11} + a_{22})\lambda + a_{11}a_{22} - a_{12}a_{21} = 0. \end{aligned} \tag{2.90}$$

This quadratic equation is called the <u>characteristic equation</u> of A. Its solution are the eigenvalues λ_1 and λ_2. The process of solution is the following:

(i) Determine λ_1 and λ_2 from Eq. (2.90).
(ii) From Eq. (2.89) with $\lambda = \lambda_1$, determine the eigenvector $x^{(1)}$, and then with $\lambda = \lambda_2$, the second eigenvector $x^{(2)}$.

Note that if x is an eigenvector of A, then so is kx (for $k \neq 0$).

Example 2.4.1. Find the eigenvalues and eigenvectors of $A = \begin{bmatrix} -2 & 1 \\ 2 & -1 \end{bmatrix}$.

Solution.

(i) The characteristic equation is $\det(A - \lambda I) = \begin{vmatrix} -2 - \lambda & 1 \\ 2 & -1 - \lambda \end{vmatrix} = (2 + \lambda)(1 + \lambda) - 2 = \lambda(\lambda + 3) = 0$. This yields $\lambda_1 = 0$ and $\lambda_2 = -3$. For $\lambda = \lambda_1$ in the first equation of Eq. (2.89), $-2x_1 + x_2 = 0$, which gives $x_1 = 1$ and $x_2 = 2$. Also from the second equation, the result is $2x_1 - x_2 = 0$ (however, this latter equation gives the same result).

Hence, an eigenvector to $\lambda_1 = 0$ is $x^{(1)} = \begin{bmatrix} 1 \\ 2 \end{bmatrix}$. Similarly, $x^{(2)} = \begin{bmatrix} 1 \\ -1 \end{bmatrix}$ is an eigenvector to $\lambda_2 = -3$. [answer]

2.4.1 Basic concepts and theory

Many problems of engineering lead to systems of differential equations. Of central interest are first-order systems,

$$
\begin{aligned}
y_1' &= f_1(t, y_1, ..., y_n), \\
y_2' &= f_2(t, y_1, ..., y_n), \\
&\quad \cdots \\
y_n' &= f_n(t, y_1, ..., y_n).
\end{aligned}
\tag{2.91}
$$

In addition, an nth-order differential equation

$$
y^{(n)} = F(t, y, y', ..., y^{(n-1)})
\tag{2.92}
$$

can always be reduced to a system of first-order differential equations by setting

$$
y_1 = y, \; y_2 = y', \; y_3 = y'', ..., y_n = y^{(n-1)}.
\tag{2.93}
$$

Therefore,

$$
y_1' = y_2, \; y_2' = y_3, ..., y_{(n-1)}' = y_n, \; \text{and, from Eq. (2.92), } y_n' = F(t, y_1, y_2...).
\tag{2.94}
$$

Extending the notion of a system of equations, Eq. (2.91) is a <u>linear system</u> if it can be written as

$$
\begin{aligned}
y_1' &= a_{11}(t)y_1 + ... + a_{1n}(t)y_n + g_1(t), \\
&\quad \cdots \\
y_n' &= a_{n1}(t)y_1 + ... + a_{nn}(t)y_n + g_n(t),
\end{aligned}
\tag{2.95}
$$

or in vector form,

$$
y' = Ay + g,
\tag{2.96}
$$

$$\text{where } A = \begin{bmatrix} a_{11} & a_{12} & \cdots & a_{1n} \\ a_{21} & a_{22} & \cdots & a_{2n} \\ & & \cdots & \\ a_{n1} & a_{n2} & \cdots & a_{nn} \end{bmatrix}, \; y = \begin{bmatrix} y_1 \\ y_2 \\ \vdots \\ y_n \end{bmatrix}, \text{ and } g = \begin{bmatrix} g_1 \\ g_2 \\ \vdots \\ g_n \end{bmatrix}.$$

If $g = 0$, the system is called <u>homogeneous</u>, that is,

$$y' = Ay; \tag{2.97}$$

otherwise it is nonhomogeneous.

The general theory of linear systems is similar to that of a single linear equation (see Section 2.2.1, Section 2.2.4, and Section 2.2.5). A <u>basis</u> of solutions of Eq. (2.96) on some interval **I** is a linearly independent set of n solutions $y^{(1)}, \ldots, y^{(n)}$. A <u>general solution</u> of Eq. (2.96) consists of the corresponding linear combination:

$$y = c_1 y^{(1)} + \ldots + c_n y^{(n)} \qquad (c_1, \ldots, c_n \text{ are arbitrary constants}). \tag{2.98}$$

The <u>Wronskian</u> of $y^{(1)}, \ldots, y^{(n)}$ is defined as

$$W(y^{(1)}, \ldots, y^{(n)}) = \begin{vmatrix} y_1^{(1)} & y_1^{(2)} & \cdots & y_1^{(n)} \\ y_2^{(1)} & y_2^{(2)} & \cdots & y_2^{(n)} \\ \cdot & \cdot & \cdots & \cdot \\ y_n^{(1)} & y_n^{(2)} & \cdots & y_n^{(n)} \end{vmatrix}. \tag{2.99}$$

The Wronskian of two solutions y and z of a single second-order equation is (see Eq. (2.56) of Section 2.2.4)

$$W(y, z) = \begin{vmatrix} y & z \\ y' & z' \end{vmatrix}. \tag{2.100}$$

Using Eq. (2.56) and writing the equation as a first-order system, $y_1 = y$, $y_2 = y'$, $z_1 = z$, and $z_2 = z'$, $W(y, z)$ takes the form of Eq. (2.99) with $n = 2$.

2.4.2 Homogeneous linear systems with constant coefficients

Consider the homogeneous linear system:

$$\boxed{y' = Ay,} \tag{2.101}$$

where A has constant $n \times n$ entries. Substituting into Eq. (2.101)

$$y = x e^{\lambda t} \tag{2.102}$$

and dividing by $e^{\lambda t}$ yields the eigenvalue problem

$$\boxed{Ax = \lambda x.}$$
(2.103)

If the constant matrix A in Eq. (2.101) has a linearly independent set of n eigenvectors $x^{(1)}, ..., x^{(n)}$, corresponding to eigenvalues $\lambda_1, ..., \lambda_n$, the general solution is

$$y = c_1 x^{(1)} e^{\lambda_1 t} + ... + c_n x^{(n)} e^{\lambda_n t}.$$
(2.104)

Example 2.4.2. Solve the system $\begin{array}{l} y_1' = -2y_1 + y_2 \\ y_2' = 2y_1 - y_2 \end{array}$ with initial conditions $y_1(0) = 3$ and $y_2(0) = 0$.

Solution.
From Example 2.4.1 for this system, the eigenvalues are $\lambda_1 = 0$ and $\lambda_2 = -3$. The corresponding eigenvectors are, respectively, $x^{(1)T} = [1 \quad 2]$ and $x^{(2)T} = [1 \quad -1]$.

Hence, the general solution is $y = c_1 y^{(1)} + c_2 y^{(2)} = c_1 \begin{bmatrix} 1 \\ 2 \end{bmatrix} + c_2 \begin{bmatrix} 1 \\ -1 \end{bmatrix} e^{-3t}.$

For the initial conditions $y(0)^T = [3 \quad 0]$, one obtains $y(0) = c_1 \begin{bmatrix} 1 \\ 2 \end{bmatrix} + c_2 \begin{bmatrix} 1 \\ -1 \end{bmatrix} = \begin{bmatrix} 3 \\ 0 \end{bmatrix}$. Thus, $\begin{array}{l} c_1 + c_2 = 3 \\ 2c_1 - c_2 = 0 \end{array}$, which yields $c_1 = 1$ and $c_2 = 2$.

Therefore, the particular solution is $y = \begin{bmatrix} 1 \\ 2 \end{bmatrix} + 2 \begin{bmatrix} 1 \\ -1 \end{bmatrix} e^{-3t}$ or $\begin{array}{l} y_1 = 1 + 2e^{-3t} \\ y_2 = 2 - 2e^{-3t} \end{array}.$

[answer]

Example 2.4.3. Consider the chemical kinetics problem in Fig. 2.3 with coupled forward and backward rate equations of first order for a closed system. In this process, there is desorption from a surface into the surrounding atmosphere with a back deposition onto the surface. The first-order rate constants are $k_1 = 2 \times 10^{-2}$ s^{-1} and $k_2 = 1 \times 10^{-2}$ s^{-1}. The inventories on the surface, $N_1(t)$ (mg), and in the surrounding atmosphere, $N_2(t)$ (mg), are obtained from the following mass balance equations for the system:

$$\frac{dN_1}{dt} = k_2 N_2 - k_1 N_1,$$

$$\frac{dN_2}{dt} = k_1 N_1 - k_2 N_2.$$

Initially, there is 300 mg on the surface and no inventory in the atmosphere, so that this system of equations is subject to the initial conditions $N_1(0) = 300$ mg and $N_2(0) = 0$ mg.

(a) Solve this homogeneous linear system for $N_1(t)$ and $N_2(t)$ (this system of equations is also solved by Laplace transform methods as in Problem 3.14).

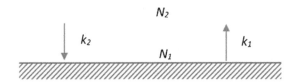

FIGURE 2.3

Diagram of the chemical kinetics system.

(b) What is the equilibrium value for N_1 and N_2? Explain the physical significance of this result.

Solution.

(a) The system can be written as $\begin{aligned} N_1' &= -.02N_1 + .01N_2 \\ N_2' &= .02N_1 - .01N_2 \end{aligned}$, with initial conditions

$N_1(0) = 300$ and $N_2(0) = 0$, or $N' = AN$ so that $\begin{bmatrix} \dot{N}_1 \\ \dot{N}_2 \end{bmatrix} = \begin{bmatrix} -.02 & .01 \\ .02 & -.01 \end{bmatrix} \begin{bmatrix} N_1 \\ N_2 \end{bmatrix}$ and

$N(0)^T = [300 \ \ 0]$.

The characteristic equation is $\det(A - \lambda I) = \begin{vmatrix} -.02 - \lambda & .01 \\ .02 & -.01 - \lambda \end{vmatrix} = (.02 +$

$\lambda)(.01 + \lambda) - (.02)(.01) = \lambda(\lambda + .03) = 0$. The eigenvalues are $\lambda_1 = 0$ and $\lambda_2 = -.03$. Eigenvectors are obtained from $-(.02 + \lambda)x_1 + .01x_2 = 0$. For $\lambda_1 = 0$, this substitution gives $-.02x_1 + .01x_2 = 0$, which implies $x_2 = 2x_1$. One can therefore take $x^{(1)T} = [1 \ \ 2]$. For $\lambda_2 = -.03$, this substitution gives $-(.02 - .03)x_1 + .01x_2 = 0$, which implies $x_1 = -x_2$, and $x^{(2)T} = [1 \ \ -1]$.

Hence, the general solution is

$$N = \begin{bmatrix} N_1 \\ N_2 \end{bmatrix} = c_1 N^{(1)} + c_2 N^{(2)} = c_1 \begin{bmatrix} 1 \\ 2 \end{bmatrix} + c_2 \begin{bmatrix} 1 \\ -1 \end{bmatrix} e^{-.03t}.$$

Using the initial conditions $N(0) = c_1 \begin{bmatrix} 1 \\ 2 \end{bmatrix} + c_2 \begin{bmatrix} 1 \\ -1 \end{bmatrix} = \begin{bmatrix} 300 \\ 0 \end{bmatrix}$, $\begin{aligned} c_1 + c_2 &= 300 \\ 2c_1 - c_2 &= 0 \end{aligned}$,

which yields $c_1 = 100$ and $c_2 = 200$.

Therefore, the particular solution is

$$N = 100 \begin{bmatrix} 1 \\ 2 \end{bmatrix} + 200 \begin{bmatrix} 1 \\ -1 \end{bmatrix} e^{-.03t} \quad \text{or} \quad \begin{aligned} N_1 &= 100 \left(1 + 2e^{-.03t}\right) \\ N_2 &= 200 \left(1 - e^{-.03t}\right) \end{aligned}. \quad \text{[answer]}$$

This physical problem is analogous to the system of equations solved in Example 2.4.2.

(b) At equilibrium as $t \to \infty$, $N_1 = 100$ and $N_2 = 200$. A double inventory results in the atmosphere compared to the surface because the rate constant $k_1 = 2k_2$ (twice as much material per unit time is leaving the surface than returning to it). [answer]

No basis of eigenvectors available

Suppose μ is a double eigenvalue of A (that is, $\det(A - \lambda I)$ has a factor $(\lambda - \mu)^2$), for which there is only one eigenvector x. Hence, only one solution is implied: $y^{(1)} = xe^{\mu t}$. A second solution of Eq. (2.101) can be obtained by substituting

$$y^{(2)} = xte^{\mu t} + ue^{\mu t} \tag{2.105}$$

into Eq. (2.101), yielding $y^{(2)'} = xe^{\mu t} + \mu xte^{\mu t} + \mu ue^{\mu t} = Ay^{(2)} = Axte^{\mu t} + Aue^{\mu t}$. Since $\mu x = Ax$, two terms cancel and dividing by $e^{\mu t}$ gives $x + \mu u = Au$. Thus, one obtains

$$(A - \mu I)u = x. \tag{2.106}$$

Example 2.4.4. Solve $y' = Ay = \begin{bmatrix} 3 & 1 \\ -1 & 1 \end{bmatrix} y$.

Solution.

(i) The characteristic equation is $\det(A - \lambda I) = \begin{vmatrix} 3 - \lambda & 1 \\ -1 & 1 - \lambda \end{vmatrix} = \lambda^2 - 4\lambda + 4 = (\lambda - 2)^2 = 0$. This equation has a double root $\lambda = 2$. Eigenvectors are obtained from $(3 - \lambda)x_1 + x_2 = 0$. For $\lambda = 2$, this implies $x_1 + x_2 = 0$. Thus, taking $x^{(1)T} = [-1 \quad 1]$, Eq. (2.106) gives $(A - 2I)u = \begin{bmatrix} 1 & 1 \\ -1 & -1 \end{bmatrix} u = \begin{bmatrix} -1 \\ 1 \end{bmatrix}$. Thus, $\begin{matrix} u_1 + u_2 = -1 \\ -u_1 - u_2 = 1 \end{matrix}$ and one can take $u^T = [-1 \quad 0]$. Therefore, the general solution is $y = c_1 y^{(1)} + c_2 y^{(2)} = c_1 \begin{bmatrix} -1 \\ 1 \end{bmatrix} e^{2t} + c_2 \left(\begin{bmatrix} -1 \\ 1 \end{bmatrix} t + \begin{bmatrix} -1 \\ 0 \end{bmatrix} \right) e^{2t}$. [answer]

If A has a triple eigenvalue μ and only a single linearly independent eigenvector x corresponding to it, one obtains a second solution Eq. (2.105) with u satisfying Eq. (2.106) and a third solution of the form

$$y^{(3)} = \frac{1}{2} xt^2 e^{\mu t} + ute^{\mu t} + ve^{\mu t}, \tag{2.107}$$

where v is determined from

$$(A - \mu I)v = u. \tag{2.108}$$

If A has a triple eigenvalue and two linearly independent eigenvectors $x^{(1)}$ and $x^{(2)}$, then three linearly independent solutions are

$$y^{(1)} = x^{(1)} e^{\mu t}, \, y^{(2)} = x^{(2)} e^{\mu t}, \, y^{(3)} = xte^{\mu t} + ue^{\mu t}, \tag{2.109}$$

where x is a linear combination of $x^{(1)}$ and $x^{(2)}$ such that

$$(A - \mu I)u = x \tag{2.110}$$

is solvable for u.

2.4.3 Nonhomogeneous linear systems

For the nonhomogeneous linear system

$$y' = Ay + g, \tag{2.111}$$

the general solution in which the $n \times n$ matrix $A(t)$ and $g(t)$ are continuous on some interval **J** is given by

$$y = y^{(h)} + y^{(p)}, \tag{2.112}$$

where $y^{(h)}(t)$ is a general solution of the homogeneous system in Eq. (2.101) and $y^{(p)}(t)$ is a particular solution of Eq. (2.111) (that is, a solution containing arbitrary constants). The particular solution is obtained as follows.

Method of undetermined coefficients

This method is applicable if the components of g are integer powers of t, exponential functions, or sine and cosine functions.

Example 2.4.5. Solve

$$y' = Ay + g = \begin{bmatrix} 1 & -4 \\ -2 & 3 \end{bmatrix} y + \begin{bmatrix} 5t^2 \\ 6t - 6 \end{bmatrix}. \tag{2.113}$$

Solution.
The form of g suggests $y^{(p)} = u + vt + wt^2$. Therefore, by substitution into Eq. (2.113), $y^{(p)'} = v + 2wt = Au + Avt + Awt^2 + g$. In components, one has

$$\begin{bmatrix} v_1 \\ v_2 \end{bmatrix} + \begin{bmatrix} 2w_1 \\ 2w_2 \end{bmatrix} t = \begin{bmatrix} u_1 - 4u_2 \\ -2u_1 + 3u_2 \end{bmatrix} + \begin{bmatrix} v_1 - 4v_2 \\ -2v_1 + 3v_2 \end{bmatrix} t + \begin{bmatrix} w_1 - 4w_2 \\ -2w_1 + 3w_2 \end{bmatrix} t^2 + \begin{bmatrix} 5t^2 \\ 6t - 6 \end{bmatrix}.$$

Thus:

- Equating t^2 terms: $0 = w_1 - 4w_2 + 5$, $0 = -2w_1 + 3w_2$. These relations yield $w_1 = 3$ and $w_2 = 2$.
- Equating t terms: $2w_1 = v_1 - 4v_2$, $2w_2 = -2v_1 + 3v_2 + 6$. These relations yield $v_1 = -2$ and $v_2 = -2$.
- Equating constant terms: $v_1 = u_1 - 4u_2$, $v_2 = -2u_1 + 3u_2 - 6$. These relations yield $u_1 = -2$ and $u_2 = 0$.

From these results and the general solution of the homogeneous system in Section 2.4.2, the general solution is

$$y = y^{(h)} + y^{(p)} = c_1 \begin{bmatrix} 2 \\ 1 \end{bmatrix} e^{-t} + c_2 \begin{bmatrix} 1 \\ -1 \end{bmatrix} e^{5t} + \begin{bmatrix} 3t^2 - 2t - 2 \\ 2t^2 - 2t \end{bmatrix}. \qquad \text{[answer]}$$

A modification is required if a term g involves $e^{\lambda t}$, where λ is an eigenvalue of A. Then instead of assuming $y^{(p)} = ue^{\lambda t}$, one lets

$$y^{(p)} = ute^{\lambda t} + ve^{\lambda t}. \qquad (2.114)$$

Example 2.4.6. Solve

$$y' = Ay + g = \begin{bmatrix} -2 & 1 \\ 2 & -1 \end{bmatrix} y + \begin{bmatrix} -1 \\ -2 \end{bmatrix} e^{-3t}. \qquad (2.115)$$

Solution.
From Section 2.4.2, the general solution of the homogeneous system is $y^{(h)} = c_1 \begin{bmatrix} 1 \\ 2 \end{bmatrix} + c_2 \begin{bmatrix} 1 \\ -1 \end{bmatrix} e^{-3t}$. Since $\lambda = -3$ is an eigenvalue of A, using Eq. (2.114), $y^{(p)} = ute^{-3t} + ve^{-3t}$, and substituting this equation into Eq. (2.115) yields $y^{(p)'} = ue^{-3t} - 3ute^{-3t} - 3ve^{-3t} = Aute^{-3t} + Ave^{-3t} + g$. Equating the te^{-2t} terms implies $-3u = Au$ (that is, u is an eigenvector of A with eigenvalue $\lambda = -3$). Therefore, $u^T = a[1 \ -1]$. Equating the other terms gives

$$u - 3v = Av + \begin{bmatrix} -1 \\ -2 \end{bmatrix} \text{ or } (A + 3I)v = u - \begin{bmatrix} -1 \\ -2 \end{bmatrix} = a \begin{bmatrix} 1 \\ -1 \end{bmatrix} + \begin{bmatrix} 1 \\ 2 \end{bmatrix}.$$

Alternatively, in components, $\begin{matrix} v_1 + v_2 = a + 1 \\ 2v_1 + 2v_2 = -a + 2 \end{matrix}$, implying $a = 0$. Therefore, $v_2 = 1 - v_1$. One can choose $v^T = [0 \ 1]$. The general solution is

$$y = c_1 \begin{bmatrix} 1 \\ 2 \end{bmatrix} + c_2 \begin{bmatrix} 1 \\ -1 \end{bmatrix} e^{-3t} + \begin{bmatrix} 0 \\ 1 \end{bmatrix} e^{-3t}. \qquad \text{[answer]}$$

Method of variation of parameters
This method is applicable to the nonhomogeneous linear system

$$y' = A(t)y + g(t), \qquad (2.116)$$

with variable $A = A(t)$ and general $g(t)$. The general solution of the homogeneous system

$$y^{(h)} = c_1 y_1^{(1)} + \cdots + c_n y_1^{(n)} \qquad (2.117)$$

can be written in components

$$y^{(h)} = \begin{bmatrix} c_1 y_1^{(1)} + \cdots + c_n y_1^{(n)} \\ \vdots \\ c_1 y_n^{(1)} + \cdots + c_n y_n^{(n)} \end{bmatrix} = \begin{bmatrix} y_1^{(1)} \cdots y_1^{(n)} \\ \vdots \\ y_n^{(1)} \cdots y_n^{(n)} \end{bmatrix} \begin{bmatrix} c_1 \\ \vdots \\ c_n \end{bmatrix} = Y(t)c, \qquad (2.118)$$

where $Y(t)$ is the matrix with the columns $y^{(1)}, ..., y^{(n)}$, the vectors of the basis in Eq. (2.117), and $c^{(T)} = [c_1...c_n]$ is constant. Thus Y obeys the homogeneous matrix equation

$$Y' = AY. \tag{2.119}$$

Letting

$$y^{(p)} = Y(t)u(t) \tag{2.120}$$

and substituting this equation into Eq. (2.116) yields

$$Y'u + Yu' = AYu + g. \tag{2.121}$$

Thus, using Eq. (2.119) in Eq. (2.121) gives

$$Yu' = g \text{ or } u' = Y^{-1}g. \tag{2.122}$$

Integrating Eq. (2.122)

$$u(t) = \int_{t_o}^{t} Y^{-1}(\tilde{t})g(\tilde{t})d\tilde{t} + C. \tag{2.123}$$

Equivalently, the general solution is

$$y = Yu = YC + Y\int_{t_o}^{t} Y^{-1}(\tilde{t})g(\tilde{t})d\tilde{t}. \tag{2.124}$$

The particular solution $y^{(p)}$ is obtained by setting $C = 0$, that is,

$$\boxed{y^{(p)} = Y\int_{t_o}^{t} Y^{-1}(\tilde{t})g(\tilde{t})d\tilde{t}.} \tag{2.125}$$

Example 2.4.7. For the system in Example 2.4.6,

$$y^{(h)} = c_1\begin{bmatrix}1\\2\end{bmatrix} + c_2\begin{bmatrix}e^{-3t}\\-e^{-3t}\end{bmatrix}. \tag{2.126}$$

Solution.

We have, $Y = [y^{(1)} \quad y^{(2)}] = \begin{bmatrix}1 & e^{-3t}\\2 & -e^{-3t}\end{bmatrix}$ and $g = \begin{bmatrix}-1\\-2\end{bmatrix}e^{-3t}$.

From Section 2.4, $Y^{-1} = \dfrac{1}{\det Y}\begin{bmatrix}y_{22} & -y_{12}\\-y_{21} & y_{11}\end{bmatrix} = \dfrac{1}{-3e^{-3t}}\begin{bmatrix}-e^{-3t} & -e^{-3t}\\-2 & 1\end{bmatrix} =$

$\dfrac{1}{3}\begin{bmatrix}1 & 1\\2e^{3t} & -e^{3t}\end{bmatrix}$. Therefore, $Y^{-1}g = \dfrac{1}{3}\begin{bmatrix}1 & 1\\2e^{3t} & -e^{3t}\end{bmatrix}\begin{bmatrix}-e^{-3t}\\-2e^{-3t}\end{bmatrix} = \begin{bmatrix}-e^{-3t}\\0\end{bmatrix}$. Inte-

grating and choosing $C = 0$ yields $u(t) = \int_0^t \begin{bmatrix}-e^{-3\tilde{t}}\\0\end{bmatrix}d\tilde{t} = \begin{bmatrix}\frac{1}{3}(e^{-3t} - 1)\\0\end{bmatrix}$.

Thus,

$$y^{(p)} = Yu = \begin{bmatrix} 1 & e^{-3t} \\ 2 & -e^{-3t} \end{bmatrix} \begin{bmatrix} \frac{1}{3}\left(e^{-3t} - 1\right) \\ 0 \end{bmatrix} = \begin{bmatrix} \frac{1}{3}\left(e^{-3t} - 1\right) \\ \frac{2}{3}\left(e^{-3t} - 1\right) \end{bmatrix}$$

$$= -\frac{1}{3}\begin{bmatrix} 1 \\ 2 \end{bmatrix} + \frac{1}{3}\begin{bmatrix} 1 \\ 2 \end{bmatrix}e^{-3t}.$$

The last term can be rewritten as $\frac{1}{3}\begin{bmatrix} 1 \\ 2 \end{bmatrix}e^{-3t} = \left(\frac{1}{3}\begin{bmatrix} 1 \\ -1 \end{bmatrix} + \begin{bmatrix} 0 \\ 1 \end{bmatrix}\right)e^{-3t}$. Hence, the general solution is

$$y = y^{(h)} + y^{(p)} = c_1'\begin{bmatrix} 1 \\ 2 \end{bmatrix} + c_2'\begin{bmatrix} 1 \\ -1 \end{bmatrix}e^{-3t} + \begin{bmatrix} 0 \\ 1 \end{bmatrix}e^{-3t}. \qquad \text{[answer]}$$

Here the constants have been incorporated into the homogeneous solution of Example 2.4.2, such that $c_1' = c_1 - 1/3$ and $c_2' = c_2 + 1/3$. This general solution is in agreement with Example 2.4.6.

Method of diagonalization

This method can be applied to the system

$$y' = Ay + g(t), \qquad (2.127)$$

for which A has a basis of vectors $x^{(1)}, ..., x^{(n)}$. It can be shown that then

$$D = X^{-1}AX \qquad (2.128)$$

is a diagonal matrix with eigenvalues $\lambda^{(1)}, ..., \lambda^{(n)}$ of A on the main diagonal. Therefore, X is an $n \times n$ matrix with columns $x^{(1)}, ..., x^{(n)}$. Note that X^{-1} exists because the columns are linearly independent. Thus, for the eigenvectors in Example 2.4.6,

$$X = [x^{(1)}x^{(2)}] = \begin{bmatrix} 1 & 1 \\ 2 & -1 \end{bmatrix} \text{ and } D = \begin{bmatrix} \frac{1}{3} & \frac{1}{3} \\ \frac{2}{3} & -\frac{1}{3} \end{bmatrix}\begin{bmatrix} -2 & 1 \\ 2 & -1 \end{bmatrix}\begin{bmatrix} 1 & 1 \\ 2 & -1 \end{bmatrix} = \begin{bmatrix} 0 & 0 \\ 0 & -3 \end{bmatrix}.$$

Define $z = X^{-1}y$ and $y = Xz$, and substituting this into Eq. (2.127) (where X is a constant), one obtains

$$Xz' = AXz + g. \qquad (2.129)$$

Multiplying Eq. (2.129) on the left by X^{-1} gives

$$z' = X^{-1}AXz + h, \qquad (2.130)$$

where $h = X^{-1}g$. Using Eq. (2.128), one obtains $z' = Dz + h$ or in components $z_j' = \lambda_j z_j + h_j$, where $j = 1, ..., n$. Each of these "decoupled" n linear equations can be solved as (see Section 2.1)

$$z_j = c_j e^{\lambda_j t} + e^{\lambda_j t} \int e^{-\lambda_j t} h_j(t) \, dt. \tag{2.131}$$

These quantities are the components of $z(t)$, and therefore $y = Xz$.

Example 2.4.8. For Example 2.4.6, one has $h = X^{-1}g = \dfrac{1}{3} \begin{bmatrix} 1 & 1 \\ 2 & -1 \end{bmatrix} \begin{bmatrix} -e^{-3t} \\ -2e^{-3t} \end{bmatrix} =$

$\dfrac{1}{3} \begin{bmatrix} -3e^{-3t} \\ 0 \end{bmatrix} = \begin{bmatrix} -e^{-3t} \\ 0 \end{bmatrix}$. Since the eigenvalues are $\lambda_1 = 0$ and $\lambda_2 = -3$, the diag-

onalized system is $z' = \begin{bmatrix} 0 & 0 \\ 0 & -3 \end{bmatrix} z + h$. Thus, $\begin{matrix} z_1' = -e^{-3t} \\ z_2' = -3z_2 \end{matrix}$. From Eq. (2.131), the

solutions are $z_1 = c_1 + \dfrac{e^{-3t}}{3}$ and $z_2 = c_2 e^{-3t}$. The general solution is

$$y = Xz = \begin{bmatrix} 1 & 1 \\ 2 & -1 \end{bmatrix} \begin{bmatrix} c_1 + \frac{e^{-3t}}{3} \\ c_2 e^{-3t} \end{bmatrix} = \begin{bmatrix} c_1 + \frac{e^{-3t}}{3} + c_2 e^{-3t} \\ 2c_1 + \frac{2}{3}e^{-3t} - c_2 e^{-3t} \end{bmatrix}.$$

Hence, $y = c_1 \begin{bmatrix} 1 \\ 2 \end{bmatrix} + c_2 \begin{bmatrix} 1 \\ -1 \end{bmatrix} e^{-3t} + \dfrac{1}{3} \begin{bmatrix} 1 \\ 2 \end{bmatrix} e^{-3t}$. The last term can again be rewritten

as $\dfrac{1}{3} \begin{bmatrix} 1 \\ 2 \end{bmatrix} e^{-3t} = \left(\dfrac{1}{3} \begin{bmatrix} 1 \\ -1 \end{bmatrix} + \begin{bmatrix} 0 \\ 1 \end{bmatrix} \right) e^{-3t}$, so that the general solution is

$$y = c_1 \begin{bmatrix} 1 \\ 2 \end{bmatrix} + c_2 \begin{bmatrix} 1 \\ -1 \end{bmatrix} e^{-3t} + \begin{bmatrix} 0 \\ 1 \end{bmatrix} e^{-3t}. \qquad \text{[answer]}.$$

This result is again identical to Example 2.4.6 and Example 2.4.7.

2.5 Series solutions and special functions

The power series method is the standard method for solving linear differential equations with variable coefficients as

$$y'' + p(x)y' + q(x)y = 0. \tag{2.132}$$

A power series is of the form

$$\sum_{m=0}^{\infty} a_m (x - x_0)^m = a_0 + a_1(x - x_0) + a_2(x - x_0)^2 + \dots, \tag{2.133}$$

where $a_0, a_1, a_2...$ are the constant <u>coefficients</u> of the series, x_0 is a constant (called the center of the series), and x is a variable. Letting $x_0 = 0$, consider a series to Eq. (2.133) of the form

$$y = \sum_{m=0}^{\infty} a_m x^m = a_0 + a_1 x + a_2 x^2 + a_3 x^3 ...$$ (2.134)

with

$$y' = \sum_{m=0}^{\infty} m a_m x^{m-1} = a_1 + 2a_2 x + 3a_3 x^2 + ...,$$ (2.135)

$$y'' = \sum_{m=0}^{\infty} m(m-1) a_m x^{m-2} = 2a_2 + 3 \cdot 2a_3 + 4 \cdot 3a_4 x^2 + ...$$ (2.136)

Example 2.5.1. Solve $y'' + y = 0$.

Solution. Letting $y = \sum_{n=0}^{\infty} a_n x^n$, the derivatives are $y' = \sum_{n=1}^{\infty} n a_n x^{n-1}$ and $y'' = \sum_{n=2}^{\infty} n(n-1) a_n x^{n-2}$. Substituting these expressions into the differential equation

yields $\sum_{n=2}^{\infty} n(n-1) a_n x^{n-2} + \sum_{n=0}^{\infty} a_n x^n = 0$. Relabeling so that both expressions have the same power of x (where $n \to n + 2$) gives

$$\sum_{n=0}^{\infty} [(n+2)(n+1) a_{n+2} + a_n] x^n = 0.$$

Equating each coefficient to zero gives

$(n=0)$ $2 \cdot 1 a_2 = -a_0 \Rightarrow a_2 = -\dfrac{a_0}{2!},$ $(n=1)$ $3 \cdot 2 a_3 = -a_1 \Rightarrow a_3 = -\dfrac{a_1}{3!},$

$(n=2)$ $4 \cdot 3 a_4 = -a_2 \Rightarrow a_4 = \dfrac{a_0}{4!},$ $(n=3)$ $5 \cdot 4 a_5 = -a_3 \Rightarrow a_5 = \dfrac{a_1}{5!},$

$(n=4)$ $6 \cdot 5 a_6 = -a_4 \Rightarrow a_6 = -\dfrac{a_0}{6!},$ $(n=5)$ $7 \cdot 6 a_5 = -a_3 \Rightarrow a_7 = -\dfrac{a_1}{7!}.$

Thus, $a_{2k} = \dfrac{(-1)^k}{(2k)!} a_0$ and $a_{2k+1} = \dfrac{(-1)^k}{(2k+1)!} a_1$. The solution is therefore

$$y(x) = a_0 \sum_{k=0}^{\infty} \frac{(-1)^k}{(2k)!} x^{2k} + a_1 \sum_{k=0}^{\infty} \frac{(-1)^k}{(2k+1)!} x^{2k+1}$$

$$= a_0 \cos x + a_1 \sin x.$$

The analytical result is also easily obtained from Section 2.2.2.

2.5.1 Legendre equation

Consider Legendre's differential equation:

$$(1 - x^2)y'' - 2xy' + n(n+1)y = 0. \qquad (2.137)$$

Assuming a power series solution

$$y = \sum_{m=0}^{\infty} a_m x^m \qquad (2.138)$$

and substituting this equation and its derivative into Eq. (2.137) yields

$$y = (1 - x^2) \sum_{m=2}^{\infty} m(m-1)a_m x^{m-2} - 2x \sum_{m=1}^{\infty} m a_m x^{m-1} + k \sum_{m=0}^{\infty} a_m x^m = 0, \quad (2.139)$$

where $k = n(n+1)$. Thus, expanding

$$y = \sum_{m=2}^{\infty} m(m-1)a_m x^{m-2} - \sum_{m=2}^{\infty} m(m-1)a_m x^m - 2 \sum_{m=1}^{\infty} m a_m x^m + k \sum_{m=0}^{\infty} a_m x^m = 0,$$
$$(2.140)$$

or written out,

$$
\begin{array}{llll}
2 \cdot 1 a_2 & +3 \cdot 2 a_3 x & +4 \cdot 3 a_4 x^2 + \cdots & +(s+2)(s+1)a_{s+2} x^s + \cdots \\
& & -2 \cdot 1 a_2 x^2 - \cdots & -s(s-1)a_s x^s - \cdots \\
& -2 \cdot 1 a_1 x & -2 \cdot 2 a_2 x^2 - \cdots & -2s a_s x^s - \cdots \\
k a_0 & +k a_1 x & +k a_2 x^2 + \cdots & +k a_s x^s + \cdots \quad = 0.
\end{array}
$$

Equating the coefficients of each power of x to zero,

$$x^0 : 2a_2 + n(n+1)a_0 = 0,$$
$$x^1 : 6a_3 + [-2 + n(n+1)]a_1 = 0,$$
$$x^s (s = 2, 3...) : (s+2)(s+1)a_{s+2} + \underbrace{[-s(s-1) - 2s + n(n+1)]}_{= (n-s)(n+s+1)} a_s = 0.$$

Therefore,

$$a_{s+2} = -\frac{(n-s)(n+s+1)}{(s+2)(s+1)} a_s \qquad (s = 0, 1, ...). \qquad (2.141)$$

From the <u>recurrence relation</u> in Eq. (2.141) one obtains

$$a_2 = -\frac{n(n+1)}{2!}a_0, \qquad a_3 = -\frac{(n-1)(n+2)}{3!}a_1,$$

$$a_4 = -\frac{(n-2)(n+3)}{4 \cdot 3}a_2, \qquad a_5 = -\frac{(n-3)(n+4)}{5 \cdot 4}a_3,$$

$$= \frac{(n-2)n(n+1)(n+3)}{4!}a_0, \qquad = \frac{(n-3)(n-1)(n+2)(n+4)}{5!}a_1.$$

Inserting the values for the coefficients into Eq. (2.138),

$$y(x) = a_0 y_1(x) + a_1 y_2(x), \tag{2.142}$$

where

$$y_1(x) = 1 - \frac{n(n+1)}{2!}x^2 + \frac{(n-2)n(n+1)(n+3)}{4!}x^4 - +\ldots,$$

$$y_2(x) = x - \frac{(n-1)(n+2)}{3!}x^3 + \frac{(n-3)(n-1)(n+2)(n+4)}{5!}x^5 - +\ldots.$$

Hence, Eq. (2.142) is a general solution of Eq. (2.137) for $-1 < x < 1$.

Legendre polynomials

If n in Eq. (2.137) is a nonnegative integer, then the right-hand side of Eq. (2.141) is zero when $s = n$, that is, $a_{n+2} = a_{n+4} = a_{n+6} = \ldots = 0$. If n is even, $y_1(x)$ reduces to a polynomial of degree n. If n is odd, the same is true for $y_2(x)$. These polynomials multiplied by some constants are called <u>Legendre polynomials</u>.

From Eq. (2.141), $a_s = -\frac{(s+2)(s+1)}{(n-s)(n+s+1)}a_{s+2}$ $(s \leqslant n-2)$ so that $a_{n-2} = -\frac{n(n-1)}{2(2n-1)}a_n$. Choosing $a_n = \frac{(2n)!}{2^n(n!)^2}$ so that the polynomial will have a value of unity when $x = 1$ gives

$$a_{n-2} = -\frac{n(n-1)(2n)!}{2(2n-1)2^n(n!)^2}$$

$$= -\frac{\cancel{n}(\cancel{n}-1)2n(2\cancel{n}-1)(2n-2)!}{\cancel{2}(2\cancel{n}-1)2^n \underbrace{\cancel{n}(n-1)!\cancel{n}(\cancel{n}-1)(n-2)!}_{n!}} = -\frac{(2n-2)!}{2^n(n-1)!(n-2)!}.$$

Similarly, $a_{n-4} = -\frac{(n-2)(n-3)}{4(2n-3)}a_{n-2} = \frac{(2n-4)!}{2^n 2!(n-2)!(n-4)!}$. In general, when $n - 2m \geqslant 0$,

$$a_{n-2m} = (-1)^m \frac{(2n-2m)!}{2^n m!(n-m)!(n-2m)!}. \tag{2.143}$$

From Eq. (2.143), the solution of Eq. (2.137) is the Legendre polynomial of degree n:

$$P_n(x) = \sum_{m=0}^{M}(-1)^m \frac{(2n-2m)!}{2^n m!(n-m)!(n-2m)!}x^{n-2m},$$ (2.144)

where $M = \dfrac{n}{2}$ or $\dfrac{n-1}{2}$ (whichever is an integer). For example,

$$P_0(x) = 1, \qquad\qquad P_1(x) = x,$$
$$P_2(x) = \frac{1}{2}(3x^2 - 1), \quad P_3(x) = \frac{1}{2}(5x^3 - 3x).$$

2.5.2 Frobenius method

The differential equation

$$y'' + \frac{b(x)}{x}y' + \frac{c(x)}{x^2}y = 0,$$ (2.145)

where $b(x)$ and $c(x)$ are analytic at $x = 0$, can be solved by a Frobenius (or extended power series) method. One solution exists for Eq. (2.145) of the form

$$y = x^r \sum_{m=0}^{\infty} a_m x^m = x^r \left(a_0 + a_1 x + a_2 x^2 + a_3 x^3 \ldots\right),$$ (2.146)

where r may be real or complex (and is chosen so that $a_0 \neq 0$). To solve Eq. (2.145), this equation can be rewritten as

$$x^2 y'' + xb(x)y' + c(x)y = 0,$$ (2.147)

where $b(x)$ and $c(x)$ are expanded in a power series: $b(x) = b_0 + b_1 x + b_2 x^2 + \ldots$ and $c(x) = c_0 + c_1 x + c_2 x^2 + \ldots$. Inserting the derivatives of Eq. (2.146),

$$y'(x) = \sum_{m=0}^{\infty}(m+r)a_m x^{m+r-1} = x^{r-1}[ra_0 + (r+1)a_1 x + \ldots],$$

$$y''(x) = \sum_{m=0}^{\infty}(m+r)(m+r-1)a_m x^{m+r-2}$$
$$= x^{r-2}[r(r-1)a_0 + (r+1)ra_1 x + \ldots],$$

into Eq. (2.147) yields

$$x^r[r(r-1)a_0 + \ldots] + (b_0 + b_1 x + \ldots)x^r(ra_0 + \ldots)$$
$$+ (c_0 + c_1 x + \ldots)x^r(a_0 + a_1 x + \ldots) = 0.$$ (2.148)

Equating the coefficients of the smallest power of x (x^r) to zero yields the quadratic indicial equation:

$$r(r-1) + b_0 r + c_0 = 0. \qquad (2.149)$$

Thus, one of the two solutions will always be in the form of Eq. (2.146), where r is a root of Eq. (2.149). The form of the other solution depends on the root yielding three possible cases:

Case I: Distinct roots not differing from an integer;

$$y_1(x) = x^{r_1}\left(a_0 + a_1 x + a_2 x^2 + \ldots\right), \qquad (2.150a)$$

$$y_2(x) = x^{r_2}\left(A_0 + A_1 x + A_2 x^2 + \ldots\right), \qquad (2.150b)$$

where the coefficients are obtained successively from Eq. (2.148) with $r = r_1$ and $r = r_2$.

Case II: Double root $r_1 = r_2 = r$;

$$y_1(x) = x^r\left(a_0 + a_1 x + a_2 x^2 + \ldots\right) \qquad \left[r = \frac{1}{2}(1 - b_0)\right], \qquad (2.151a)$$

$$y_2(x) = y_1(x)\ln x + x^r\left(A_1 x + A_2 x^2 + \ldots\right) \qquad [x > 0], \qquad (2.151b)$$

Case III: Roots differing by an integer;

$$y_1(x) = x^{r_1}\left(a_0 + a_1 x + a_2 x^2 + \ldots\right), \qquad (2.152a)$$

$$y_2(x) = k y_1(x)\ln x + x^{r_2}\left(A_0 + A_1 x + A_2 x^2 + \ldots\right), \qquad (2.152b)$$

where $r_1 - r_2 > 0$ and k may be zero.

Example 2.5.2. (*Case I*) Solve $2xy'' + y' + y = 0$.

Solution. Substitute Eq. (2.146) and its derivatives into this differential equation, obtaining

$$2\sum_{m=0}^{\infty}(m+r)(m+r-1)a_m x^{m+r-1} + \sum_{m=0}^{\infty}(m+r)a_m x^{m+r-1} + \sum_{m=0}^{\infty}a_m x^{m+r} = 0.$$

$$\qquad (2.153)$$

Equating the sum of the coefficients of x^{r-1} (that is, the smallest power of x) to zero yields the indicial equation $2r(r-1) + r = 0$, which implies that $2r^2 - r = 0$. Hence, $r_1 = \frac{1}{2}$ and $r_2 = 0$. Equating the sum of the coefficients of x^{r+s} to zero (that is, taking $m + r - 1 = r + s$, which implies $m = s + 1$ in the first two series and $m = s$

in the last series), $2(s+r+1)(s+r)a_{s+1}+(s+r+1)a_{s+1}+a_s=0$, or simplifying, $(s+r+1)(2s+2r+1)a_{s+1}+a_s=0$. This latter relation implies

$$a_{s+1} = -\frac{a_s}{(s+r+1)(2s+2r+1)} \qquad (s=0,1,\dots). \qquad (2.154)$$

(i) <u>First solution:</u> For $r=\frac{1}{2}$, Eq. (2.154) gives $a_{s+1} = -\dfrac{a_s}{(2s+3)(s+1)}$ and thus $a_1 = -\dfrac{a_0}{3\cdot1}$, $a_2 = -\dfrac{a_1}{5\cdot2}$, $a_3 = -\dfrac{a_2}{7\cdot3}$, or by successive substitution, $a_1 = -\dfrac{a_0}{3\cdot1}$, $a_2 = \dfrac{a_0}{5\cdot3\cdot2\cdot1}$, $a_3 = -\dfrac{a_0}{7\cdot5\cdot3\cdot3\cdot2\cdot1}$, or in general (with $a_0=1$), $a_m = \dfrac{(-1)^m 2^m}{(2m+1)!}$ $(m=0,1,\dots)$. Therefore,

$$y_1(x) = x^{1/2}\sum_{m=0}^{\infty}\frac{(-1)^m 2^m}{(2m+1)!}x^m = \frac{1}{\sqrt{2}}\sum_{m=0}^{\infty}\frac{(-1)^m (2x)^{(2m+1)/2}}{(2m+1)!} = \frac{1}{\sqrt{2}}\sin(\sqrt{2x}).$$

[answer]

(ii) <u>Second solution:</u> For $r=0$, Eq. (2.154) gives $A_{s+1} = -\dfrac{A_s}{(s+1)(2s+1)}$ $(s=0,1,\dots)$ so that $A_1 = -\dfrac{A_0}{1\cdot1}$, $A_2 = -\dfrac{A_1}{2\cdot3}$, $A_3 = -\dfrac{A_2}{3\cdot5}$, or $A_1 = -\dfrac{A_0}{1\cdot1}$, $A_2 = \dfrac{A_0}{1\cdot2\cdot3}$, $A_3 = -\dfrac{A_0}{1\cdot2\cdot3\cdot3\cdot5}$. In general (with $A_0=1$), $A_m = \dfrac{(-1)^m 2^m}{(2m)!}$. Therefore, $y_2(x) = \displaystyle\sum_{m=0}^{\infty}\frac{(-1)^m 2^m}{(2m)!}x^m = \sum_{m=0}^{\infty}\frac{(-1)^m (\sqrt{2x})^{2m}}{(2m)!} = \cos(\sqrt{2x})$. [answer]

Hence, the general solution is $y = c_1\sin(\sqrt{2x})+c_2\cos(\sqrt{2x})$, which is in agreement with the form of Eq. (2.150).

Example 2.5.3. *(Case II)* Solve $x^2y'' - (2x^2)y' + \left(x^2+\dfrac{1}{4}\right)y=0$.

Solution. Substituting Eq. (2.146) and its derivatives,

$$\sum_{m=0}^{\infty}(m+r)(m+r-1)a_m x^{m+r} - 2\sum_{m=0}^{\infty}(m+r)a_m x^{m+r+1} + \sum_{m=0}^{\infty}a_m x^{m+r+2}$$

$$+ \frac{1}{4}\sum_{m=0}^{\infty}a_m x^{m+r} = 0.$$

Equating the sum of the coefficients of the smallest power x^{r-1} to zero yields $[r(r-1)+1/4]=0$, which implies that $r_1=r_2=1/2$ (there is a double root).

(i) <u>First solution:</u> Equating the sum of coefficients of x^{r+s} to zero with $m=s$ in the first and fourth terms, $m=s-1$ in the second term, and $m=s-2$ in the third term,

$(s+r)(s+r-1)a_s - 2(s+r-1)a_{s-1} + a_{s-2} + \frac{1}{4}a_s = 0$ or $[(s+r)(s+r-1) + \frac{1}{4}]a_s - 2(s+r-1)a_{s-1} + a_{s-2} = 0$. For $r = 1/2$, this expression reduces to $s^2 a_s = 2(s - \frac{1}{2})a_{s-1} - a_{s-2}$. For $s = 1$, $a_1 = a_0$ and for $s \geq 2$, $a_s = \dfrac{(2s-1)a_{s-1} - a_{s-2}}{s^2}$.

Hence, $a_2 = \dfrac{3a_1 - a_0}{4} = \dfrac{a_0}{2!}$, $a_3 = \dfrac{5a_2 - a_1}{9} = \dfrac{5a_0/2 - a_0}{9} = \dfrac{3a_0}{18} = \dfrac{a_0}{3!}$, and $a_4 = \dfrac{a_0}{4!}$, and so on. By choosing $a_0 = 1$,

$$y_1(x) = x^{1/2}\left[1 + \frac{x}{1!} + \frac{x^2}{2!} + \frac{x^3}{3!} + \frac{x^4}{4!} + \dots\right] = x^{1/2}e^x. \qquad \text{[answer]}$$

(ii) <u>Second solution</u>: Using the method of reduction of order (Section 2.2.4), $y_2 = y_1 \displaystyle\int U\,dx$, where $U = \dfrac{1}{y_1^2}e^{-\int p\,dx}$ for the differential equation $y'' + p(x)y' + q(x)y = 0$. Thus, for the equation $y'' - 2y' + \left(1 + \dfrac{1}{4x^2}\right)y = 0$, one has $p = -2$ and $-\displaystyle\int(-2)\,dx = 2x$. Hence, $U = \dfrac{e^{2x}}{xe^{2x}} = \dfrac{1}{x}$ and $y_2(x) = x^{1/2}e^x\displaystyle\int\frac{1}{x}\,dx = x^{1/2}e^x(\ln x + c)$. [answer]

Thus, the first and second solutions are in agreement with the form of Eq. (2.151). The general solution is therefore $y(x) = x^{1/2}e^x[c_1 + c_2\ln x]$.

Example 2.5.4. (*Case III*) Solve $xy'' + 2y' + xy = 0$.

Solution. Substituting Eq. (2.146) and its derivatives,

$$\sum_{m=0}^{\infty}(m+r)(m+r-1)a_m x^{m+r-1} + 2\sum_{m=0}^{\infty}(m+r)a_m x^{m+r-1} + \sum_{m=0}^{\infty}a_m x^{m+r+1} = 0,$$

or simplifying,

$$\sum_{m=0}^{\infty}(m+r)(m+r+1)a_m x^{m+r-1} + \sum_{m=0}^{\infty}a_m x^{m+r+1} = 0.$$

The lowest power of x implies the indicial equation is $r(r+1) = 0$. The roots are $r_1 = 0$ and $r_2 = -1$, which differ by an integer.

Set $m = s + 1$ in the first series and $m = s - 1$ in the second series so that

$$(s+r+1)(s+r+2)a_{s+1} + a_{s-1} = 0. \qquad (2.155)$$

(i) <u>First solution</u>:

Inserting $r = 0$ in Eq. (2.155), $(s + 1)(s + 2)a_{s+1} + a_{s-1} = 0$, yields $a_{s+1} = -\dfrac{a_{s-1}}{(s+1)(s+2)}$ for ($s \geq 1$) and $a_1 = 0$ for $s = 0$. Hence, $a_1 = 0$, $a_2 = \dfrac{-a_0}{2 \cdot 3} = \dfrac{-a_0}{3!}$, $a_3 = \dfrac{-a_1}{3 \cdot 4} = 0$, $a_4 = \dfrac{-a_2}{4 \cdot 5} = \dfrac{a_0}{5!}$, and so on. Hence $a_{2m} = (-1)^m \dfrac{a_0}{(2m+1)!}$ and $a_{2m+1} = 0$. Taking $a_0 = 1$ yields

$$y_1(x) = \left[1 - \frac{x^2}{3!} + \frac{x^4}{5!} - \cdots \right] = \left(\frac{1}{x} \right) \left[x - \frac{x^3}{3!} + \frac{x^5}{5!} - \cdots \right] = \frac{\sin x}{x}. \qquad \text{[answer]}$$

(ii) Second solution: Using the method of reduction of order (Section 2.2.4), $y_2 = y_1 \int U \, dx$, where $U = \dfrac{1}{y_1^2} e^{-\int p \, dx}$ for the differential equation $y'' + p(x)y' + q(x)y = 0$. Thus, for the equation $y'' + \dfrac{2}{x} y' + y$, one has $p = 2/x$ and $- \int (2/x) \, dx = \ln x^{-2}$. Hence, $U = \dfrac{x^2}{\sin^2 x} \cdot \dfrac{1}{x^2} = \dfrac{1}{\sin^2 x}$ and $y_2(x) = \dfrac{\sin x}{x} \int \dfrac{dx}{\sin^2 x} = \dfrac{\sin x \cot x}{x} = \dfrac{\cos x}{x}$. [answer]

One obtains the same result using the second root of the indicial equation, $r = -1$ in Eq. (2.155).

Thus, for this case $k = 0$ in Eq. (2.152) and the general solution is given by $y(x) = c_1 \dfrac{\sin x}{x} + c_2 \dfrac{\cos x}{x}$.

Example 2.5.5. Sometimes a differential equation can be transformed into a simpler one using a suitable transformation. For instance, solve $x^2 y'' + x y' + 4y = 0$.

Solution.

Let $x = e^z$, so that $\dfrac{dx}{dz} = e^z$. Hence, using this transformation with the chain rule, the first and second terms of the equation can be rewritten. Given that $y' = \dfrac{dy}{dx} = \dfrac{dy}{dz} \cdot \dfrac{dz}{dx} = e^{-z} \dfrac{dy}{dz}$, the second term becomes

$$xy' = e^z e^{-z} \frac{dy}{dz} = \frac{dy}{dz}.$$

Similarly, for the first term,

$$y'' = \frac{d^2 y}{dx^2} = \frac{d}{dx} \left[e^{-z} \frac{dy}{dz} \right] = \left\{ \frac{d}{dz} \left[e^{-z} \frac{dy}{dz} \right] \right\} e^{-z}$$

$$= e^{-z} \left\{ e^{-z} \frac{d^2 y}{dz^2} - \frac{dy}{dz} e^{-z} \right\} = e^{-2z} \left(\frac{d^2 y}{dz^2} - \frac{dy}{dz} \right),$$

so that

$$x^2 y'' = \left(\frac{d^2 y}{dz^2} - \frac{dy}{dz} \right).$$

Hence, the differential equation is transformed into a linear differential equation with constant coefficients: $\frac{d^2 y}{dz^2} + 4y = 0$. The solution to this latter differential equation is $y = A \cos 2z + B \sin 2z$. Substituting in $z = \ln x$, the final solution is

$$y = A \cos(2 \ln x) + B \sin(2 \ln x). \qquad \text{[answer]}$$

2.5.3 Bessel's equation

The Frobenius method can be used to solve Bessel's differential equation:

$$\boxed{x^2 y'' + xy' + (x^2 - v^2)y = 0,} \tag{2.156}$$

or in standard form,

$$\boxed{y'' + \frac{1}{x} y' + \left(1 - \frac{v^2}{x^2} \right) y = 0.} \tag{2.157}$$

Substituting a series of the form

$$y = \sum_{m=0}^{\infty} a_m x^{m+r} \tag{2.158}$$

and its derivatives into Eq. (2.156) gives

$$\sum_{m=0}^{\infty} (m+r)(m+r-1)a_m x^{m+r} + \sum_{m=0}^{\infty} (m+r)a_m x^{m+r} + \sum_{m=0}^{\infty} a_m x^{m+r+2}$$

$$- v^2 \sum_{m=0}^{\infty} a_m x^{m+r} = 0.$$

Thus, letting $m = s$ for the first, second, and fourth series and $m = s - 2$ for the third yields

$$r(r-1)a_0 + ra_0 - v^2 a_0 = 0 \qquad\qquad (s = 0), \quad (2.159\text{a})$$
$$(r+1)ra_1 + (r+1)a_1 - v^2 a_1 = 0 \qquad\qquad (s = 1), \quad (2.159\text{b})$$
$$(s+r)(s+r-1)a_s + (s+r)a_s + a_{s-2} - v^2 a_s = 0 \quad (s = 2, 3, ...). \quad (2.159\text{c})$$

From Eq. (2.159a), the indicial equation is

$$(r+v)(r-v) = 0 \tag{2.160}$$

with roots $r_1 = v$ and $r_2 = -v$. For $r = r_1 = v$, Eq. (2.159b) yields $a_1 = 0$ and Eq. (2.159c) may be written

$$(s + 2v)sa_s + a_{s-2} = 0. \tag{2.161}$$

Since $a_1 = 0$ (and $v \geq 0$), Eq. (2.161) implies $a_3 = 0$, $a_5 = 0$. Setting $s = 2m$ in Eq. (2.161), $a_{2m} = -\dfrac{1}{2^2 m(v + m)} a_{2m-2}$ $(m = 1, 2, ...)$, or successively, this relation gives $a_2 = -\dfrac{a_0}{2^2(v + 1)}$, $a_4 = -\dfrac{a_2}{2^2 \cdot 2(v + 2)} = \dfrac{a_0}{2^4 2!(v + 1)(v + 2)}$, or generally

$$a_{2m} = \frac{(-1)^m a_0}{2^{2m} m!(v + 1)(v + 2)...(v + m)} \quad (m = 1, 2, ...). \tag{2.162}$$

Bessel function of the first kind

(i) If $v = n =$ an integer and choosing $a_0 = \dfrac{1}{2^n n!}$ (so that $n!(n + 1)...(n + m) = (n + m)!$, Eq. (2.162) becomes

$$a_{2m} = \frac{(-1)^m}{2^{2m+n} m!(n + m)!} \quad (m = 1, 2, ...). \tag{2.163}$$

A solution to Eq. (2.156), using Eq. (2.158) and Eq. (2.163), is the Bessel function of the <u>first kind</u> of order n:

$$\boxed{J_n(x) = x^n \sum_{m=0}^{\infty} \frac{(-1)^m x^{2m}}{2^{2m+n} m!(n + m)!}.} \tag{2.164}$$

(ii) If $v \neq$ an integer, the <u>gamma function</u> arises, which is defined by the integral

$$\boxed{\Gamma(v) = \int_0^{\infty} e^{-t} t^{v-1} dt} \tag{2.165}$$

and has the property

$$\boxed{\Gamma(v + 1) = v\Gamma(v).} \tag{2.166}$$

Also, the gamma function has the property that $\Gamma\left(\frac{1}{2}\right) = \sqrt{\pi}$ and $\Gamma(1) = 1$. The gamma function generalizes the factorial function, where by Eq. (2.166)

$$\Gamma(n + 1) = n! \quad (n = 0, 1, ...). \tag{2.167}$$

Hence, the Bessel function of the first kind (of order $v \neq$ an integer) is

$$\boxed{J_v(x) = x^v \sum_{m=0}^{\infty} \frac{(-1)^m x^{2m}}{2^{2m+v} m!\Gamma(v + m + 1)}.} \tag{2.168}$$

The general solution of Eq. (2.156) for $\nu \neq$ an integer and for all $x \neq 0$ is

$$y(x) = c_1 J_\nu(x) + c_2 J_{-\nu}(x). \tag{2.169}$$

However, for integer $\nu = n$,

$$J_{-n}(x) = (-1)^n J_n(x) \quad (n = 1, 2, ...), \tag{2.170}$$

and therefore $J_{-n}(x)$ and $J_n(x)$ are not linearly independent and Eq. (2.169) (with $\nu = n$) is no longer a general solution. Therefore, a second linearly independent solution is required.

Bessel function of the second kind (or Neumann's function)

A standard second solution $Y_\nu(x)$ is defined by

$$Y_\nu(x) = \frac{1}{\sin \nu \pi} [J_\nu(x) \cos \nu \pi - J_{-\nu}(x)], \quad \nu \neq 0, 1, 2....$$

Also $Y_n(x) = \lim_{\nu \to n} Y_\nu(x)$. For the case $n = 0, 1, 2...$, the following series expansion is obtained:

$$\begin{aligned}
Y_n(x) = {}& \frac{2}{\pi} J_n(x) \left(\ln \frac{x}{2} + \gamma \right) + \frac{x^n}{\pi} \sum_{m=0}^{\infty} \frac{(-1)^{m-1}(h_m + h_{m+n})}{2^{2m+n} m!(m+n)!} x^{2m} \\
& - \frac{x^{-n}}{\pi} \sum_{m=0}^{n-1} \frac{(n-m-1)!}{2^{2m-n} m!} x^{2m},
\end{aligned} \tag{2.171}$$

where γ is Euler's constant $= 0.57721$ and $h_s = \sum_{k=1}^{s} \frac{1}{k}$ with $h_0 = 0$.

Summary

The complete general solution of Eq. (2.156) is

$$\begin{aligned}
y(x) &= c_1 J_\nu(x) + c_2 J_{-\nu}(x), \quad \nu \neq 0 \text{ and not a positive integer,} \\
y(x) &= c_1 J_n(x) + c_2 Y_n(x), \quad \nu = n = 0 \text{ or a positive integer.}
\end{aligned} \tag{2.172}$$

Modified Bessel equation

An alternative form of Eq. (2.156) is the differential equation

$$x^2 y'' + x y' - (x^2 + \nu^2) y = 0. \tag{2.173}$$

This equation is transformed into the form of Eq. (2.156) by substituting $x = iz$. The complete solutions of Eq. (2.173) are

$$y(x) = c_1 I_\nu(x) + c_2 I_{-\nu}(x) \text{ for all } \nu \text{ except } \nu = 0 \text{ or positive integer,}$$

$$y(x) = c_1 I_n(x) + c_2 K_n(x) \text{ for } n = 0 \text{ or positive integer,}$$

(2.174)

where $I_\nu(x) = i^{-\nu} J_\nu(ix)$ (modified Bessel function of the first kind of order ν) and $K_n(x) = \frac{\pi}{2} i^{n+1} [J_n(ix) + i Y_n(ix)]$ (modified Bessel function of the second kind of order n).

Generalized form of Bessel's equation
Consider

$$x^2 y'' + x(a + 2bx^r)y' + [c + dx^{2s} - b(1 - a - r)x^r + b^2 x^{2r}]y = 0.$$

(2.175)

The generalized solution of Eq. (2.175) is

$$y(x) = x^{(1-a)/2} e^{-(bx^r/r)} \left[c_1 Z_\nu \left(\frac{\sqrt{|d|}}{s} x^s \right) + c_2 Z_{-\nu} \left(\frac{\sqrt{|d|}}{s} x^s \right) \right],$$

(2.176)

where $\nu = \frac{1}{s} \sqrt{\left(\frac{1-a}{2} \right)^2 - c}$, noting that $\nu \geq 0$ and a and s are constants. Here Z_ν denotes a Bessel function as follows:

a) $\begin{rcases} Z_\nu \equiv J_\nu, \\ Z_{-\nu} \equiv J_{-\nu} \end{rcases}$ for $\dfrac{\sqrt{d}}{s}$ real, $\nu \neq 0$, $\nu \neq$ integer,

b) $\begin{rcases} Z_\nu \equiv J_n, \\ Z_{-\nu} \equiv Y_n \end{rcases}$ for $\dfrac{\sqrt{d}}{s}$ real, $\nu = 0$ or integer only,

c) $\begin{rcases} Z_\nu \equiv I_\nu, \\ Z_{-\nu} \equiv I_{-\nu} \end{rcases}$ for $\dfrac{\sqrt{d}}{s}$ imaginary, $\nu \neq 0$, $\nu \neq$ integer,

d) $\begin{rcases} Z_\nu \equiv I_n, \\ Z_{-\nu} \equiv K_n \end{rcases}$ for $\dfrac{\sqrt{d}}{s}$ imaginary, $\nu = 0$ or integer only.

All of the Bessel functions and modified Bessel functions are tabulated. Fig. 2.4 shows a plot of ordinary and modified Bessel functions of zero order.

Recurrence formulas for Bessel functions
The recurrence formulas are also valid for $Y_\nu(x)$:

1. $J_{\nu+1}(x) = \dfrac{2\nu}{x} J_\nu(x) - J_{\nu-1}(x)$;

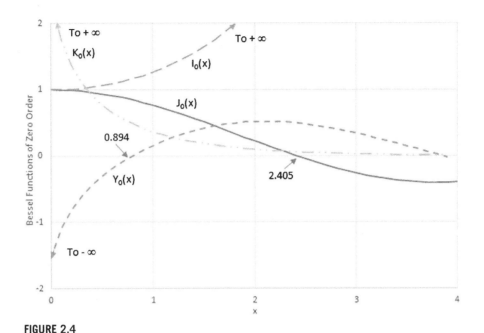

FIGURE 2.4

Ordinary and modified Bessel functions of zero order.

2. $J_\nu'(x) = \dfrac{1}{2}[J_{\nu-1}(x) - J_{\nu+1}(x)];$

3. $\dfrac{d}{dx}[x^\nu J_\nu(x)] = x^\nu J_{\nu-1}(x);$

4. $\dfrac{d}{dx}\left[x^{-\nu} J_\nu(x)\right] = -x^{-\nu} J_{\nu+1}(x).$

Functions related to Bessel functions

Hankel functions of the first and second kind are defined by

$$H_\nu^{(1)} = J_\nu(x) + i Y_\nu(x),$$
$$H_\nu^{(2)} = J_\nu(x) - i Y_\nu(x).$$

Example 2.5.6. As shown in Fig. 2.5, a cylindrical wall surrounding a heated rod radially conducts heat. The heat escapes at the end of the wall into the air of temperature T_A. The rod has a radius R with a surface temperature T_R. The wall has a length l.

Solution.

(a) Heat balance

For the volume element $dV = 2\pi r l\, dr$, the heat balance is

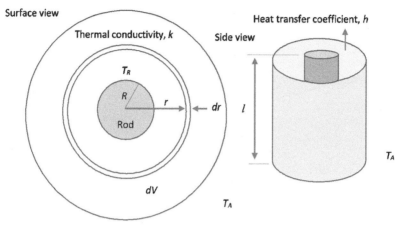

FIGURE 2.5

Cylindrical wall surrounding a heated rod.

$$\text{Input} \quad = \text{conduction (Fourier equation)} = -k(2\pi rl)\frac{dT}{dr},$$

$$\text{Output} \quad = \text{conduction} + \text{loss to air}$$

$$= -\left[k(2\pi rl)\frac{dT}{dr} + d\left(k2\pi rl\frac{dT}{dr} \right) \right] + h(2\pi r dr)[T - T_A],$$

where $(2\pi r dr)$ is the surface area of the volume element in the last term.
The rate of accumulation = Input − Output = 0 at steady state, so that

$$-k(2\pi rl)\frac{dT}{dr} + k(2\pi rl)\frac{dT}{dr} + d\left(k2\pi rl\frac{dT}{dr} \right) - h(2\pi r dr)[T - T_A] = 0.$$

Dividing by dr yields $kl\dfrac{d}{dr}\left(r\dfrac{dT}{dr} \right) = hr(T - T_A)$ or $r\dfrac{d^2T}{dr^2} + \dfrac{dT}{dr} = \dfrac{hr}{kl}(T - T_A)$.

This latter equation can be rewritten as $\dfrac{d^2T}{dr^2} + \dfrac{1}{r}\dfrac{dT}{dr} - \beta(T - T_A) = 0$, where $\beta = \dfrac{h}{kl}$.

Letting $y = T - T_A$ and multiplying through by r^2 gives

$$\boxed{r^2\frac{d^2y}{dr^2} + r\frac{dy}{dr} - r^2\beta y = 0.} \tag{2.177}$$

Comparing with Eq. (2.175), $a + 2br^{\tilde{r}} = 1$, so that $a = 1$ and $b = 0$. Also $c + dr^{2s} - b(1 - d - \tilde{r})r^{\tilde{r}} + b^2 r^{2\tilde{r}} = -r^2\beta$, which implies $c = 0$, $s = 1$, and $d = -\beta$.

From

$$v = \frac{1}{s}\sqrt{\left(\frac{1-a}{2}\right)^2 - c} = 1\sqrt{\left(\frac{1-1}{2}\right)^2 - 0} = 0,$$

$$\frac{\sqrt{d}}{s} = \sqrt{-\beta} \text{ (imaginary)},$$

condition "d" applies.

Hence, $Z_v \equiv I_n = I_0$ and $Z_{-v} \equiv K_n = K_0$. Therefore, from Eq. (2.176), the general solution (where $x \equiv r$) $y(r) = c_1 I_0(r\sqrt{\beta}) + c_2 K_0(r\sqrt{\beta})$. The constants are obtained from the following boundary conditions:

(i) $T(r \longrightarrow \infty) = T_A$ (that is, a finite air temperature);
(ii) $T(r = R) = T_R$.

Using the first boundary condition, since $I_0(r) \longrightarrow \infty$ as $r \longrightarrow \infty$ (see Fig. 2.4), this implies $c_1 = 0$. Thus, $y = c_2 K_0(r\sqrt{\beta})$ or $T - T_A = c_2 K_0(r\sqrt{\beta})$. The second boundary condition gives $T_R - T_A = c_2 K_0(R\sqrt{\beta})$ so that $c_2 = \dfrac{T_R - T_A}{K_0(R\sqrt{\beta})}$.

The final solution is therefore $\boxed{\dfrac{T(r) - T_A}{T_R - T_A} = \dfrac{K_0(r\sqrt{\beta})}{K_0(R\sqrt{\beta})}}$.

2.5.4 Sturm–Liouville problems

The Bessel and Legendre equations, as well as other engineering equations, can be put in the general form of a Sturm–Liouville equation. For example, the Bessel equation $\tilde{x}^2 \dfrac{d^2 y}{d\tilde{x}^2} + \tilde{x}\dfrac{dy}{d\tilde{x}} + (\tilde{x}^2 - n^2)y(\tilde{x}) = 0$, letting $\tilde{x} = kx$ (which implies $\dfrac{dy}{d\tilde{x}} = \dfrac{y'}{k}$ and $\dfrac{d^2 y}{d\tilde{x}^2} = \dfrac{y''}{k^2}$), reduces to $x^2 y'' + xy' + (k^2 x^2 - n^2)y = 0$, or dividing by x,

$$[xy']' + \left(-\frac{n^2}{x} + \lambda x\right)y = 0, \quad \text{where } \lambda = k^2.$$

Similarly, Legendre's equation (see Section 2.5.1),

$$(1 - x^2)y'' - 2xy' + n(n+1)y = 0,$$

can be written as $[(1 - x^2)y']' + \lambda y = 0$. Both equations are of the form of a Sturm–Liouville equation,

$$\boxed{[r(x)y']' + [q(x) + \lambda p(x)]y = 0.} \tag{2.178}$$

This Eq. (2.178) is considered on some interval, $a \le x \le b$, where p, q, r, and r' are continuous and $p(x) > 0$. The boundary conditions at the endpoints are

$$k_1 y(a) + k_2 y'(a) = 0, \tag{2.179a}$$

$$l_1 y(b) + l_2 y'(b) = 0, \qquad (2.179b)$$

where k_1 and k_2 are not both zero and l_1 and l_2 are not both zero. The solution of the Sturm–Liouville problem (that is, Eq. (2.178) and Eq. (2.179)) is the eigenfunction $y(x)$ for the associated eigenvalue λ.

Example 2.5.7. Find the eigenvalues and eigenfunctions of the Sturm–Liouville problem ($r = p = 1$ and $q = 0$) $y'' + \lambda y = 0$ with $y(0) = 0$ and $y(\pi) = 0$.

Solution. For $\lambda = v^2$, $y(x) = A\cos vx + B\sin vx$. Using the boundary conditions, $y(0) = A = 0$ and $y(\pi) = B\sin v\pi = 0$. The second boundary condition implies $v = 0, \pm 1, \pm 2\dots$. However, for the case $v = 0$, one obtains the trivial solution $y \equiv 0$. Therefore, the eigenvalues are $\lambda = v^2$, where $v = 1, 2, \dots$, with the corresponding eigenfunctions $y(x) = \sin vx$. [answer]

Definitions

(i) The functions y_1 and y_2 are orthogonal on $a \le x \le b$ with respect to a weight $p(x) > 0$ if

$$\int_a^b p(x) y_m(x) y_n(x) dx = 0 \quad \text{for } m \ne n. \qquad (2.180)$$

(ii) The norm $\|y_m\|$ of y_m is

$$\|y_m\| = \sqrt{\int_a^b p(x) y_m^2(x)\, dx}. \qquad (2.181)$$

(iii) Functions are orthonormal if (i) applies and $\|y_m\| = 1$.

Example 2.5.8. The functions $y_m(x) = \sin mx$, $m = 1, 2\dots$, form an orthonormal set on the interval $-\pi \le x \le \pi$, that is,

$$\int_{-\pi}^{\pi} \sin mx \sin nx dx = \frac{1}{2}\int_{-\pi}^{\pi} \cos(m-n)dx - \frac{1}{2}\int_{-\pi}^{\pi} \cos(m+n)\, dx = 0 \quad (m \ne n).$$

Also $\|y_m\| = \sqrt{\int_{-\pi}^{\pi} \sin^2 mx\, dx} = \sqrt{\pi}\ (m = 1, 2, \dots)$.

Therefore the orthonormal set is $\dfrac{\sin x}{\sqrt{\pi}}, \dfrac{\sin 2x}{\sqrt{\pi}}, \dfrac{\sin 3x}{\sqrt{\pi}}, \dots$ [answer]

Orthogonality of eigenfunctions

The eigenfunctions $y_m(x)$ and $y_n(x)$ of the Sturm–Liouville problem in Eq. (2.178) and Eq. (2.179) that correspond to the different eigenvalues λ_m and λ_n are orthogonal on the interval $a \le x \le b$ with respect to the weight function p.

Note that:
if $r(a) = 0$, then Eq. (2.179a) no longer applies;

if $r(b) = 0$, then Eq. (2.179b) no longer applies;

if $r(a) = r(b)$, then Eq. (2.179) is replaced by the periodic boundary conditions $y(a) = y(b)$ and $y'(a) = y'(b)$.

Example 2.5.9. Find an orthonormal set of eigenfunctions for the periodic Sturm–Liouville problem $y'' + \lambda y = 0$ with $y(\pi) = y(-\pi)$ and $y'(\pi) = y'(-\pi)$.

Solution. From Example 2.5.7, the general solution is $y = A \cos kx + B \sin kx$ $(k = \sqrt{\lambda})$. Use the boundary conditions

$$A \cos(k\pi) + B \sin(k\pi) = A \cos(-k\pi) + B \sin(-k\pi),$$
$$- kA \sin(k\pi) + kB \cos(k\pi) = -kA \sin(-k\pi) + kB \cos(-k\pi).$$

Since $\cos(-\alpha) = \cos(\alpha)$ and $\sin(-\alpha) = -\sin(\alpha)$, the above equations give $\sin k\pi = 0$, $\lambda = k^2 = n^2 = 0, 1, 4, 9....$ The eigenfunctions are

$$1, \cos x, \sin x, \cos 2x, \sin 2x, ...$$

These eigenfunctions are orthogonal, where

$$\int_{-\pi}^{\pi} \cos mx \sin mx dx = \frac{1}{2} \int_{-\pi}^{\pi} \sin 2mx \, dx = 0$$

as per Eq. (2.180). In addition, using Eq. (2.181),

$$\|1\| = \sqrt{\int_{-\pi}^{\pi} dx} = \sqrt{2\pi},$$

$$\| \cos mx \| = \sqrt{\int_{-\pi}^{\pi} \cos^2 mx \, dx} = \sqrt{\pi},$$

$$\| \sin mx \| = \sqrt{\int_{-\pi}^{\pi} \sin^2 mx \, dx} = \sqrt{\pi}.$$

The orthonormal set is $\dfrac{1}{\sqrt{2\pi}}, \dfrac{\cos x}{\sqrt{\pi}}, \dfrac{\sin x}{\sqrt{\pi}}, \dfrac{\cos 2x}{\sqrt{\pi}}, \dfrac{\sin 2x}{\sqrt{\pi}},....$ [answer]

Orthogonality of Bessel functions

Recall the Sturm–Liouville equation for Bessel's equation

$$[x J_n'(kx)]' + \left(-\frac{n^2}{x} + k^2 x \right) J_n(kx) = 0.$$

Here $p(x) = x$, $q(x) = -\dfrac{n^2}{x}$, $r(x) = x$, and $\lambda = k^2$. Since $r(0) = 0$, the orthogonality theorem implies an orthogonality on the interval $0 \le x \le R$. Thus from Eq. (2.179b),

$J_n(kR) = 0$ (n fixed), where $kR = \alpha_{mn}$ and $J_n(\alpha_{mn})$ has infinitely many zeros ($m = 1, 2, ...$). Thus, $k = k_{mn} = \alpha_{mn}/R$. For each fixed nonnegative integer n, the Bessel functions form an orthogonal set $J_n(k_{1n}x)$, $J_n(k_{2n}x)$, ... on the interval $0 \le x \le R$ with weight $p(x) = x$, such that

$$\boxed{\int_0^R x J_n(k_{mn}x) J_n(k_{jn}x) \, dx = 0} \quad \text{for } j \ne m. \tag{2.182}$$

2.5.5 Eigenfunction expansions

Eigenfunctions arising from the Sturm–Liouville problem can be used to develop series of given functions (for example, Fourier series), and provide the necessary techniques for the solution of engineering problems in heat conduction, fluid flow, etc.

Notation

For an orthonormal set with respect to $p(x) > 0$ on $a \le x \le b$,

$$(y_m, y_n) = \int_a^b p(x) y_m(x) y_n(x) dx = \delta_{mn},$$

where δ_{mn} is the Kronecker delta function defined as

$$\delta_{mn} = \begin{cases} 0 & \text{if} \quad m \ne n, \\ 1 & \text{if} \quad m = n. \end{cases}$$

For the norm, $\|y_m\| = \sqrt{(y_m, y_m)} = \sqrt{\int_a^b p(x) y_m^2(x) \, dx}$.

Now if $y_0, y_1, ...$ are an orthogonal set with respect to $p(x)$ on the interval $a \le x \le b$, then a given function can be represented by the <u>orthogonal expansion</u> or <u>generalized Fourier series</u>

$$\boxed{f(x) = \sum_{m=0}^{\infty} a_m \, y_m(x) = a_0 y_0(x) + a_1 y_1(x) + ...} \tag{2.183}$$

Moreover, if y_m are eigenfunctions of the Sturm–Liouville problem, Eq. (2.183) is called an **eigenfunction expansion**.

As a result of orthogonality, the <u>Fourier constants</u> $a_0, a_1, ...$ can be easily determined, where by multiplying Eq. (2.183) by $p(x) y_n(x)$ (n fixed) and integrating we find

$$(f, y_n) = \int_a^b p f y_n \, dx = \int_a^b p \left(\sum_{m=0}^{\infty} a_m y_m \right) y_n \, dx = \sum_{m=0}^{\infty} a_m (y_m, y_n).$$

Because of the orthogonality, the integrals (y_m, y_n) on the right-hand side are zero except when $m = n$, where $(y_n, y_n) = \|y_n\|^2$. Therefore, $(f, y_n) = a_n \|y_n\|^2$. The coefficients of Eq. (2.183) are given by

$$a_m = \frac{(f, y_m)}{\|y_m\|^2} = \frac{1}{\|y_m\|^2} \int_a^b p(x) f(x) y_m(x) dx \quad (m = 0, 1, 2...). \quad (2.184)$$

2.5.6 Fourier series

The Sturm–Liouville problem in Example 2.5.9 gave the orthogonal set

$$1, \cos x, \sin x, \cos 2x, \sin 2x, ...$$

on the interval $-\pi \le x \le \pi$ with $p(x) = 1$. The corresponding eigenfunction expansion can be written as

$$f(x) = a_0 + \sum_{m=1}^{\infty} (a_m \cos mx + b_m \sin mx). \quad (2.185)$$

Eq. (2.185) is called a _Fourier series_ of $f(x)$ and the coefficients in Eq. (2.184) are the _Fourier coefficients_ of $f(x)$. From Eq. (2.184) and the norms derived in Example 2.5.9, the Fourier coefficients are given by the so-called _Euler formulas_:

$$a_0 = \frac{1}{2\pi} \int_{-\pi}^{\pi} f(x) dx, \quad (2.186a)$$

$$a_m = \frac{1}{\pi} \int_{-\pi}^{\pi} f(x) \cos mx \, dx \quad (m = 1, 2, ...), \quad (2.186b)$$

$$b_m = \frac{1}{\pi} \int_{-\pi}^{\pi} f(x) \sin mx \, dx \quad (m = 1, 2, ...). \quad (2.186c)$$

Example 2.5.10. Find the Fourier series of the periodic function in Fig. 2.6 (where $f(x + 2\pi) = f(x)$):

$$f(x) = \begin{cases} -x & \text{if } -\pi < x < 0, \\ x & \text{if } 0 < x < \pi. \end{cases}$$

Solution. Using Eq. (2.186), the Fourier coefficients can be evaluated as

$$a_0 = \frac{1}{2\pi} \left[\int_{-\pi}^{0} (-x) \, dx + \int_{0}^{\pi} (x) \, dx \right] = \frac{\pi}{2},$$

$$a_m = \frac{1}{\pi} \left[\int_{-\pi}^{0} (-x) \cos mx \, dx + \int_{0}^{\pi} (x) \cos mx \, dx \right]$$

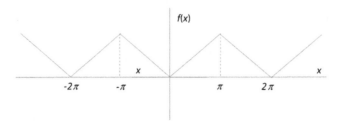

FIGURE 2.6

Periodic function $f(x)$.

$$= -\frac{1}{\pi}\left[\frac{\cos mx}{m^2} + \frac{x\sin mx}{m}\right]\Big|_{-\pi}^{0} + \frac{1}{\pi}\left[\frac{\cos mx}{m^2} + \frac{x\sin mx}{m}\right]\Big|_{0}^{\pi}$$

$$= \frac{2}{\pi m^2}[-1 + \cos m\pi] = \frac{2}{\pi m^2}[-1 + (-1)^m] = \begin{cases} -\dfrac{4}{\pi m^2} & m\text{ odd,} \\ 0 & m\text{ even,} \end{cases}$$

$$b_m = \frac{1}{\pi}\left[\int_{-\pi}^{0}(-x)\sin mx\,dx + \int_{0}^{\pi}(x)\sin mx\,dx\right] = 0,$$

where $(m = 1, 2, 3, ...)$. The Fourier series is

$$f(x) = \frac{\pi}{2} - \frac{4}{\pi}\left(\cos x + \frac{1}{3^2}\cos 3x + \frac{1}{5^2}\cos 5x + ...\right). \qquad \text{[answer]}$$

Example 2.5.11. (Fourier–Legendre series) As shown, the Legendre equation is a Sturm–Liouville equation:

$$[(1 - x^2)y']' + \lambda y = 0,$$

with $\lambda = n(n+1), r = 1 - x^2, q = 0$, and $p = 1$. Here $r(-1) = r(1) = 0$ is a periodic boundary condition for the Sturm–Liouville problem on the interval $-1 \le x \le 1$. For $n = 0, 1, 2...$ (that is, $\lambda = 0 \cdot 1, 1 \cdot 2, 2 \cdot 3, ...$), the Legendre polynomials $P_n(x)$ are the eigenfunction solutions. Also, by the orthogonality theorem,

$$\int_{-1}^{1} P_m(x)P_n(x)\,dx = 0,$$

and the norm is

$$\|P_m\| = \sqrt{\int_{-1}^{1} P_m^2(x)\,dx} = \sqrt{\frac{2}{2m+1}} \qquad (m = 0, 1, 2, ...).$$

The eigenfunction expansion for a Fourier–Legendre series is

$$f(x) = \sum_{m=0}^{\infty} a_m P_m(x) = a_0 P_0 + a_1 P_1 + \dots = a_0 + a_1 x + a_2 \left(\frac{3}{2}x^2 - \frac{1}{2} \right) + \dots,$$

where using the norm in Eq. (2.184), the coefficients are

$$a_m = \frac{2m+1}{2} \int_{-1}^{1} f(x) P_m(x) \, dx.$$

Example 2.5.12. (Fourier–Bessel series) From Section 2.5.4, the Bessel functions form an orthogonal set, $J_n(k_{1n}x)$, $J_n(k_{2n}x)$, $J_n(k_{3n}x)$, ..., where n is fixed and $k_{mn} = \alpha_{mn}/R$ on the interval $0 \leq x \leq R$. From Eq. (2.183), the corresponding Fourier–Bessel series is

$$f(x) = \sum_{m=0}^{\infty} a_m J_n(k_{mn}x) = a_1 J_n(k_{1n}x) + a_2 J_n(k_{2n}x) + \dots.$$

The norm is given by

$$\| J_n(k_{mn}x) \| = \sqrt{\int_{0}^{R} x J_n^2(k_{mn}x) \, dx} = \sqrt{\frac{R^2}{2} J_{n+1}^2(k_{mn}R)}.$$

The coefficients evaluated by Eq. (2.184) are

$$a_m = \frac{2}{R^2 J_{n+1}^2(\alpha_{mn})} \int_{0}^{R} x f(x) J_n(k_{mn}x) \, dx.$$

Fourier series for functions of any period

A periodic function $f(x)$ is said to have a period p if for all x,

$$f(x + p) = f(x),$$

where p is a positive constant. Eq. (2.185) and Eq. (2.186) are applicable to functions with a period $p = 2\pi$ (e.g., see Example 2.5.10). If a function $f(x)$ has a period $p = 2L$, then the Fourier series corresponding to $f(x)$ is

$$f(x) = a_0 + \sum_{n=1}^{\infty} \left(a_n \cos \frac{n\pi}{L} x + b_n \sin \frac{n\pi}{L} x \right), \qquad (2.187)$$

where the Fourier coefficients are

$$a_0 = \frac{1}{2L} \int_{-L}^{L} f(x) \, dx, \qquad (2.188a)$$

$$a_n = \frac{1}{L}\int_{-L}^{L} f(x)\cos\frac{n\pi}{L}x\,dx \qquad (n = 1, 2, ...), \qquad (2.188b)$$

$$b_n = \frac{1}{L}\int_{-L}^{L} f(x)\sin\frac{n\pi}{L}x\,dx \qquad (n = 1, 2, ...). \qquad (2.188c)$$

Example 2.5.13. Find the Fourier series of the periodic function in Fig. 2.7 with the period $p = 2L = 4$ (that is, $L = 2$), where

$$f(x) = \begin{cases} 1 + \dfrac{x}{2} & \text{if } -2 < x < 0, \\[2mm] 1 - \dfrac{x}{2} & \text{if } 0 < x < 2. \end{cases}$$

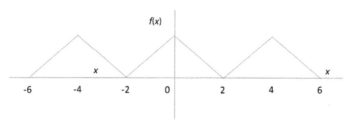

FIGURE 2.7

Function with a period $p = 4$.

Solution. Using Eq. (2.188), the Fourier coefficients can be evaluated as

$$a_0 = \frac{1}{4}\int_{-2}^{2} f(x)\,dx = \frac{1}{4}\int_{-2}^{0}\left(1+\frac{x}{2}\right)dx + \frac{1}{4}\int_{0}^{2}\left(1-\frac{x}{2}\right)dx = \frac{1}{2},$$

$$a_n = \frac{1}{2}\int_{-2}^{2} f(x)\cos\frac{n\pi x}{2}\,dx = \frac{1}{2}\int_{-2}^{0}\left(1+\frac{x}{2}\right)\cos\frac{n\pi x}{2}\,dx$$

$$+ \frac{1}{2}\int_{0}^{2}\left(1-\frac{x}{2}\right)\cos\frac{n\pi x}{2}\,dx$$

$$= -\frac{2}{(n\pi)^2}\left[\cos\left(\frac{n\pi x}{2}\right)\Big|_{0}^{2}\right] = \frac{2}{n^2\pi^2}(1-\cos(n\pi)) = \frac{2}{n^2\pi^2}\left(1-(-1)^n\right)$$

$$= \begin{cases} \dfrac{4}{(n\pi)^2} & n \text{ odd}, \\[3mm] 0 & n \text{ even}, \end{cases}$$

$$b_n = 0.$$

Therefore, the Fourier series is given by

$$f(x) = \frac{1}{2} + \frac{4}{\pi^2}\left(\cos\frac{\pi x}{2} + \frac{1}{3^2}\cos\frac{3\pi x}{2} + \frac{1}{5^2}\cos\frac{5\pi x}{2} + ...\right). \qquad \text{[answer]}$$

Even and odd functions

A function is called even if $f(-x) = f(x)$ (for example, $\cos x$). Similarly, a function is called odd if $f(-x) = -f(x)$ (for example, $\sin x$). The function in Example 2.5.13 was even and had only cosine terms in its Fourier series. This result, in fact, is general where the Fourier series of an <u>even</u> function of period $2L$ is called a "Fourier cosine series," that is,

$$f(x) = a_0 + \sum_{n=1}^{\infty} a_n \cos\frac{n\pi}{L}x \qquad (f \text{ even}), \qquad (2.189)$$

with coefficients

$$a_0 = \frac{1}{L}\int_0^L f(x)\,dx, \; a_n = \frac{2}{L}\int_0^L f(x)\cos\frac{n\pi}{L}x\,dx \qquad (n = 1, 2, 3, ...). \quad (2.190)$$

The Fourier series of an <u>odd</u> function of period $2L$ is a "<u>Fourier sine series</u>," that is,

$$f(x) = \sum_{n=1}^{\infty} b_n \sin\frac{n\pi}{L}x \qquad (f \text{ odd}), \qquad (2.191)$$

with the coefficients

$$b_n = \frac{2}{L}\int_0^L f(x)\sin\frac{n\pi}{L}x\,dx. \qquad (2.192)$$

Half-range Fourier sine or cosine series

A half-range Fourier sine or cosine series is a series in which only sine or cosine terms are present, respectively. When a half-range series corresponding to a given function is desired, the function is generally defined in the interval $(0, L)$ (which is half of the interval $(-L, L)$, hence the name half-range). The function then specified as odd or even is clearly defined in the other half of the interval $(-L, 0)$. As such, the <u>half-range cosine series</u> is given by Eq. (2.189) and Eq. (2.190) (for the even periodic extension), and the <u>half-range sine series</u> is given by Eq. (2.191) and Eq. (2.192) (for the odd periodic extension).

Example 2.5.14. (Triangle and its half-range expansions for the function in Fig. 2.8)

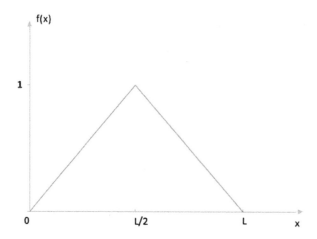

FIGURE 2.8

Half-range triangular wave.

Find the half-range expansions of the function

$$
f(x) = \begin{cases} \dfrac{2}{L}x & \text{if } 0 < x < L/2, \\[2mm] \dfrac{2}{L}(L - x) & \text{if } L/2 < x < L. \end{cases}
$$

Solution.

(a) <u>Even periodic extension</u>

From Eq. (2.190),

$$
a_0 = \frac{1}{L}\left[\frac{2}{L}\int_0^{L/2} x\,dx + \frac{2}{L}\int_{L/2}^L (L - x)\,dx\right] = \frac{1}{2},
$$

$$
a_n = \frac{2}{L}\left[\frac{2}{L}\int_0^{L/2} x\cos\frac{n\pi x}{L}\,dx + \frac{2}{L}\int_{L/2}^L (L - x)\cos\frac{n\pi x}{L}\,dx\right].
$$

Integrating the last expression by parts gives $a_n = \dfrac{4}{n^2\pi^2}\left(2\cos\dfrac{n\pi}{2} - \cos n\pi - 1\right)$.

Therefore, $a_2 = -\dfrac{16}{2^2\pi^2}$, $a_6 = -\dfrac{16}{6^2\pi^2}$, $a_{10} = -\dfrac{16}{10^2\pi^2}$, ..., and $a_n = 0$ if $n \neq$ 2, 6, 10, 14, ...

Hence, the first half-range expansion using Eq. (2.189) is

$$
f(x) = \frac{1}{2} - \frac{16}{\pi^2}\left(\frac{1}{2^2}\cos\frac{2\pi x}{L} + \frac{1}{6^2}\cos\frac{6\pi x}{L} + \ldots\right). \qquad \text{[answer]}
$$

This series represents the even periodic extension of the given function $f(x)$ of period $2L$ (see Fig. 2.9(a)).

(b) Odd periodic extension

Similarly from Eq. (2.192), $b_n = \dfrac{8}{n^2\pi^2} \sin \dfrac{n\pi}{2}$.

Hence, the second half-range extension is

$$f(x) = \frac{8}{\pi^2}\left(\frac{1}{1^2}\sin\frac{\pi x}{L} - \frac{1}{3^2}\sin\frac{3\pi x}{L} + \frac{1}{5^2}\sin\frac{5\pi x}{L} - \ldots\right). \qquad \text{[answer]}$$

This series represents the odd periodic extension of $f(x)$ of period $2L$ (see Fig. 2.9(b)).

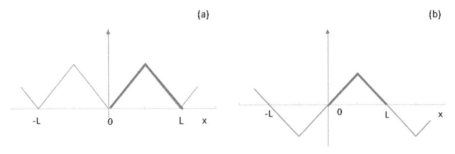

(a)   (b)

FIGURE 2.9

Periodic extensions of $f(x)$ in Example 2.5.14 for an (a) even and (b) odd extension.

Dirichlet conditions

Given are the following conditions:

(i) $f(x)$ is defined and single-valued except possibly at a finite number of points in $(-L, L)$;
(ii) $f(x)$ is periodic outside $(-L, L)$, with period $2L$;
(iii) $f(x)$ and $f'(x)$ are piecewise continuous in $(-L, L)$.

Then the series in Eq. (2.187) with coefficients in Eq. (2.188) converges to

(i) $f(x)$ if x is a point of continuity;
(ii) $\dfrac{f(x+0) + f(x-0)}{2}$ if x is a point of discontinuity.

In this theorem, $f(x+0)$ and $f(x-0)$ are the right- and left-hand limits of $f(x)$ and represent $\lim_{\epsilon \to 0} f(x+\epsilon)$ and $\lim_{\epsilon \to 0} f(x-\epsilon)$, respectively, where $\epsilon > 0$. The conditions in (i) to (iii) imposed on $f(x)$ are <u>sufficient</u> but not necessary, and are generally satisfied in practice.

Parseval's identity

Parseval's identity states that

$$\frac{1}{L} \int_{-L}^{L} \{f(x)\}^2 \, dx = a_0^2 + \sum_{n=1}^{\infty} \left(a_n^2 + b_n^2 \right)$$

if a_0, a_n, and b_n are the Fourier coefficients corresponding to $f(x)$ and if $f(x)$ satisfies the Dirichlet conditions.

Complex notation for Fourier series

Using Euler identities,

$$e^{i\theta} = \cos\theta + i \sin\theta,$$
$$e^{-i\theta} = \cos\theta - i \sin\theta,$$

(2.193)

where $i = \sqrt{-1}$, the Fourier series for $f(x)$ can be written as

$$f(x) = \sum_{n=-\infty}^{\infty} c_n e^{in\pi x/L},$$

(2.194)

where

$$c_n = \frac{1}{2L} \int_{-L}^{L} f(x) e^{-in\pi x/L} \, dx.$$

(2.195)

Eq. (2.194) supposes that the Dirichlet conditions are satisfied and that $f(x)$ is continuous at x. If $f(x)$ is discontinuous at x, the left-hand side of Eq. (2.194) should be replaced by $[f(x+0) + f(x-0)]/2$.

Example 2.5.15. Find the complex Fourier series of $f(x) = e^{-\pi x}$ over the range $-\pi < x < \pi$ with the periodicity $f(x + 2\pi) = f(x)$.

Solution. Using Eq. (2.195) with Maple in Appendix A,

$$c_n = \frac{1}{2\pi} \int_{-\pi}^{\pi} e^{-\pi x} e^{-inx} \, dx = \frac{1}{2\pi} \left(\frac{e^{\pi^2} e^{in\pi} - e^{-\pi^2} e^{-in\pi}}{(\pi + in)} \right).$$

Multiplication by $\dfrac{(\pi - in)}{(\pi - in)}$ and noting that $e^{in\pi} = e^{-in\pi} = (-1)^n$ gives

$$c_n = \frac{1}{2\pi} \left(\frac{\pi - in}{\pi^2 + n^2} \right) (-1)^n \left[e^{\pi^2} - e^{-\pi^2} \right] = \frac{\sinh \pi^2}{\pi} \left(\frac{\pi - in}{\pi^2 + n^2} \right) (-1)^n.$$

Hence, the complex Fourier series using Eq. (2.194) is

$$e^{-\pi x} = \frac{\sinh \pi^2}{\pi} \sum_{n=-\infty}^{\infty} (-1)^n \left(\frac{\pi - in}{\pi^2 + n^2} \right) e^{inx} \qquad (-\pi < x < \pi). \qquad \text{[answer]}$$

Note that from the above complex series, one can obtain the usual Fourier series as follows. Using the Euler identities in Eq. (2.193) (with $\theta = nx$), the term $(\pi - in)e^{inx}$ in the series is

$$(\pi - in)(\cos nx + i \sin nx) = \pi(\cos nx + n \sin nx) - i(n \cos nx - \pi \sin nx)$$

$$(n \text{ positive}),$$

$$(\pi + in)(\cos nx - i \sin nx) = \pi(\cos nx + n \sin nx) + i(n \cos nx - \pi \sin nx)$$

$$(n \text{ negative}),$$

$$\pi \quad (n = 0).$$

Hence, the imaginary terms cancel out in the series, and the real Fourier series is

$$e^{-\pi x}$$

$$= 2 \sinh \pi^2 \left[\frac{1}{2} - \frac{1}{\pi^2 + 1^2} (\cos x + \sin x) + \frac{1}{\pi^2 + 2^2} (\cos 2x + \sin 2x) - + \ldots \right],$$

where $-\pi < x < \pi$.

In summary, Fourier series are series of cosine and sine terms and arise in the important task of representing general periodic functions. They constitute a very important tool in solving problems that involve ordinary and partial differential equations (see Chapter 5).

Problems

2.1 Consider the nonhomogeneous system of first-order, linear differential equations $y_1' = y_2 + \cosh t$ and $y_2' = y_1$ with boundary conditions $y_1(0) = 0$ and $y_2(0) = -\frac{1}{2}$. Solve this initial value problem by the following three methods:

(a) Matrix methods. Use the method of variation of parameters to determine the particular solution.

(b) Laplace transform methods.

(c) Convert the two first-order differential equations into a single second-order differential equation and solve this latter equation.

2.2 Consider the following problems:

(a) State the differential equation for the Sturm–Liouville problem where $r(x) = x$, $p(x) = x^{-1}$, and $q(x) = 0$.

(b) Using the transformation $x = e^t$, show that this differential equation reduces to $\dfrac{d^2 y}{dt^2} + \lambda y = 0$.

(c) Given the boundary conditions $y(1) = 0$ and $y(e) = 0$ for the differential equation in (a), what are the corresponding eigenvalues and eigenfunctions?

2.3 Using the transformation $x = \cosh t$, show that the differential equation $\dfrac{d^2 y}{dt^2} + \coth t \dfrac{dy}{dt} - 20y = 0$ reduces to Legendre's differential equation and give the expression for the Legendre polynomial which is a solution to this transformed equation.

2.4 Consider the hypergeometric equation $x(1 - x)y'' + [b - (2 + b)x]y' - by = 0$, where b is a constant larger than unity. Using a Frobenius method, show that the series solution about $x = 0$ is the geometric series such that $y_1(x) = 1 + x + x^2 + \ldots$.

2.5 Find a general solution, in terms of Bessel functions, for the differential equation $4x^2 y'' + 4xy' + (x - n^2)y = 0$, where n is an integer, by using either the substitution or the generalized form for the solution of the Bessel equation. What is the solution that satisfies the condition $|y(0)| < \infty$?

2.6 Using J_o for a series expansion, show that the coefficients of the Fourier–Bessel series for $f(x) = 1$ over the interval $0 \leq x \leq 1$ are given as $a_m = 2/\{\alpha_{m0} J_1(\alpha_{m0})\}$, where α_{m0} are the zeros of J_0.

2.7 Consider the function $f(x) = \begin{cases} 0 & \text{if } -1 \leq x \leq 0, \\ 2 & \text{if } 0 \leq x \leq 1. \end{cases}$

Calculate the first two terms of the Fourier–Legendre series of $f(x)$ over the interval $-1 \leq x \leq 1$. What is the value of the complete Fourier–Legendre series of $f(x)$ at $x = 0$?

2.8 Consider the differential equation $16x^2 y'' + 3y = 0$.

(a) Calculate the two basis functions using a Frobenius method and give the general solution.

(b) Derive the second basis function from $y_1(x)$ using a method of reduction of order.

2.9 Give the Legendre polynomial $P_n(x)$ which is a solution to the differential equation $\dfrac{1}{2}(1 - x^2) y'' - xy' + 3y = 0$.

2.10 Find a general solution, in terms of Bessel functions, for the differential equation $x^2 \dfrac{d^2 y}{dx^2} + x \dfrac{dy}{dx} - (1 + 4x^4) y = 0$ by (i) using the transformation $z = x^2$ and (ii) employing the inspection method for the generalized form of Bessel's equation.

2.11 The steady-state temperature distribution for a fin of cross-sectional area A, constant perimeter P, constant conductivity k, and length L can be determined from the following differential equation:

$$\frac{d}{dx}\left(kA\frac{dT}{dx}\right) - hP(T - T_\infty) = 0,$$

where h is the heat transfer coefficient for the fin surrounded by a fluid with a constant temperature T_∞.

(a)   Is this differential equation homogeneous or nonhomogeneous? With a change of variable $\theta = T - T_\infty$, find the general solution for the variable $\theta(x)$ by putting $m^2 = hP/(kA)$.

(b)   Find a solution to the boundary value problem for the boundary conditions $\theta(0) = \theta_0$ and $\theta'(L) = 0$.

2.12   Given the periodic function $f(x) = x$ for $-2 < x < 2$ and $f(x + 4) = f(x)$.

(a)   Sketch this function over the interval $-6 \leq x \leq 6$.

(b)   Find the Fourier series of this function.

(c)   What is the value of $f(x)$ that one would compute from the Fourier series solution at the discontinuity of $x = 2$.

2.13   Given is the differential equation $xy'' + (1 - x)y' + ny = 0$.

(a)   Determine one of the basis solutions for this differential equation, when $n = 2$, using a Frobenius method.

(b)   The Laguerre polynomial $L_n(x)$ is in fact a solution of this differential equation as given by the Rodrigues formula $L_n(x) = e^x \dfrac{d^n}{dx^n}\left(x^n e^{-x}\right)$. Using this formula, specify the Laguerre polynomial $L_2(x)$. What is the value of the arbitrary constant in part (a) to obtain the Laguerre polynomial solution $L_2(x)$?

2.14   Consider the differential equation $\dfrac{d^2 y}{dx^2} - xy = 0$. Using the transformations $y = u\sqrt{x}$ and $\dfrac{2}{3}ix^{3/2} = z$, where $i = \sqrt{-1}$, show that this differential equation reduces to the Bessel differential equation

$$\frac{d^2 u}{dz^2} + \frac{1}{z}\frac{du}{dz} + u\left(1 - \frac{1}{(3z)^2}\right) = 0.$$

What is the solution for $u(z)$ and $y(x)$?

2.15   Determine the particular solution for the nonhomogeneous, second-order, ordinary differential equation $y'' + y' - 2y = 4\sin 2x$ using the following methods:

(a)   undetermined coefficients,

(b)   variation of parameters,

(c)   Fourier transforms (see Problem 3.10).

Laplace and Fourier transforms

The Fourier transform and the Laplace transform are related quantities involving integral transformations to simplify the solution of problems involving differential equations. Both methods are used for solving differential and integral problems. Fourier transforms can be used for signal design; Laplace transforms for control theory. Both methods arise in engineering to simplify calculations in system modeling or, under certain circumstances, converting partial differential equations into ordinary differential equations. The properties of these transforms are discussed.

3.1 Laplace transform methods

3.1.1 Definition of a Laplace transform

The Laplace transform of a function $f(t)$ is defined as

$$\mathscr{L}\{f(t)\} = F(s) = \int_0^t e^{-st} f(t)\, dt \tag{3.1}$$

and exists according to whether the integral in Eq. (3.1) exists (converges).

In practice, there will be a real number s_o such that Eq. (3.1) exists for $s > s_o$, and does not exist for $s \leqslant s_o$. The set of values $s > s_o$ for which Eq. (3.1) exists is called the range of convergence or existence of $\mathscr{L}\{f(t)\}$. The original function $f(t)$ is called the inverse Laplace transform of $F(s)$ and is denoted by

$$f(t) = \mathscr{L}^{-1}\{F(s)\}. \tag{3.2}$$

The symbol $\mathscr{L}$ in Eq. (3.1) is called the Laplace transform operator, and is a linear operator (as well as $\mathscr{L}^{-1}$), that is,

$$\mathscr{L}\{c_1 f_1(t) + c_2 f_2(t)\} = c_1 \mathscr{L}\{f_1(t)\} + c_2 \mathscr{L}\{f_2(t)\}. \tag{3.3}$$

Example 3.1.1. Find the Laplace transform of $f(t) = 1$.

Advanced Mathematics for Engineering Students. https://doi.org/10.1016/B978-0-12-823681-9.00011-3

Solution. We have

$$\mathcal{L}\{f(t)\} = \int_0^\infty e^{-st}\,dt = \lim_{T\to\infty}\left[-\frac{1}{s}e^{-sT}\right]_0^T$$

$$= \lim_{T\to\infty}\left[-\frac{1}{s}e^{-sT} + \frac{1}{s}e^0\right] = \frac{1}{s} \quad (s > 0). \qquad \text{[answer]}$$

The Laplace transforms of other elementary functions are listed in Table 3.1.

Table 3.1 Laplace transforms of some elementary functions.

$f(t)$	$\mathcal{L}\{f(t)\} = F(s)$	$f(t)$	$\mathcal{L}\{f(t)\} = F(s)$
1	$\dfrac{1}{s}, \quad s > 0$	$\cos \omega t$	$\dfrac{s}{s^2 + \omega^2}, \quad s > 0$
$t^n, \quad n = 1, 2, 3...$	$\dfrac{n!}{s^{n+1}}, \quad s > 0$	$\sin \omega t$	$\dfrac{\omega}{s^2 + \omega^2}, \quad s > 0$
$t^a, \quad a > -1$	$\dfrac{\Gamma(a+1)}{s^{a+1}}, \quad s > 0$	$\cosh at$	$\dfrac{s}{s^2 - a^2}, \quad s > |a|$
e^{at}	$\dfrac{1}{s-a}, \quad s > a$	$\sinh at$	$\dfrac{a}{s^2 - a^2}, \quad s > |a|$
$\dfrac{t^{n-1}}{\Gamma(n)}$	$\dfrac{1}{s^n}, \quad n > 0$	$e^{at}\cos bt$	$\dfrac{(s-a)}{(s-a)^2 + b^2}, \quad s > a$
$\dfrac{t^{n-1}e^{at}}{(n-1)!} \quad 0! = 1$	$\dfrac{1}{(s-a)^n}, \quad n = 1, 2, 3...$	$e^{at}\sin bt$	$\dfrac{b}{(s-a)^2 + b^2}, \quad s > a$
$t^n e^{at}$	$\dfrac{n!}{(s-a)^{n+1}}, \quad s > a, n = 0, 1, 2...$	$e^{at}\cosh bt$	$\dfrac{(s-a)}{(s-a)^2 - b^2}, \quad s - a > |b|$
$\dfrac{t^{n-1}e^{at}}{\Gamma(n)}$	$\dfrac{1}{(s-a)^n}, \quad n > 0$	$e^{at}\sinh bt$	$\dfrac{b}{(s-a)^2 - b^2}, \quad s - a > |b|$

Existence of Laplace transforms

If $f(t)$ is piecewise continuous on every finite interval in the range $t \geq 0$ and satisfies

$$\boxed{|f(t)| \leq Me^{\gamma t}} \tag{3.4}$$

for some constants γ and M, then $\mathcal{L}\{f(t)\}$ exists for $s > \gamma$.

Uniqueness

If the Laplace transform of a given function exists, it is uniquely determined. Conversely, if two functions have the same transform, these functions cannot differ over an interval of positive length, although they may differ at various isolated points, that is, the inverse of a given transform is essentially unique.

3.1.2 **Laplace transforms of derivatives**

Laplace transforms are useful for solving linear differential equations, and therefore it is useful to find Laplace transforms of derivatives.

Laplace transforms of first derivatives of $f(t)$

Let $f(t)$ be continuous, have a piecewise continuous derivative $f'(t)$ on every interval in the range $t \geq 0$, and satisfy Eq. (3.4). Then $\mathcal{L}\{f'(t)\}$ exists when $s > \gamma$ and

$$\boxed{\mathcal{L}\{f'(t)\} = s\mathcal{L}\{f(t)\} - f(0).} \tag{3.5}$$

Laplace transforms of $f^{(n)}(t)$

The above theorem can be generalized if $f^{(n-1)}(t)$ is continuous, $f^{(n)}(t)$ is piecewise continuous on every finite interval, and $f(t), f'(t), ..., f^{(n-1)}(t)$ satisfy Eq. (3.4):

$$\boxed{\mathcal{L}\{f^{(n)}(t)\} = s^n \mathcal{L}\{f(t)\} - s^{n-1}f(0) - s^{n-2}f'(0) - ... - f^{(n-1)}(0).} \tag{3.6}$$

Example 3.1.2. Let $f(t) = t^2$ and find the Laplace transform of $\mathcal{L}\{t^2\}$.

Solution. Since $f(0) = 0$, $f'(0) = 0$, $f''(t) = 2$, and $\mathcal{L}\{2\} = 2\mathcal{L}\{1\} = \dfrac{2}{s}$ (see Table 3.1), from Eq. (3.6), $\mathcal{L}\{f''(t)\} = s^2\mathcal{L}\{f(t)\} - sf(0) - f'(0)$. Hence, $\mathcal{L}\{f''(t)\} = \mathcal{L}\{2\} = \dfrac{2}{s} = s^2\mathcal{L}\{f(t)\} - s \cdot 0 - 0$. Therefore, $\mathcal{L}\{t^2\} = \dfrac{2}{s^3}$ (which is in agreement with Table 3.1).   [answer]

Example 3.1.3. Solve $y'' - y = t$ with $y(0) = 1$ and $y'(0) = 1$.

Solution. From Eq. (3.6) and Table 3.1, the subsidiary equation is $s^2 Y(s) - s\, \underbrace{y(0)}_{=1}$

$-\underbrace{y'(0)}_{=1} - Y(s) = \dfrac{1}{s^2}$ so that $(s^2 - 1)Y(s) = s + 1 + \dfrac{1}{s^2}$.

Hence, on solving the subsidiary equation, $Y(s) = \dfrac{s+1}{s^2-1} + \dfrac{1}{s^2(s^2-1)} = \dfrac{1}{s-1} +$

$\left(\dfrac{1}{s^2-1} - \dfrac{1}{s^2} \right)$.

The problem solution is therefore $y(t) = \mathcal{L}^{-1}\{(Y(s)\} = \mathcal{L}^{-1}\{\dfrac{1}{s-1}\} +$

$\mathcal{L}^{-1}\{\dfrac{1}{s^2-1}\} - \mathcal{L}^{-1}\{\dfrac{1}{s^2}\} = e^t + \sinh t - t$.   [answer]

The general methodology for solving this problem with a Laplace transform method is shown below:

General methodology

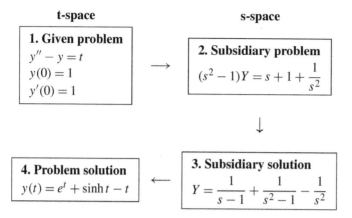

Example 3.1.4. Simultaneous first-order rate equations.

Consider the set of first-order rate equations describing the consecutive reactions $A \xrightarrow{k_1} B$, $B \xrightarrow{k_2} C$. The rate equations are

$$\frac{dn_a}{dt} = -k_1 n_a \quad \text{and} \quad \frac{dn_b}{dt} = k_1 n_a - k_2 n_b$$

and the material balance between the initial and final products is given by

$$n_{ao} + n_{bo} + n_{co} = n_a + n_b + n_c.$$

Solution.

(a) <u>Direct integration</u>

The solution of the first rate equation is directly obtained by separation of variables (see Section 2.1): $\int_0^t \frac{dn_a}{n_a} = - \int_0^t k_1\, dt$. Thus, on integrating,

$$\ln[n_a(t)] - \ln\underbrace{[n_a(0)]}_{n_{ao}} = -k_1 t \quad \text{or} \quad \ln\left[\frac{n_a(t)}{n_{ao}}\right] = -k_1 t.$$

Hence, the solution is $n_a(t) = n_{ao} e^{-k_1 t}$. Substitution of this result into the other rate equation results in a linear differential equation of first order:

$$\frac{dn_b}{dt} + k_2 n_b = k_1 n_{ao} e^{-k_1 t}.$$

Multiplying through by the integrating factor $e^{k_2 t}$ (see Section 2.1):

$$e^{k_2 t}\frac{dn_b}{dt} + k_2 n_b e^{k_2 t} = k_1 n_{ao} e^{-(k_1-k_2)t} \quad \text{or} \quad \frac{\left(dn_b e^{k_2 t}\right)}{dt} = k_1 n_{ao} e^{-(k_1-k_2)t}.$$

Integrating both sides of the equation, $\displaystyle\int_0^t \left(dn_b e^{k_2 t}\right) = \int_0^t k_1 n_{ao} e^{-(k_1-k_2)t}\, dt$ yields

$$n_b(t)e^{k_2 t} - n_{bo} = -\frac{k_1 n_{ao}}{(k_1 - k_2)} e^{-(k_1-k_2)t}\Big|_0^t.$$ Hence,

$$n_b(t)e^{k_2 t} - n_{bo} = \frac{k_1 n_{ao}}{(k_2 - k_1)}\left[e^{-(k_1-k_2)t} - 1\right].$$

Thus, simplifying, $n_b(t) = n_{bo}e^{-k_2 t} + \dfrac{k_1 n_{ao}}{(k_2 - k_1)}\left[e^{-k_1 t} - e^{-k_2 t}\right].$

Also by the mass balance, $n_c = n_{a0} + n_{bo} + n_{co} - n_a - n_b.$

Therefore, $n_c(t) = n_{a0} + n_{bo} + n_{co} - n_{ao}e^{-k_1 t} - n_{bo}e^{-k_2 t} - \dfrac{k_1 n_{ao}}{(k_2 - k_1)}\left[e^{-k_1 t} - e^{-k_2 t}\right].$

[answer]

(b) <u>Laplace transform</u>

The subsidiary equations become

$$s N_a - n_{ao} = -k_1 N_a,$$
$$s N_b - n_{bo} = k_1 N_a - k_2 N_b.$$

Solving these two equations yields

$$N_a(s) = \frac{n_{ao}}{k_1 + s},$$

$$N_b(s) = \frac{n_{bo} + k_1 N_a}{k_2 + s} = \frac{n_{bo}}{k_2 + s} + \frac{k_1 n_{ao}}{(k_1 + s)(k_2 + s)}$$

$$= \frac{n_{bo}}{k_2 + s} + \frac{k_1 n_{ao}}{(k_2 - k_1)}\left[\frac{1}{(k_1 + s)} - \frac{1}{(k_2 + s)}\right].$$

From Table 3.1,

$$n_a(t) = \mathcal{L}^{-1}\{N_a(s)\} = n_{ao}e^{-k_1 t},$$

$$n_b(t) = \mathcal{L}^{-1}\{N_b(s)\} = n_{bo}e^{-k_2 t} + \frac{k_1 n_{ao}}{(k_2 - k_1)}\left[e^{-k_1 t} - e^{-k_2 t}\right]. \qquad \text{[answer]}$$

Recall $\mathcal{L}^{-1}\left\{\dfrac{1}{s+a}\right\} = e^{-at}$ from Table 3.1. This result is the same as in part (a).

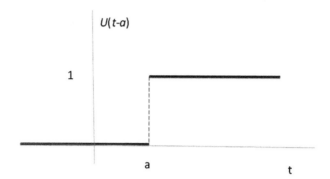

FIGURE 3.1

Unit step function.

3.1.3 The unit step function

The unit step function, also called the Heaviside unit step function (see Fig. 3.1), is defined as

$$U(t-a) = \begin{cases} 0 \ \text{if} \ \ t < a, \\ 1 \ \text{if} \ \ t > a. \end{cases}$$

It is possible to express various discontinuous functions in terms of the unit step function. The Laplace transform of the unit step function is

$$\mathcal{L}\{U(t-a)\} = \frac{e^{-as}}{s} \qquad s > 0$$

and

$$\mathcal{L}^{-1}\left\{\frac{e^{-sa}}{s}\right\} = U(t-a).$$

3.1.4 Special theorems on Laplace transforms

Some important results involving Laplace transforms (and corresponding inverse Laplace transforms) are detailed below. In all cases, it is assumed that $f(t)$ satisfies the existence theorem.

First shifting theorem (s-shifting)

We have

$$\mathcal{L}\{e^{at} f(t)\} = F(s-a)$$

and

$$\mathcal{L}^{-1}\{F(s-a)\} = e^{at} f(t).$$

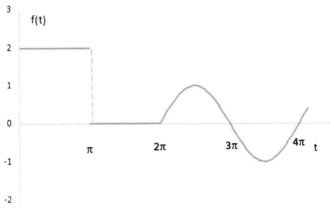

FIGURE 3.2

Example of a second shifting function.

For example, by applying the theorem and using Table 3.1,

$$\mathscr{L}\{e^{at}\cos\omega t\} = \frac{s - a}{(s - a)^2 + \omega^2}.$$

Second shifting theorem (t-shifting)

We have

$$\mathscr{L}\{U(t - a)f(t - a)\} = e^{-as}F(s)$$
and
$$\mathscr{L}^{-1}\{e^{-as}F(s)\} = U(t - a)f(t - a).$$

Example 3.1.5. Find the Laplace transform of $\mathscr{L}\{f(t)\}$, where $f(t)$ is given by Fig. 3.2 such that

$$f(t) = \begin{cases} 2 & \text{if} \quad 0 < t < \pi, \\ 0 & \text{if} \quad \pi < t < 2\pi, \\ \sin t & \text{if} \quad t > 2\pi. \end{cases}$$

Solution. In terms of step functions, $f(t) = 2U(t) - 2U(t - \pi) + U(t - 2\pi)\sin t$.

By applying the theorem and using Table 3.1, $\mathscr{L}\{f(t)\} = \dfrac{2}{s} - \dfrac{2e^{-\pi s}}{s} + \dfrac{e^{-2\pi s}}{s^2 + 1}.$

[answer]

Integration of a function

We have

$$\mathscr{L}\left\{\int_0^t f(u)\,du\right\} = \frac{F(s)}{s}$$

and

$$\mathscr{L}^{-1}\left\{\frac{F(s)}{s}\right\} = \int_0^t f(u)\,du.$$

Example 3.1.6. Given the Laplace transform of $\mathscr{L}\{f(t)\} = \dfrac{1}{s(s^2 + \omega^2)}$, find the function $f(t)$.

Solution. From Table 3.1, $\mathscr{L}^{-1}\left\{\dfrac{1}{s^2 + \omega^2}\right\} = \dfrac{1}{\omega}\sin\omega t$. Therefore, using the theorem,

$$\mathscr{L}^{-1}\left\{\frac{1}{s}\left(\frac{1}{s^2 + \omega^2}\right)\right\} = \frac{1}{\omega}\int_0^t \sin\omega u\,du = \frac{1}{\omega^2}(1 - \cos\omega t). \quad \text{[answer]}$$

Integration of a transform

We have

$$\mathscr{L}\left\{\frac{f(t)}{t}\right\} = \int_s^\infty F(u)\,du$$

and

$$\mathscr{L}^{-1}\left\{\int_s^\infty F(u)\,du\right\} = \frac{f(t)}{t}.$$

if $\lim\limits_{t\to 0} \dfrac{f(t)}{t}$ exists.

Example 3.1.7. Find $\mathscr{L}\left\{\dfrac{1 - e^{-t}}{t}\right\}$.

Solution. Since $\lim\limits_{t\to 0}\dfrac{1 - e^{-t}}{t} = \lim\limits_{t\to 0}\dfrac{e^{-t}}{1} = 1$ from l'Hopital's rule, $1 - e^{-t}$ is continuous, and Eq. (3.4) applies, the above theorem can be used. By Table 3.1, $\mathscr{L}\{1 - e^{-t}\} = \dfrac{1}{s} - \dfrac{1}{s+1}$. As such, by the theorem,

$$\mathscr{L}\left\{\frac{1 - e^{-t}}{t}\right\} = \int_s^\infty \left(\frac{1}{u} - \frac{1}{u+1}\right)du = \lim_{K\to\infty}\int_s^K \left(\frac{1}{u} - \frac{1}{u+1}\right)du$$

$$= \lim_{K\to\infty} [\ln u - \ln(u+1)]\Big|_s^K = \lim_{K\to\infty}\left[\ln\left(1 + \frac{1}{s}\right) - \ln\left(1 + \frac{1}{K}\right)\right]$$

$$= \ln\left(1 + \frac{1}{s}\right). \quad \text{[answer]}$$

Differentiation of a transform

We have

$$\mathscr{L}\{t^n f(t)\} = (-1)^n \frac{d^n F}{ds^n} = (-1)^n F^{(n)}(s), \qquad n = 1, 2, 3, \ldots,$$

and

$$\mathscr{L}^{-1}\{F^{(n)}(s)\} = (-1)^n t^n f(t).$$

For instance, for $n = 1$, $\mathscr{L}\{tf(t)\} = -F'(s)$.

Example 3.1.8. Find the Laplace transform of $\mathscr{L}\{t \sin \beta t\}$.

Solution. From Table 3.1, $\mathscr{L}\{\sin \omega t\} = \dfrac{\omega}{s^2 + \omega^2}$. Thus, setting $\omega = \beta$ and using the above theorem for $n = 1$, $\mathscr{L}\{t \sin \beta t\} = -\dfrac{d}{ds}\left(\dfrac{\beta}{s^2 + \beta^2}\right) = \dfrac{2\beta s}{(s^2 + \beta^2)^2}$.   [answer]

Periodic functions

If $f(t)$ has a period $p > 0$, such that $f(t + p) = f(t)$, then

$$\mathscr{L}\{f(t)\} = \frac{1}{1 - e^{-ps}} \int_0^p e^{-st} f(t)\, dt.$$

Example 3.1.9. Given the square wave function $f(t)$ in Fig. 3.3, find $\mathscr{L}\{f(t)\}$.

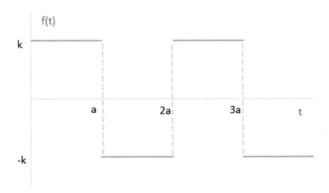

FIGURE 3.3

Example of a square wave function with a period of $2a$.

Solution. Using the above result,

$$\mathcal{L}\{f(t)\} = \frac{1}{1 - e^{-2as}} \left(\int_0^a k e^{-st}\, dt + \int_a^{2a} (-k) e^{-st}\, dt \right)$$

$$= \frac{k}{s} \frac{1 - 2e^{-as} + e^{-2as}}{(1 + e^{-as})(1 - e^{-as})} = \frac{k}{s} \left(\frac{1 - e^{-as}}{1 + e^{-as}} \right) \times \frac{e^{\frac{as}{2}}}{e^{\frac{as}{2}}}$$

$$= \frac{k}{s} \tanh \frac{as}{2}. \qquad \text{[answer]}$$

Convolution theorem

We have

$$\boxed{\mathcal{L}\left\{ \int_0^t f(u)g(t-u)\, du \right\} = \mathcal{L}\left\{ \int_0^t f(t-u)g(u)\, du \right\} = F(s)G(s)}$$

and

$$\mathcal{L}^{-1}\{F(s)G(s)\} = \int_0^t f(u)g(t-u)\, du = \int_0^t f(t-u)g(u)\, du.$$

The above integral, called the convolution of f and g, is written as

$$f * g = \int_0^t f(u)g(t-u)\, du.$$

Example 3.1.10. Given $H(s) = \dfrac{1}{(s^2 + 1)^2} = \dfrac{1}{(s^2 + 1)} \cdot \dfrac{1}{(s^2 + 1)}$, find $h(t) = \mathcal{L}^{-1}\{H(s)\}$.

Solution. From Table 3.1, $\mathcal{L}^{-1}\left\{ \dfrac{1}{s^2 + 1} \right\} = \sin t.$

Hence, using the convolution theorem,

$$h(t) = \mathcal{L}^{-1}\{H(s)\} = \sin t * \sin t = \int_0^t \sin u \, \sin(t - u)\, du$$

$$= \frac{1}{2} \int_0^t -\cos t \, du + \frac{1}{2} \int_0^t \cos(2u - t)\, du = -\frac{1}{2} t \cos t + \frac{1}{2} \sin t. \qquad \text{[answer]}$$

3.1.5 Partial fraction method

In many problems, the subsidiary equation occurs as a quotient of two polynomials:

$$Y(s) = \frac{P(s)}{Q(s)},$$

where $P(s)$ and $Q(s)$ have real coefficients and no common factors. The degree of $P(s)$ is lower than $Q(s)$.

Example 3.1.11. (unrepeated factors) Find $\mathscr{L}^{-1}\left\{\dfrac{s+1}{s(s-2)(s+3)}\right\}$.

Solution. One can set $\dfrac{s+1}{s(s-2)(s+3)} = \dfrac{A}{s} + \dfrac{B}{(s-2)} + \dfrac{C}{(s+3)}$.
To determine the constants A, B, and C, multiply by $s(s-2)(s+3)$ so that

$$s+1 = A(s-2)(s+3) + Bs(s+3) + Cs(s-2).$$

This relation must be an identity and thus it must hold for all values of s. Then, letting $s = 0, 2$, and -3 in succession, one finds $A = -\frac{1}{6}$, $B = \frac{3}{10}$, and $C = -\frac{2}{15}$. Thus, from Table 3.1,

$$\mathscr{L}^{-1}\left\{\frac{s+1}{s(s-2)(s+3)}\right\} = \mathscr{L}^{-1}\left\{\frac{-1/6}{s} + \frac{3/10}{(s-2)} + \frac{-2/15}{(s+3)}\right\}$$

$$= -\frac{1}{6} + \frac{3}{10}e^{2t} - \frac{2}{15}e^{-3t}. \qquad \text{[answer]}$$

Further, consider the two following cases:

Case 1 (unrepeated factor)

As in the previous example, if $Y(s) = \dfrac{P(s)}{Q(s)}$ has a fraction $\dfrac{A}{s-a}$, then from Table 3.1 the inverse transform is Ae^{at}, where

$$A = \lim_{s\to a} \frac{(s-a)P(s)}{Q(s)} \quad \text{or} \quad \frac{P(a)}{Q'(a)}. \tag{3.7}$$

Case 2 (repeated factor)

If $Y(s)$ has a sum of m repeated fractions, $\dfrac{A_m}{(s-a)^m} + \dfrac{A_{m-1}}{(s-a)^{m-1}} + \cdots + \dfrac{A_1}{(s-a)}$, then from Table 3.1, using the first shifting theorem, the inverse transform is

$$e^{at}\left[A_m\frac{t^{m-1}}{(m-1)!} + A_{m-1}\frac{t^{m-2}}{(m-2)!} + \cdots + A_2\frac{t}{1!} + A_1\right],$$

where

$$A_m = \lim_{s\to a}\frac{(s-a)^m P(s)}{Q(s)}, \tag{3.8a}$$

and the other constants are given by

$$A_k = \frac{1}{(m-k)!}\lim_{s\to a}\frac{d^{m-k}}{ds^{m-k}}\left[\frac{(s-a)^m P(s)}{Q(s)}\right], \quad k = 1, 2, \ldots m-1. \tag{3.8b}$$

Example 3.1.12. (repeated factor) Find $\mathscr{L}^{-1}\left\{Y(s) = \dfrac{P(s)}{Q(s)} = \dfrac{s^3-4s^2+4}{s^2(s-2)(s-1)}\right\}$.

Solution. We have $\dfrac{P(s)}{Q(s)} = \dfrac{s^3 - 4s^2 + 4}{s^2(s-2)(s-1)} = \dfrac{A_2}{s^2} + \dfrac{A_1}{s} + \dfrac{B}{(s-2)} + \dfrac{C}{(s-1)}.$

The constants A_2 and A_1 are first determined where $a = 0$ (Case 2). Using Eq. (3.8a) with $m = 2$ gives

$$A_2 = \lim_{s \to 0} \frac{s^2(s^3 - 4s^2 + 4)}{s^2(s-2)(s-1)} = \frac{s^3 - 4s^2 + 4}{(s-2)(s-1)}\bigg|_{s=0} = 2.$$

From Eq. (3.8b) with $m = 2$ and $k = 1$,

$$A_1 = \lim_{s \to 0} \frac{d}{ds}\left[\frac{(s^3 - 4s^2 + 4)}{(s-2)(s-1)}\right] = 3.$$

From Eq. (3.7), the constants B and C can be determined as

$$B = \frac{P(s)}{s^2(s-1)}\bigg|_{s=2} = -1 \quad \text{and} \quad C = \frac{P(s)}{s^2(s-2)}\bigg|_{s=1} = -1.$$

The inverse transform is $y(t) = \mathcal{L}^{-1}\{Y(s)\} = \mathcal{L}^{-1}\left\{\dfrac{2}{s^2} + \dfrac{3}{s} - \dfrac{1}{(s-2)} - \dfrac{1}{(s-1)}\right\},$ which yields the final result:

$$y(t) = 2t + 3 - e^{2t} - e^t. \qquad \text{[answer]}.$$

Finally, as a generalization of Case 1, where $Y(s) = \dfrac{P(s)}{Q(s)}$ and $P(s) = $ a polynomial of degree less than n, $Q(s) = (s - \alpha_1)(s - \alpha_2)...(s - \alpha_n)$, where $\alpha_1, \alpha_2, ..., \alpha_n$ are all distinct, then

$$\boxed{\mathcal{L}^{-1}\left\{\frac{P(s)}{Q(s)}\right\} = \sum_{k=1}^{n} \frac{P(\alpha_k)}{Q'(\alpha_k)} e^{\alpha_k t}.}$$

3.1.6 Laplace inversion formula

Using the theory of complex variables presented in Chapter 10, one can find the inverse Laplace transform with the complex inversion formula for certain cases where the inverse transform is not listed in standard tables.

If $F(s) = \mathcal{L}\{f(t)\}$, then $\mathcal{L}^{-1}\{F(s)\}$ is given by

$$\boxed{f(t) = \frac{1}{2\pi i} \int_{\gamma - i\infty}^{\gamma + i\infty} e^{st} F(s)\, ds \quad t > 0,} \qquad (3.9)$$

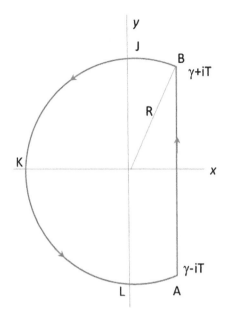

FIGURE 3.4

Contour integral for the Laplace inversion formula.

where s is a complex number, that is, $s = x + iy$. In practice, the integral in Eq. (3.9) is evaluated by considering the contour integral

$$\frac{1}{2\pi i} \oint_C e^{st} F(s) \, ds, \tag{3.10}$$

where C is the contour shown in Fig. 3.4 of the line AB and the arc $BJKLA$ of a circle of radius R. If the arc $BJKLA$ is represented by Γ and $T = \sqrt{R^2 - \gamma^2}$, then Eq. (3.9) can be given as

$$\lim_{R \to \infty} \frac{1}{2\pi i} \int_{\gamma-iT}^{\gamma+iT} e^{st} F(s) \, ds = \lim_{R \to \infty} \left\{ \frac{1}{2\pi i} \oint_C e^{st} F(s) \, ds - \frac{1}{2\pi i} \int_{\Gamma} e^{st} F(s) \, ds \right\}. \tag{3.11}$$

If constants $M > 0$ and $k > 0$ can be found such that on Γ (where $s = Re^{i\theta}$)

$$|F(s)| < \frac{M}{R^k}, \tag{3.12}$$

then the integral around Γ of $e^{st} F(s)$ approaches zero as $R \longrightarrow \infty$. In fact, condition Eq. (3.12) holds if $F(s) = P(s)/Q(s)$, where $P(s)$ and $Q(s)$ are polynomials and the degree of $P(s)$ is less than the degree of $Q(s)$.

Thus, for the case $\lim\limits_{R\to\infty}\int_\Gamma e^{st}F(s)\,ds = 0$, Eq. (3.11) can be evaluated with the use of the residue theorem (see Section 10.5). This theorem states that if $f(z)$ is a function that is analytic inside a simple closed path C and on C, except for finitely many singular points $z_1, z_2, ..., z_k$ inside C, then

$$\oint_C f(z)\,ds = 2\pi i \sum_{j=1,z=z_j}^{k} \operatorname{Res} f(z), \tag{3.13}$$

where the integral is taken counterclockwise around C. The integral in Eq. (3.13) is simply equal to the sum of the residues of $f(z)$. Thus, comparing Eq. (3.11) and Eq. (3.13),

$$f(t) = \sum \text{residues of } e^{st}F(s) \text{ at poles of } F(s). \tag{3.14}$$

A pole is a value at which $F(s)$ becomes infinite. For example, if

$$e^{st}F(s) = e^{st}\frac{P(s)}{Q(s)}, \tag{3.15}$$

a pole is a root of $Q(s) = 0$. For a simple pole (that is, one which is not repeated), the residue when $s = a$ is a root of the denominator of Eq. (3.15):

$$\operatorname{Res}_a = \lim_{s\to a}(s-a)e^{st}\frac{P(s)}{Q(s)}. \tag{3.16}$$

An equivalent formula that is also useful with transcendental functions is

$$\operatorname{Res}_a = \left[e^{st}\frac{P(s)}{dQ(s)/ds}\right]_{s=a}. \tag{3.17}$$

When a pole is repeated m times, the residue is

$$\operatorname{Res}_a = \frac{1}{(m-1)!}\left[\frac{d^{m-1}}{ds^{m-1}}(s-a)^m\frac{P(s)}{Q(s)}e^{st}\right]_{s=a}. \tag{3.18}$$

In fact, Eq. (3.16), Eq. (3.17), and Eq. (3.18) are identical to the results given in Section 3.1.5 (Cases 1 and 2).

Example 3.1.13. Evaluate $\mathcal{L}^{-1}\left\{\dfrac{1}{(s+1)(s-2)^2}\right\}$ by using the method of residues.

Solution. Since the function satisfies the condition in Eq. (3.12), we have

$$\mathcal{L}^{-1}\left\{\frac{1}{(s+1)(s-2)^2}\right\} = \frac{1}{2\pi i}\int_{\gamma-i\infty}^{\gamma+i\infty}\frac{e^{st}\,ds}{(s+1)(s-2)^2} = \frac{1}{2\pi i}\oint_C\frac{e^{st}\,ds}{(s+1)(s-2)^2}$$

$$= \sum \text{residues of } \frac{e^{st}\,ds}{(s+1)(s-2)^2} \text{ at poles } s = -1 \text{ and } s = 2.$$

The residue at simple pole $s = -1$ is $\lim\limits_{s \to -1} (s\cancel{+}1)\left\{\dfrac{e^{st}}{(s\cancel{+}1)(s-2)^2}\right\} = \dfrac{e^{-t}}{9}$, where Eq. (3.16) has been employed.

The residue at the double pole $s = 2$ is evaluated using Eq. (3.18):

$$\lim_{s \to 2} \frac{1}{(1)!} \frac{d}{ds}\left[(s\cancel{-}2)^2 \frac{e^{st}}{(s+1)(s\cancel{-}2)^2}\right] = \lim_{s \to 2} \frac{d}{ds}\left[\frac{e^{st}}{(s+1)}\right]$$

$$= \lim_{s \to 2}\left[\frac{te^{st}(s+1) - e^{st}(1)}{(s+1)^2}\right] = \frac{te^{2t}}{3} - \frac{e^{2t}}{9}.$$

Then $\mathscr{L}^{-1}\left\{\dfrac{1}{(s+1)(s-2)^2}\right\} = \sum \text{residues} = \dfrac{e^{-t}}{9} + \dfrac{te^{2t}}{3} - \dfrac{e^{2t}}{9}$.   [answer]

Example 3.1.14. (Process control)
Consider the linear equation

$$\frac{d^2 x}{dt^2} + a\frac{dx}{dt} + bx = c\frac{dy}{dt} + dy. \tag{3.19}$$

In control theory y is called the input and x is called the response or output function. A process is made up of elements, each of which has input and output signals. When the output signal is used to modify the input to an earlier element of a series, the system possesses feedback.

All signals become known as a function of time by solving the system of differential equations. Usually, however, only the relation between external signals is necessary. It is also easier to work with deviations of signals from their steady-state values. Therefore, the deviations (and all of their derivatives) will be zero at zero time and the Laplace transform of a derivative (Section 3.1.2) reduces to

$$\mathscr{L}\left\{\frac{d^n f}{dt^n}\right\} = s^n F(s).$$

For instance, the transform of Eq. (3.19) is

$$(s^2 + as + b)X(s) = (cs + d)Y(s) \text{ or } X(s) = \left(\frac{cs + d}{s^2 + as + b}\right)Y(s) = G(s)Y(s),$$

where the coefficient $G(s) = \left(\dfrac{cs + d}{s^2 + as + b}\right)$ is called a transfer function. An output transform $X(s)$, accordingly, is found by multiplying the input transform $Y(s)$ by the transfer function $G(s)$. Suppose that a particular input is constant, $y(t) = k$ with transform $Y(s) = k/s$ (see Table 3.1). Then

$$X(s) = \frac{k(cs + d)}{s(s^2 + as + b)}.$$

Finally $x(t)$ can be found by inversion with the aid of partial fraction expansion or with standard Laplace tables.

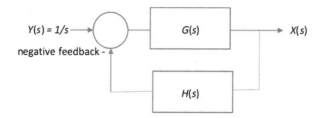

FIGURE 3.5

Block diagram with negative feedback.

Moreover, in the case of feedback, the transforms of an input y and an output x are related by the block diagram in Fig. 3.5. In equation form, the direct relation between transforms is obtained from

$$[Y(s) - H(s)X(s)] \, G(s) = X(s),$$

whence

$$X(s) = \frac{G(s)}{1 + G(s)H(s)} Y(s) = \frac{G(s)}{1 + G(s)H(s)} \left(\frac{1}{s}\right).$$

Thus, similarly, the output response can be obtained by inversion of $X(s)$.

3.2 Fourier transform methods

Fourier series are used for periodic functions. However, the method of Fourier series can be extended to nonperiodic functions with the use of integral representations.

3.2.1 Fourier integrals

The Fourier integral theorem states that if

(i) $f(x)$ satisfies the Dirichlet conditions (Section 2.5.6) in every finite interval $(-L, L)$, and

(ii) $\int_{-\infty}^{\infty} \left| f(x) \right| dx$ converges, then

$$f(x) = \int_{0}^{\infty} [A(\omega) \cos \omega x + B(\omega) \sin \omega x] \, d\omega, \qquad (3.20)$$

where

$$
\boxed{
\begin{aligned}
A(\omega) &= \frac{1}{\pi} \int_{-\infty}^{\infty} f(x) \cos \omega x \, dx, \\
B(\omega) &= \frac{1}{\pi} \int_{-\infty}^{\infty} f(x) \sin \omega x \, dx.
\end{aligned}
}
\tag{3.21}
$$

The Fourier integral expansion of $f(x)$ in Eq. (3.20) holds if x is a point of continuity of $f(x)$; otherwise if x is a point of discontinuity, $f(x)$ is replaced by $\dfrac{f(x+0) + f(x-0)}{2}$. The above conditions are sufficient but not necessary.

The similarity of Eq. (3.20) and Eq. (3.21) with corresponding results for Fourier series is apparent.

Equivalent forms of Fourier integrals

Substituting Eq. (3.21) into Eq. (3.20) gives

$$
f(x) = \frac{1}{\pi} \int_{\omega=0}^{\infty} \int_{v=-\infty}^{\infty} f(v) \{ \cos \omega v \cos \omega x + \sin \omega v \sin \omega x \} \, dv \, d\omega.
$$

Therefore,

$$
\boxed{
f(x) = \frac{1}{\pi} \int_{\omega=0}^{\infty} \int_{v=-\infty}^{\infty} f(v) \cos \omega (x - v) \, dv \, d\omega.
}
\tag{3.22}
$$

Since $\cos \omega (x - v) = \dfrac{e^{i\omega(x-v)} + e^{-i\omega(x-v)}}{2}$, Eq. (3.22) becomes

$$
\begin{aligned}
f(x) &= \frac{1}{2\pi} \int_{v=-\infty}^{\infty} f(v) \left[\int_{\omega=0}^{\infty} \left(e^{i\omega(x-v)} + e^{-i\omega(x-v)} \right) d\omega \right] dv \\
&= \frac{1}{2\pi} \int_{v=-\infty}^{\infty} f(v) \left[\int_{\omega=0}^{\infty} e^{i\omega(x-v)} \, d\omega + \int_{\omega=-\infty}^{0} e^{i\omega(x-v)} \, d\omega \right] dv.
\end{aligned}
$$

Therefore, this latter result can be written as

$$
\boxed{
\begin{aligned}
f(x) &= \frac{1}{2\pi} \int_{-\infty}^{\infty} \int_{-\infty}^{\infty} f(v) e^{i\omega(x-v)} \, dv \, d\omega \\
&= \frac{1}{2\pi} \int_{-\infty}^{\infty} e^{i\omega x} \left[\int_{-\infty}^{\infty} f(v) e^{-i\omega v} \, dv \right] d\omega.
\end{aligned}
}
\tag{3.23}
$$

3.2.2 Fourier transforms

From Eq. (3.23), it follows that if

$$F(\omega) = \frac{1}{\sqrt{2\pi}} \int_{-\infty}^{\infty} f(v)e^{-i\omega v}\, dv, \qquad (3.24)$$

then

$$f(x) = \frac{1}{\sqrt{2\pi}} \int_{-\infty}^{\infty} F(\omega)e^{i\omega x}\, d\omega. \qquad (3.25)$$

The function $F(\omega) = \mathscr{F}\{f(x)\}$ is called the <u>Fourier transform</u> of $f(x)$. The function $f(x)$ is the <u>inverse Fourier transform</u> of $F(\omega)$, that is, $f(x) = \mathscr{F}^{-1}\{F(\omega)\}$.

Example 3.2.1. Find the Fourier transform of the function in Fig. 3.6:

$$f(x) = \begin{cases} 1 & \text{if} \quad |x| < a, \\ 0 & \text{if} \quad |x| > a. \end{cases}$$

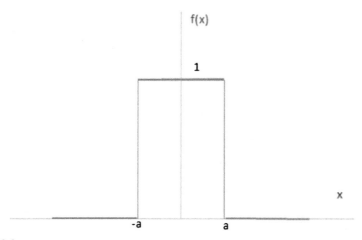

FIGURE 3.6

Square wave function.

Solution. From Eq. (3.24),

$$F(\omega) = \frac{1}{\sqrt{2\pi}} \int_{-\infty}^{\infty} f(v)e^{-i\omega v}\, dv = \frac{1}{\sqrt{2\pi}} \int_{-a}^{a} (1)e^{-i\omega v}\, dv = \frac{1}{\sqrt{2\pi}} \left. \frac{e^{-i\omega v}}{-i\omega} \right|_{-a}^{a}$$

$$= \frac{1}{\sqrt{2\pi}} \left(\frac{e^{i\omega a} - e^{-i\omega a}}{i\omega} \right) = \sqrt{\frac{2}{\pi}} \frac{\sin \omega a}{\omega}, \quad \omega \neq 0. \qquad \text{[answer]}$$

For $\omega = 0$, $F(\omega) = \sqrt{\dfrac{2}{\pi}} a$.

Example 3.2.2. Evaluate the definite integral $\displaystyle\int_{-\infty}^{\infty} \dfrac{\sin \omega a \cos \omega x}{\omega} \, d\omega$.

Solution. From Fourier's integral theorem, if $F(\omega) = \dfrac{1}{\sqrt{2\pi}} \displaystyle\int_{-\infty}^{\infty} f(v) e^{-i\omega v} \, dv$, then

$$f(x) = \frac{1}{\sqrt{2\pi}} \int_{-\infty}^{\infty} F(\omega) e^{i\omega x} \, d\omega.$$

From Example 3.2.1, $\dfrac{1}{\sqrt{2\pi}} \displaystyle\int_{-\infty}^{\infty} \sqrt{\dfrac{2}{\pi}} \dfrac{\sin \omega a}{\omega} e^{i\omega x} \, d\omega = \begin{cases} 1 & \text{if } |x| < a, \\ \frac{1}{2} & \text{if } |x| = a, \\ 0 & \text{if } |x| > a. \end{cases}$

The left-hand side of the above equation is equal to

$$\frac{1}{\pi} \int_{-\infty}^{\infty} \frac{\sin \omega a \cos \omega x}{\omega} \, d\omega + \frac{i}{\pi} \int_{-\infty}^{\infty} \frac{\sin \omega a \sin \omega x}{\omega} \, d\omega.$$

The second integral is zero since the integrand is odd. Thus,

$$\int_{-\infty}^{\infty} \frac{\sin \omega a \cos \omega x}{\omega} \, d\omega = \begin{cases} \pi & \text{if } |x| < a, \\ \frac{\pi}{2} & \text{if } |x| = a, \\ 0 & \text{if } |x| > a. \end{cases} \qquad \text{[answer]}$$

Moreover, if $x = 0$ and $a = 1$, then the result yields $\displaystyle\int_{-\infty}^{\infty} \dfrac{\sin \omega a}{\omega} \, d\omega = \pi$ or

$$\int_{0}^{\infty} \frac{\sin \omega a}{\omega} \, d\omega = \frac{\pi}{2}.$$

Fourier cosine and sine transforms

If $f(x)$ is an <u>even</u> function in Eq. (3.21), $A(\omega) = \dfrac{2}{\pi} \displaystyle\int_{0}^{\infty} f(x) \cos \omega x \, dx$ and $B(\omega) = 0$. Hence, Eq. (3.20) yields the <u>Fourier cosine transforms</u>

$$\boxed{F_C(\omega) = \sqrt{\frac{2}{\pi}} \int_{0}^{\infty} f(x) \cos \omega x \, dx} \qquad (3.26)$$

and

$$\boxed{f(x) = \sqrt{\frac{2}{\pi}} \int_{0}^{\infty} F_C(\omega) \cos \omega x \, d\omega.} \qquad (3.27)$$

Likewise, if $f(x)$ is an <u>odd</u> function, $A(\omega) = 0$ and $B(\omega) = \dfrac{2}{\pi} \displaystyle\int_0^\infty f(x) \sin \omega x \, dx$.
Hence, Eq. (3.20) yields the <u>Fourier sine transforms</u>

$$F_S(\omega) = \sqrt{\frac{2}{\pi}} \int_0^\infty f(x) \sin \omega x \, dx \tag{3.28}$$

and

$$f(x) = \sqrt{\frac{2}{\pi}} \int_0^\infty F_S(\omega) \sin \omega x \, d\omega. \tag{3.29}$$

3.2.3 Special theorems on Fourier transforms

Some important results involving Fourier transforms (and corresponding inverse Fourier transforms) are detailed below.

Linearity

If a and b are constants,

$$\mathscr{F}\{af(x) + bg(x)\} = a\mathscr{F}\{f(x)\} + b\mathscr{F}\{g(x)\}. \tag{3.30}$$

Fourier transform of derivatives

Fourier transforms are useful for solving linear differential equations, and therefore it is useful to find Fourier transforms of derivatives.

Let $f(x)$ be continuous and $f(x) \longrightarrow 0$ as $|x| \longrightarrow \infty$. Furthermore, let $f'(x)$ be absolutely integrable on the x axis. Then

$$\mathscr{F}\{f'(x)\} = i\omega \mathscr{F}\{f(x)\} \tag{3.31}$$

and for higher derivatives

$$\mathscr{F}\{f''(x)\} = -\omega^2 \mathscr{F}\{f(x)\}. \tag{3.32}$$

Cosine and sine transforms of derivatives

Let $f(x)$ be continuous and absolutely integrable on the x axis. Also let $f'(x)$ be piecewise continuous on each finite interval, and $f(x) \longrightarrow 0$ as $|x| \longrightarrow \infty$. Then

$$\begin{aligned} \mathscr{F}_C\{f'(x)\} &= \omega \mathscr{F}_S\{f(x)\} - \sqrt{\frac{2}{\pi}} f(0), \\ \mathscr{F}_S\{f'(x)\} &= -\omega \mathscr{F}_C\{f(x)\} \end{aligned} \tag{3.33}$$

and for higher derivatives

$$\mathscr{F}_C\{f''(x)\} = -\omega^2 \mathscr{F}_C\{f(x)\} - \sqrt{\frac{2}{\pi}} f'(0),$$

$$\mathscr{F}_S\{f''(x)\} = -\omega^2 \mathscr{F}_S\{f(x)\} + \sqrt{\frac{2}{\pi}} \omega f(0). \tag{3.34}$$

Convolution theorem

Define the convolution $f * g$ of functions f and g as

$$(f * g)(x) = \int_{-\infty}^{\infty} f(p)g(x-p)\,dp = \int_{-\infty}^{\infty} f(x-p)g(p)\,dp. \tag{3.35}$$

If $f(x)$ and $g(x)$ are piecewise continuous, bounded, and absolutely integrable on the x axis, then

$$\mathscr{F}\{f * g\} = \sqrt{2\pi}\,\mathscr{F}\{f\}\mathscr{F}\{g\}. \tag{3.36}$$

Taking the inverse transform of Eq. (3.36) yields

$$(f * g)(x) = \int_{-\infty}^{\infty} F(\omega)G(\omega)e^{i\omega x}\,d\omega. \tag{3.37}$$

Fourier series and integrals are important tools in solving boundary value problems as shown in Chapter 5.

Table 3.2 Special Fourier transforms.

$f(x)$	$\mathscr{F}\{f(x)\} = F(w)$	$f(x)$	$\mathscr{F}\{f(x)\} = F(w)$
$\begin{cases} 1 & \text{if } \|x\| < a, \\ 0 & \text{if } \|x\| > a \end{cases}$	$\sqrt{\dfrac{2}{\pi}} \dfrac{\sin aw}{w}$ a	$\dfrac{1}{x^2 + a^2}$	$\sqrt{\dfrac{\pi}{2}} \dfrac{e^{-aw}}{a}$
$\dfrac{x}{x^2 + a^2}$	$-i\sqrt{\dfrac{\pi}{2}} e^{-aw}$	$f^{(n)}(x)$	$i^n w^n F(w)$
$x^n f(x)$	$i^n \dfrac{d^n F}{dw^n}$	$f(ax)e^{itx}$	$\dfrac{1}{a} F\left(\dfrac{w-t}{a}\right)$

a *See Example 3.2.1.*

Special Fourier transform pairs are listed in Table 3.2 for the given definition in Eq. (3.24) and Eq. (3.25) as follows from Spiegel (1973). The listings differ slightly as a different normalization constant $\left(\dfrac{1}{\sqrt{2\pi}}\right)$ was used in Spiegel (1973), such that $\mathscr{F}\{f(x)\} = F(w) = \displaystyle\int_{-\infty}^{\infty} f(x)e^{-iwx}\,dx$ and $\mathscr{F}^{-1}\{F(w)\} = f(x) = \dfrac{1}{2\pi}\displaystyle\int_{-\infty}^{\infty} F(w)e^{iwx}\,dw$.

Table 3.3 Fourier sine transforms.

$f(x)$	$\mathscr{F}_s\{f(x)\} = F_s(w)$	$f(x)$	$\mathscr{F}_s\{f(x)\} = F_s(w)$
$\begin{cases} 1, & 0 < x < a, \\ 0, & x > a \end{cases}$	$\sqrt{\dfrac{2}{\pi}}\left(\dfrac{1-\cos aw}{w}\right)$	x^{-1}	$\sqrt{\dfrac{\pi}{2}}$
$\dfrac{x}{x^2+a^2}$	$\sqrt{\dfrac{\pi}{2}}\, e^{-aw}$	e^{-ax}	$\sqrt{\dfrac{2}{\pi}}\left(\dfrac{w}{w^2+a^2}\right)$
$x^{n-1}e^{-ax}$	$\sqrt{\dfrac{2}{\pi}}\dfrac{\Gamma(n)\sin(n\tan^{-1}\frac{w}{a})}{(w^2+a^2)^{n/2}}$	xe^{-ax^2}	$\dfrac{\sqrt{2}}{4a^{3/2}}we^{-\frac{w^2}{4a}}$
$x^{-1/2}$	$\dfrac{1}{\sqrt{w}}$	x^{-n}	$\sqrt{\dfrac{\pi}{2}}\dfrac{w^{n-1}\csc(\frac{n\pi}{2})}{\Gamma(n)},\quad 0<n<2$
$\dfrac{\sin ax}{x}$	$\dfrac{1}{\sqrt{2\pi}}\ln\left(\dfrac{w+a}{w-a}\right)$	$\dfrac{\sin ax}{x^2}$	$\begin{cases} \sqrt{\frac{\pi}{2}}w, & w<a, \\ \sqrt{\frac{\pi}{2}}a, & w>a \end{cases}$
$\dfrac{\cos ax}{x}$	$\begin{cases} 0, & w<a, \\ \sqrt{2\pi}/4, & w=a, \\ \sqrt{\pi/2}, & w>a \end{cases}$	$\tan^{-1}\dfrac{x}{a}$	$\sqrt{\dfrac{\pi}{2}}\dfrac{e^{-aw}}{w}$
$\csc ax$	$\sqrt{\dfrac{\pi}{2}}\dfrac{\tanh\left(\frac{\pi w}{2a}\right)}{a}$	$\dfrac{1}{e^{2x}-1}$	$\sqrt{\dfrac{2}{\pi}}\left[\dfrac{\pi}{4}\coth\left(\dfrac{\pi w}{2}\right)-\dfrac{1}{2w}\right]$

Fourier sine and cosine transform pairs are also listed in Table 3.3 and Table 3.4, respectively, given the definitions of Eq. (3.26) to Eq. (3.29). Again, account has been taken of the fact that the normalization constant used in Spiegel (1973) for the transform definitions differ.

Fourier sine transforms: We have $\mathscr{F}_s\{f(x)\} = F_s(w) = \displaystyle\int_0^\infty f(x)\sin wx\, dx$ and

$$\mathscr{F}_s^{-1}\{F_s(w)\} = f(x) = \frac{2}{\pi}\int_0^\infty F_s(w)\sin wx\, dw.$$

Fourier cosine transforms: We have $\mathscr{F}_c\{f(x)\} = F_c(w) = \displaystyle\int_0^\infty f(x)\cos wx\, dx$ and

$$\mathscr{F}_c^{-1}\{F_c(w)\} = f(x) = \frac{2}{\pi}\int_0^\infty F_c(w)\cos wx\, dw.$$

3.3 Discrete Fourier transforms and fast Fourier transforms

The discrete Fourier transform (DFT) is equivalent to the continuous Fourier transform in Section 3.2 but instead relies on an individual sampling of a finite sequence of data where each value is separated by a sampling time T. This method is very useful in a wide number of fields that include, for instance, engineering, science, mathematics, and music. For example, DFT applications are used in the processing of digital

Table 3.4 Fourier cosine transforms.

$f(x)$	$\mathscr{F}_c\{f(x)\} = F_c(w)$	$f(x)$	$\mathscr{F}_c\{f(x)\} = F_c(w)$
$\begin{cases} 1, & 0 < x < a, \\ 0, & x > a \end{cases}$	$\sqrt{\dfrac{2}{\pi}}\,\dfrac{\sin aw}{w}$	$\dfrac{1}{x^2+a^2}$	$\sqrt{\dfrac{\pi}{2}}\,\dfrac{e^{-aw}}{a}$
e^{-ax}	$\sqrt{\dfrac{2}{\pi}}\,\dfrac{a}{w^2+a^2}$	$x^{n-1}e^{-ax}$	$\sqrt{\dfrac{2}{\pi}}\,\dfrac{\Gamma(n)\cos(n\tan^{-1}\frac{w}{a})}{(w^2+a^2)^{n/2}}$
e^{-ax^2}	$\dfrac{1}{\sqrt{2a}}e^{-\frac{w^2}{4a}}$	x^{-n}	$\sqrt{\dfrac{\pi}{2}}\,\dfrac{w^{n-1}\sec(\frac{n\pi}{2})}{\Gamma(n)},\quad 0 < n < 1$
$x^{-1/2}$	$\dfrac{1}{\sqrt{w}}$	$\ln\left(\dfrac{x^2+a^2}{x^2+b^2}\right)$	$\dfrac{\sqrt{2}(e^{-bw}-e^{-aw})}{\pi^{3/2}w}$
$\dfrac{\sin ax}{x}$	$\begin{cases} \sqrt{\pi/2}, & w < a, \\ \sqrt{2\pi}/4, & w = a, \\ 0, & w > a \end{cases}$	$\sin ax^2$	$\dfrac{1}{2\sqrt{a}}\left(\cos\frac{w^2}{4a} - \sin\frac{w^2}{4a}\right)$
$\mathrm{sech}\,ax$	$\sqrt{\dfrac{\pi}{2}}\,\dfrac{\mathrm{sech}\left(\frac{\pi w}{2a}\right)}{a}$	$\cos ax^2$	$\dfrac{1}{2\sqrt{a}}\left(\cos\frac{w^2}{4a} + \sin\frac{w^2}{4a}\right)$
$\dfrac{\cosh(\sqrt{\pi}x/2)}{\cosh(\sqrt{\pi}x)}$	$\dfrac{\cosh(\sqrt{\pi}w/2)}{\cosh(\sqrt{\pi}w)}$	$\dfrac{e^{-a\sqrt{x}}}{\sqrt{x}}$	$\dfrac{1}{\sqrt{w}}\left(\cos(2a\sqrt{w}) - \sin(2a\sqrt{w})\right)$

signals such as for sound or radio waves that are collected over a finite interval of time. It is also used for the processing of images where pixels are used to raster an image or for the solution of partial differential equations with Dirichlet and Neumann boundary conditions. In addition, it can be applied for more precise mathematical operations such as for convolutions or with the multiplication of large integer values. To improve the computing speed of the DFT analysis, a fast Fourier transform (FFT) method is able to factor the resulting DFT matrices into very sparse matrices. This approach results in an enormous increase in speed in the analysis of long data sets. The method is also able to avoid round-off errors for more accurate evaluations.

The following discussion is based on the analysis in Oxford (2020), Smith (1997), and Press et al. (1986).

3.3.1 Discrete Fourier transforms

With a continuous Fourier transform in Section 3.2, a Fourier transform of an original signal $f(t)$ with a suitable normalization is defined by

$$F(w) = \frac{1}{\sqrt{2\pi}}\int_{-\infty}^{\infty} f(t)e^{-iwt}\,dt.$$

Consider the case of a sampling of data $f(k)$ that gives rise to N individual values of $f[0], f[1], f[2], ..., f[N-1]$ that are separated by a sample time T. One can think of each sample as an impulse with an area $f[k]$, with a transform integral that now has finite integration limits. The integral can be further replaced by a summation over the sampling points:

$$F(w) = \int_0^{(N-1)T} f(t)\,e^{-iwt}\,dt$$

$$= f[0]e^{-i\cdot 0} + f[1]e^{-iwT} + \ldots + f[k]e^{-iwkT} + \ldots f[N-1]e^{-iw[N-1]T}$$

$$= \sum_{k=0}^{N-1} f[k]e^{-iwkT}.$$

With only N data points, there are only N significant outputs. A continuous Fourier transform is evaluated over the integral limits of $-\infty$ to ∞ for a *periodic* function. Similarly, with only a finite number of data points, the DFT treats data in a periodic manner as well, where the interval $f(N)$ to $f(2N-1)$ is identical to the sequence $f(0)$ to $f(N-1)$. The fundamental frequency for one cycle per sequence is $w = \dfrac{1}{NT}$ Hz or $w = \dfrac{2\pi}{NT}$ rad/s, including $w = 0$ (for a nonoscillating or average component of the signal), as well as higher-order harmonics.

Hence, in general, the DFT $F[n]$ of the sequence $f[k]$ is defined as (Oxford, 2020)

$$F[n] = \sum_{k=0}^{N-1} f[k]e^{-i\frac{2\pi}{N}nk} \qquad \text{for } n = 0, \ldots, N-1. \qquad (3.38)$$

This equation can be equivalently written in the matrix form

$$\begin{bmatrix} F[0] \\ F[1] \\ F[2] \\ \vdots \\ F[N-1] \end{bmatrix} = \begin{bmatrix} 1 & 1 & 1 & \cdots & 1 \\ 1 & W & W^2 & \cdots & W^{N-1} \\ 1 & W^2 & W^4 & \cdots & W^{N-2} \\ \vdots & \cdot & \cdots & & \cdot \\ 1 & W^{N-1} & W^{N-2} & \cdots & W \end{bmatrix} \begin{bmatrix} f[0] \\ f[1] \\ f[2] \\ \vdots \\ f[N-1] \end{bmatrix},$$

where $W = e^{-i2\pi/N}$ and $W = W^{2N}$. An inverse transform, as analogously defined for the continuous transform, is

$$f[k] = \frac{1}{N} \sum_{n=0}^{N-1} F[n]e^{i\frac{2\pi}{N}nk} \qquad \text{for } k = 0, \ldots, N-1, \qquad (3.39)$$

where $F[n]$ coefficients are complex and the resulting spectrum is symmetrical about $N/2$. The inverse matrix is therefore derived as $1/N$ times the complex conjugate of the original matrix. For the inverse transform, the terms $F[n]$ and $F[N-n]$ give rise to two frequency components of which only the lower frequency component at $n \cdot \frac{2\pi}{T}$ Hz for $n \le N/2$ is valid, while the other one is extraneous for $n > N/2$. This latter component is termed an *aliasing frequency* and is an artifact of the signal processing that can be avoided by applying a low-pass filter. Thus, for these two contributions, one has

$$f_n[k] = \frac{1}{N}\left\{ F[n]e^{i\frac{2\pi}{N}nk} + F[N-n]e^{i\frac{2\pi}{N}(N-n)k} \right\}, \tag{3.40}$$

where for $f[k]$ real

$$F[N-n] = \sum_{k=0}^{N-1} f[k]e^{-i\frac{2\pi}{N}(N-n)k}$$

$$= \sum_{k=0}^{N-1} f[k]\underbrace{e^{-i2\pi k}}_{=1\ \forall\ k} e^{i\frac{2\pi}{N}nk} = F^*[n], \tag{3.41}$$

where $F^*[n]$ is the complex conjugate. Thus, substituting, Eq. (3.41) into Eq. (3.40), noting that $e^{i2\pi k} = 1$, and applying Euler's formula for the complex exponential, one obtains

$$\begin{aligned} f_n[k] &= \frac{1}{N}\left\{ F[n]e^{i\frac{2\pi}{N}nk} + F^*[n]e^{-i\frac{2\pi}{N}nk} \right\} \\ &= \frac{2}{N}\left\{ \text{Re}\{F[n]\}\cos\frac{2\pi}{N}nk - \text{Im}\{F[n]\}\sin\frac{2\pi}{N}nk \right\} \\ &= \frac{2}{N}\left\{ |F[n]|\cos\left\{ \left(\frac{2\pi n}{NT}\right)kT + \arg(F[n]) \right\} \right\}. \end{aligned} \tag{3.42}$$

This formula gives a sampled sinusoidal wave at a frequency of $\frac{2\pi n}{NT}$ and a signal amplitude of $\frac{2}{N}|F[n]|$.

Example 3.3.1. Given a continuous signal that is composed of one nonoscillating component and two oscillating components, $f(t) = 3 + 2\cos(2\pi t - \pi/2) + \cos 4\pi t$, evaluate the DFT for this signal sampled at four times per second (that is, 4 Hz) from $t = 0$ to $t = 3/4$.

Solution. Putting the time $t = kT = \frac{k}{4}$, the values for the discrete sampling of the continuous signal in Fig. 3.7 become

$$f[k] = 3 + 2\cos\left(\frac{\pi}{2}k - \frac{\pi}{2}\right) + \cos\pi k.$$

Thus, with $N = 4$

$$f[0] = 3 + 2\cos\left(-\frac{\pi}{2}\right) + \cos(0) = 4,$$

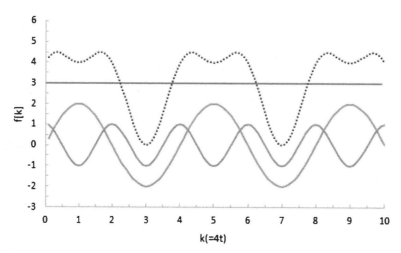

FIGURE 3.7

Continuous signal and individual components of the signal.

$$f[1] = 3 + 2\cos(0) + \cos(\pi) = 4,$$

$$f[2] = 3 + 2\cos\left(\frac{\pi}{2}\right) + \cos(2\pi) = 4,$$

$$f[3] = 3 + 2\cos\left(\frac{3\pi}{2} - \frac{\pi}{2}\right) + \cos(3\pi) = 0.$$

From Eq. (3.38), with $N = 4$

$$F[n] = \sum_{k=0}^{3} f[k]e^{-i\frac{2\pi}{4}nk} = \sum_{k=0}^{3} f[k](-i)^{nk}.$$

Hence,

$$\begin{bmatrix} F[0] \\ F[1] \\ F[2] \\ F[3] \end{bmatrix} = \begin{bmatrix} 1 & 1 & 1 & 1 \\ 1 & -i & -1 & i \\ 1 & -1 & 1 & -1 \\ 1 & i & -1 & -i \end{bmatrix} \begin{bmatrix} f[0] \\ f[1] \\ f[2] \\ f[3] \end{bmatrix} = \begin{bmatrix} 12 \\ -4i \\ 4 \\ 4i \end{bmatrix}. \qquad \text{[answer]}$$

Example 3.3.2. For Example 3.3.1, evaluate the contribution of $f[k]$ from the inverse transform results for $F[0]$, $F[1]$, and $F[2]$.

Solution.

First component: Since $F[0] = 12$, from Eq. (3.39) as a special case that $n = 0$,
$$f_0[k] = \frac{1}{N}F[0] = \frac{12}{4} = 3 \text{ (that is, a constant).} \quad \text{[answer]}$$

Second component: Since $F[1] = -4i = F^*[3]$, the peak amplitude for the fundamental component $f_1[k]$ is $\frac{2}{N}|F[1]| = \frac{2}{4} \times 4 = 2$. The phase for $-4i$ from polar coordinates for this point in a complex x-y diagram is $\arg(F[1]) = -\frac{\pi}{2}$. Hence, from Eq. (3.42), $f_1[k] = 2\cos\left(\frac{2\pi}{NT}kT - \frac{\pi}{2}\right) = 2\cos(\frac{\pi}{2}k - \frac{\pi}{2})$.   [answer]

Third component: Since $F[2] = 4$, where $n = \frac{N}{2}$, the second term in Eq. (3.40) is extraneous so that there are no $N - n$ components such that $f_2[k] = \frac{1}{N}F[2]e^{i\frac{2\pi}{4}2\cdot k} = e^{i\pi k} = \cos\pi k$ since $\sin\pi k = 0$ for all k.   [answer]

As expected, all $f_n[k]$ (for $n = 1, 2, 3$) match the three components for the discrete sampling of the continuous signal $f[k]$ in Example 3.3.1.

In addition to aliasing where different signals can become indistinguishable when sampled, spectral leakage can also occur when a noninteger number of periods of a signal is transformed. This phenomenon results in a spread of the signal among several frequencies after the DFT analysis.

Boundary value problems

The DFT technique is also used to numerically solve boundary value problems for partial differential equations in Chapter 5 and Chapter 6. For instance, consider heat conduction in a flat plate as described by a Poisson partial differential equation, $\nabla^2 u = f(x, y)$, in Eq. (6.13). To solve this problem, one can apply a discrete inverse Fourier transform with a double sum for the temperature in the plate $u(x, y)$ at the position x and y with an equivalent relation to Eq. (3.39), such that (see Press et al., 1986)

$$u_{jl} = \frac{1}{JL}\sum_{m=0}^{J-1}\sum_{n=0}^{L-1}\hat{u}_{mn}e^{-2\pi ijm/J}e^{-2\pi iln/L}.$$

Here the function $u(x, y)$ is represented by values at the discrete set of points $x_j = x_0 + jh$, $j = 0, 1, ..., J$, and $y_l = y_0 + lh$, $l = 0, 1, ..., L$, and h is the common grid spacing. Similarly,

$$f_{jl} = \frac{1}{JL}\sum_{m=0}^{J-1}\sum_{n=0}^{L-1}\hat{f}_{mn}e^{-2\pi ijm/J}e^{-2\pi iln/L}.$$

Note that the nomenclature of Press et al. (1986) is adopted in this derivation, which differs only with a sign change in the exponential for the transform pairs. One just needs to be consistent on taking the transform and its inverse, depending on which definition is followed. Substituting these expressions into Eq. (6.17), one obtains

$$\hat{u}_{mn}\left(e^{2\pi im/J} + e^{-2\pi in/J} + e^{2\pi in/L} + e^{-2\pi in/L} - 4\right) = \hat{f}_{mn}h^2.$$

This latter expression can be simplified and solved for $\hat{u}_{mn}$:

$$\hat{u}_{mn} = \frac{\hat{f}_{mn}h^2}{2\left(\cos\dfrac{2\pi m}{J} + \cos\dfrac{2\pi n}{L} - 2\right)}.$$

Thus, for the solution of the heat conduction equation, one computes $\hat{f}_{mn}$ as the Fourier transform

$$\hat{f}_{mn} = \sum_{j=0}^{J-1}\sum_{l=0}^{L-1} f_{jl}e^{2\pi imj/J}e^{2\pi inl/L}.$$

This transform is used in the expression for $\hat{u}_{mn}$ and then an inverse transform of this subsequent result is taken to obtain the final solution for u_{jl}. This analysis is only applicable for periodic boundary conditions: $u_{jl} = u_{j+J,l} = u_{j,l+L}$.

However, the methodology can be easily extended to a Dirichlet boundary condition, where $u = 0$ on the rectangular boundaries. In this case,

$$u_{jl} = \frac{2}{J}\frac{2}{L}\sum_{m=1}^{J-1}\sum_{n=1}^{L-1}\hat{u}_{mn}\sin\frac{\pi jm}{J}\sin\frac{\pi ln}{L}$$

and one computes analogously the sine transform for $\hat{f}_{mn}$,

$$\hat{f}_{mn} = \sum_{j=1}^{J-1}\sum_{l=1}^{L-1} f_{jl}\sin\frac{\pi jm}{J}\sin\frac{\pi ln}{L}.$$

This expression is used in the similar expression for $\hat{u}_{mn}$,

$$\hat{u}_{mn} = \frac{\hat{f}_{mn}h^2}{2\left(\cos\dfrac{\pi m}{J} + \cos\dfrac{\pi n}{L} - 2\right)}.$$

Again, an inverse sine transform is applied to the resulting expression for $\hat{u}_{mn}$ to obtain the final solution for u_{jl}.

In the case of an inhomogeneous boundary condition, such that $u = 0$ on all boundaries except at $x = Jh$, where $u = g(y)$, one simply adds the above solution to the solution for the homogeneous equation $\nabla^2 u = \dfrac{\partial^2 u}{\partial x^2} + \dfrac{\partial^2 u}{\partial y^2} = 0$ that satisfies the required boundary condition. For instance in the continuum for a flat plate from Example 5.2.6, the solution is given by Eq. (5.28):

$$u^H = \sum_n A_n\sinh\frac{n\pi x}{Jh}\sin\frac{n\pi y}{Lh}.$$

Here A_n is found by imposing the boundary condition such that $u = g(y)$ at $x = Jh$. Analogously, for the discrete case

$$u_{jl}^H = \frac{2}{L} \sum_{n=1}^{L-1} A_n \sinh \frac{n\pi j}{J} \sin \frac{n\pi l}{L},$$

where A_n is obtained from the inverse formula

$$A_n = \frac{1}{\sinh \pi n} \sum_{l=1}^{L-1} g_l \sin \frac{n\pi l}{L},$$

where $g_l = g(y = lh)$. The complete solution is

$$u = u_{jl} + u_{jl}^H.$$

Algorithms using FFT methods (see Section 3.3.2) are given by Press et al. (1986) to efficiently solve these transforms numerically.

3.3.2 Fast Fourier transforms

The number of multiplications to evaluate a DFT grows significantly with the length of a transform as N^2, resulting in a loss of computational speed. This result occurs due to the matrix and vector multiplication shown for the matrix form of Eq. (3.38) in Section 3.3.1. The FFT method was developed, subsequently, recognizing and eliminating the redundancy in the DFT calculations. Here $F[n]$ in Eq. (3.38) can be rewritten as

$$F[n] = \sum_{k=0}^{N-1} f[k] W_N^{nk},$$

where $W_N^{nk} = e^{-i\frac{2\pi}{N}nk}$. It is recognized that W_N^{nk} is evaluated many times and is periodic, as demonstrated in Example 3.3.3 for $N = 8$ (which is an integral power of 2).

Example 3.3.3. Evaluate W_N^k for $N = 8$ and $k = 0$ to 7.

Solution. We have $W_8^1 = e^{-i\frac{2\pi}{8}} = \frac{1-i}{\sqrt{2}} = a$. Thus, it follows that $a^2 = -i$, $a^3 = -ai = -a^*$, $a^4 = -1$, $a^5 = -a$, $a^6 = i$, $a^7 = ai = a^*$ and $a^8 = 1$. Hence, it can be seen that $W_8^4 = -W_8^0$, $W_8^5 = -W_8^1$, $W_8^6 = -W_8^2$, and $W_8^7 = -W_8^3$. [answer]

To take advantage of this symmetry and periodicity, one splits the N samples into two summations of size $N/2$, with one summation having an even value for k and the other an odd value, so that

$$F[n] = \sum_{m=0}^{N/2-1} f[2m] W_N^{2mn} + \sum_{m=0}^{N/2-1} f[2m+1] W_N^{(2m+1)n}.$$

Given that $W_N^{2mn} = e^{-i\frac{2\pi}{N}(2mn)} = e^{-i\frac{2\pi}{N/2}mn} = W_{N/2}^{mn}$, one obtains

$$F[n] = \sum_{m=0}^{N/2-1} f[2m]W_{N/2}^{mn} + W_N^m \sum_{m=0}^{N/2-1} f[2m+1]W_{N/2}^{mn} \tag{3.43}$$
$$= G[n] + W_N^m H[n].$$

Consequently, the N-point DFT can be split into two single $N/2$-point transforms, where one contains even input data $G[n]$ and the other odd input data $H[n]$. Even though the frequency index n ranges over N values, only $N/2$ computations are needed as both $G[n]$ and $H[n]$ are periodic in n having a period $N/2$. This results in important computational savings, as demonstrated in Example 3.3.4 for $N = 8$.

Example 3.3.4. For the case of N samples containing even input data $f[0]$, $f[2]$, $f[4]$, $f[6]$ and odd input data $f[1]$, $f[3]$, $f[5]$, $f[7]$, evaluate the N-point DFT.

Solution. We have

$$F[0] = G[0] + W_8^0 H[0],$$
$$F[1] = G[1] + W_8^1 H[1],$$
$$F[2] = G[2] + W_8^2 H[2],$$
$$F[3] = G[3] + W_8^3 H[3],$$
$$F[4] = G[4] + W_8^4 H[4] = G[0] - W_8^0 H[0],$$
$$F[5] = G[5] + W_8^5 H[5] = G[1] - W_8^1 H[1],$$
$$F[6] = G[6] + W_8^6 H[6] = G[2] - W_8^2 H[2],$$
$$F[7] = G[7] + W_8^7 H[7] = G[3] - W_8^3 H[3]. \qquad \text{[answer]}$$

This calculation is depicted in the flow diagram of Fig. 3.8. The basis of the FFT calculation is a butterfly computation shown in Fig. 3.9 because of the criss-cross nature of the calculation.

With N as a power of two, the process in Fig. 3.8 can be repeated with a breakdown of the $N/2$ signals into $N/4$ transforms, and so on (see Fig. 3.10). Each stage uses a so-called interface decomposition, which enables a separating of the even- and odd-numbered samples. There are $\log_2 N$ stages required for the decomposition. For example, in Fig. 3.10, for the eight-point signal (2^3), three stages are required. Similarly, a 16-point signal (2^4) requires four stages, a 512-point signal (2^7) needs seven stages, a 4096-point signal (2^{12}) involves 12 stages, and so on. As previously mentioned, a DFT analysis requires N^2 complex operations. On the other hand, there are $N/2$ complex multiplications to combine the results from the previous stage and $\log_2 N$ stages so that $(N/2) \cdot \log_2 N$ operations are needed to evaluate an N-point DFT with the FFT. This results in a considerable saving in the computational time.

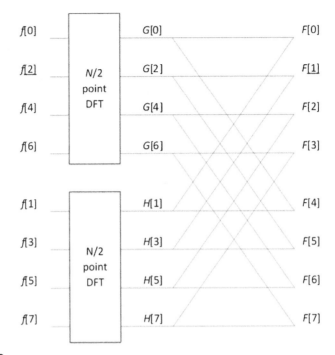

FIGURE 3.8

FFT analysis for an eight-point sampling separated into two four-point transformations.

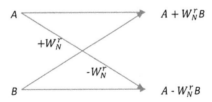

FIGURE 3.9

Basic butterfly operation in an FFT.

For instance, for $N = 256$, the DFT would require 65,536 multiplications compared to only 1,024 with the FFT. For $N = 1024$, the DFT requires a very large number of 1,048,576 multiplications compared with a much smaller number of 5,120 for the FFT.

The time domain decomposition is usually performed using a <u>bit reversal</u> sorting. Here there is a reordering of the N time domain samples employing a binary counting with the bits flipped left-for-right (see Fig. 3.11). This process allows the algorithm to effectively decimate the sequence of input samples at each stage into separate even- and odd-indexed samples.

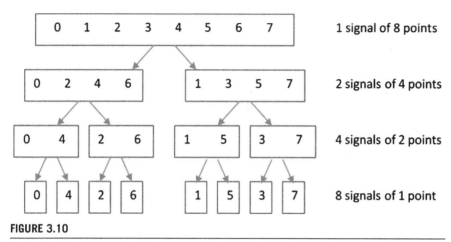

FIGURE 3.10

FFT decomposition where an N-point signal is decomposed into N signals, each containing a single point.

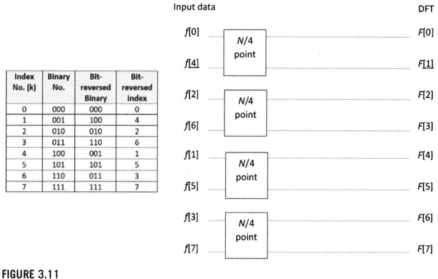

FIGURE 3.11

Evaluation of the bit-reversed counting of $N = 8$ input data separated as an $N/4$ DFT.

Thus, in the FFT process N-point time domain signals are decomposed into N separate time domain signals, each with a single point, as shown in Fig. 3.10. The next step involves a calculation of the N frequency spectra using the corresponding N time domain data. The N spectra are then synthesized into a single frequency spectrum. The algorithm for this analysis is shown in Fig. 3.12.

Time Domain Data

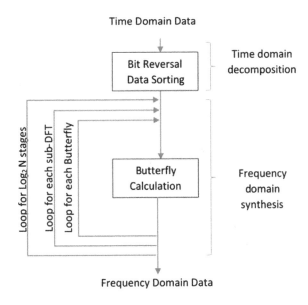

FIGURE 3.12

Algorithm for the fast Fourier transform.

Problems

3.1 Find $\mathcal{L}^{-1}\left\{\dfrac{s}{(s^2-a^2)^2}\right\}$ using the property for differentiation of a transform.

3.2 Find $\mathcal{L}\left\{\dfrac{\cos at-\cos bt}{t}\right\}$ using the property for integration of a transform.

3.3 Find $\mathcal{L}^{-1}\left\{\dfrac{1}{s^4-1}\right\}$ using the convolution theorem.

3.4 Find the inverse Laplace transform $\mathcal{L}^{-1}\left\{\dfrac{e^{-\pi s}s}{(s^2+1)}\right\}$ using the second shifting theorem, and sketch the resultant function.

3.5 Find the inverse Laplace transform $\mathcal{L}^{-1}\left\{\dfrac{1}{(s^2-1)}\right\}$ using the property for the differentiation of a transform.

3.6 Find the inverse Laplace transform $\mathcal{L}^{-1}\left\{\left(\dfrac{\Gamma(2)}{s^2}\right)\left(\dfrac{s}{s^2+1}\right)\right\}$ using the convolution theorem, where Γ is the gamma function.

3.7 Consider the function $f(x)=\begin{cases}0 & \text{if } -1\le x\le 0,\\ 2 & \text{if } 0\le x\le 1.\end{cases}$

  (a) Sketch this function and write it in terms of Heaviside step functions.

  (b) What is the Laplace transform of $f(x)$?

3.8 Find the Fourier transform $\mathscr{F}\{f(x)\}$ by direct integration, where $f(x)$ is given by $f(x) = \begin{cases} e^x & \text{if } x < 0, \\ 0 & \text{if } x > 0. \end{cases}$

3.9 Show that the Fourier sine transform pair as given in mathematical handbooks (for example, Spiegel, 1973),

$$f(x) = x^{a-1} \quad (0 < a < 1) \quad \text{and} \quad F(w) \equiv \mathscr{F}\{f(x)\} = \sqrt{\frac{2}{\pi}} \frac{\Gamma(a)}{w^a} \sin\frac{a\pi}{2},$$

reduces to the following transform pair for an appropriate choice of a:

$$f(x) = \frac{1}{\sqrt{x}} \quad \text{and} \quad F(w) \equiv \mathscr{F}\{f(x)\} = \frac{1}{\sqrt{w}}.$$

3.10 Determine the particular solution for the nonhomogeneous, second-order, ordinary differential equation $y'' + y' - 2y = 4\sin 2x$ using a Fourier transform method (compare your answer to the solution in Problem 2.15(a) and (b)).

3.11 Show that the normal distribution reduces to $f(x) = \dfrac{e^{-(x/a)^2}}{a\sqrt{\pi}}$ if it is centered at the origin, where $a = \sqrt{2}\sigma$. The maximum of this function occurs at $f(0) = \dfrac{1}{a\sqrt{\pi}}$, where a is the width of the curve at half the maximum value.

(a) Show that the Fourier transform of this function is $F(w) = \mathscr{F}\{f(x)\} = \dfrac{1}{\sqrt{2\pi}}e^{(-w/b)^2}$. This function is also a Gaussian function, where $b = 2/a$ is the corresponding width of the transformed curve at half-maximum.

(b) Recover $f(x)$ by evaluating the inverse Fourier transform $\mathscr{F}^{-1}\{F(w)\}$ for the function $F(w)$ given in part (a).

3.12 Given the function $f(x) = \dfrac{1}{a}e^{-|x|/a} = \begin{cases} \dfrac{1}{a}e^{-x/a} & \text{for } x > 0, \\ \dfrac{1}{a}e^{x/a} & \text{for } x < 0. \end{cases}$

(a) Show that the Fourier transform of this function $f(x)$ is $F(w) = \sqrt{\dfrac{2}{\pi}}\left[\dfrac{1}{1+(wa)^2}\right]$.

(b) Recover $f(x)$ by evaluating the inverse Fourier transform $\mathscr{F}^{-1}\{F(w)\}$ for the function $F(w)$ given in part (a) based on a contour integration method (see also Chapter 10). Hint: Consider the contour integral

$$\frac{1}{\pi a^2}\oint_C \frac{e^{izx}}{\left(z+\frac{i}{a}\right)\left(z-\frac{i}{a}\right)}\,dz,$$

where the contours C_1 and C_2 are given in Fig. 3.13(a) and Fig. 3.13(b) for $x > 0$ and $x < 0$, respectively, in the limit that $R \longrightarrow \infty$.

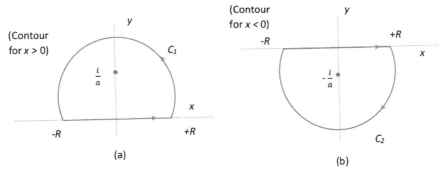

FIGURE 3.13

Contours for the integral (a) for the case that $x > 0$ and (b) for the case that $x < 0$.

3.13   Consider the Laplace transform $F(s) = \dfrac{1}{s-2}$.

(a)   Show that this transform satisfies the condition $|F(s)| < \dfrac{M}{R^k}$ in Eq. (3.12) by finding the values of M and k. Hint: Given the relation $|a+b| \le |a| + |b|$, letting $a = z_1 - z_2$ and $b = z_2$, it follows that $|z_1 - z_2| \ge |z_1| - |z_2|$ or $\dfrac{1}{|z_1 - z_2|} \le \dfrac{1}{|z_1| - |z_2|}$. Using this latter relation, with $z_1 = s = Re^{i\theta}$ and $z_2 = 2$, one obtains the required result: $\left| \dfrac{1}{s-2} \right| \le \dfrac{1}{R-2}$.

(b)   What is the residue of $\dfrac{e^{st}}{s-2}$ at the simple pole $s = 2$?

(c)   From parts (a) and (b), calculate $\mathcal{L}^{-1}\left\{ \dfrac{1}{s-2} \right\}$ using the complex inversion formula. Compare your result to that given in Table 3.1 for $a = 2$.

3.14   Solve the system of chemical kinetic equations in Example 2.4.3 using Laplace transforms.

Matrices, linear systems, and vector analysis

This chapter describes the evaluation of the determinant and the calculation of the inverse of a matrix. It also describes a solution method for systems of equations, which can be solved in a matrix form. This approach becomes important in numerical methods for solution of partial differential equations, as detailed in Chapter 5. Moreover, matrix procedures and solution of systems of equations can be applied with the Maple software package, as further detailed in Appendix A.

In addition, this chapter further details vector analysis, which is concerned with the differentiation and integration of vector fields. Vector fields arise in physics and engineering disciplines, particularly in the description of electromagnetic and gravitational phenomena as well as fluid flow.

Definition of a matrix

A matrix of order $m \times n$ is a rectangular array of numbers having m rows and n columns. It can be written in the form

$$
A = \begin{bmatrix}
a_{11} & a_{12} & \cdots & a_{1n} \\
a_{21} & a_{22} & \cdots & a_{2n} \\
\cdot & \cdot & \cdots & \cdot \\
a_{m1} & a_{m2} & \cdots & a_{mn}
\end{bmatrix}.
$$

Each number a_{jk} in this matrix is called an element. The subscripts j and k indicate respectively the row and column of the matrix.

The transpose of a matrix, denoted by A^T, is obtained by interchanging the rows and columns of a matrix A, that is, if $A = (a_{jk})$, then $A^T = (a_{kj})$ or

$$
A^T = \begin{bmatrix}
a_{11} & a_{21} & \cdots & a_{m1} \\
a_{12} & a_{22} & \cdots & a_{m2} \\
\cdot & \cdot & \cdots & \cdot \\
a_{1n} & a_{2n} & \cdots & a_{mn}
\end{bmatrix}.
$$

Advanced Mathematics for Engineering Students. https://doi.org/10.1016/B978-0-12-823681-9.00012-5

4.1 **Determinants**

If the matrix A is a square matrix, then the determinant of A of order n is given by

$$D = \det A = \begin{vmatrix} a_{11} & a_{12} & \cdots & a_{1n} \\ a_{21} & a_{22} & \cdots & a_{2n} \\ \cdot & \cdot & \cdots & \cdot \\ a_{m1} & a_{m2} & \cdots & a_{mn} \end{vmatrix}. \tag{4.1}$$

The value of $\det A$ can be obtained as follows.

(i) Given any element a_{jk} of D, a new determinant of order $(n-1)$ can be obtained by removing elements of the jth row and kth column, called the <u>minor</u> M_{jk}. For example, the minor corresponding to element 5 in the 2nd row and 3rd column of the fourth-order determinant

$$\begin{vmatrix} 3 & 0 & 1 & 2 \\ -3 & 2 & 5 & 0 \\ 1 & -1 & 2 & 2 \\ -2 & 4 & 3 & 1 \end{vmatrix}$$

is

$$\begin{vmatrix} 3 & 0 & 2 \\ 1 & -1 & 2 \\ -2 & 4 & 1 \end{vmatrix},$$

which is obtained by removing the shaded elements.

(ii) Multiplying the minor of a_{jk} by $(-1)^{j+k}$, the result is the <u>cofactor</u> of a_{jk} expressed as $C_{jk} = (-1)^{j+k} M_{jk}$. For example, the cofactor of element 5 is

$$C_{23} = (-1)^{2+3} \begin{vmatrix} 3 & 0 & 2 \\ 1 & -1 & 2 \\ -2 & 4 & 1 \end{vmatrix}.$$

The value of the determinant is then defined as the sum of the products of the elements of any row (or column) by their corresponding cofactors and is called the <u>Laplace expansion</u>,

$$\det A = \sum_{k=1}^{n} a_{jk} C_{jk}. \tag{4.2}$$

Example 4.1.1. (Determinant of second order)
We have

$$D = \det A = \begin{vmatrix} a_{11} & a_{12} \\ a_{21} & a_{22} \end{vmatrix}.$$

For $(j = 1)$ the first row is $D = a_{11}a_{22} + a_{12}(-a_{21}) = a_{11}a_{22} - a_{12}a_{21}$. The same value is obtained using the elements of the second row (or first and second columns).

Example 4.1.2. (Determinant of third order)
We have

$$D = \det A = \begin{vmatrix} a_{11} & a_{12} & a_{13} \\ a_{21} & a_{22} & a_{23} \\ a_{31} & a_{32} & a_{33} \end{vmatrix}.$$

The cofactors of the elements in the first row are $(j = 1)$

$$C_{11} = (-1)^{1+1} \begin{vmatrix} a_{22} & a_{23} \\ a_{32} & a_{33} \end{vmatrix} = a_{22}a_{33} - a_{23}a_{32},$$

$$C_{12} = (-1)^{1+2} \begin{vmatrix} a_{21} & a_{23} \\ a_{31} & a_{33} \end{vmatrix} = -(a_{21}a_{33} - a_{23}a_{31}),$$

$$C_{13} = (-1)^{1+3} \begin{vmatrix} a_{21} & a_{22} \\ a_{31} & a_{32} \end{vmatrix} = a_{21}a_{32} - a_{22}a_{31}.$$

Hence, $D = a_{11}C_{11} + a_{12}C_{12} + a_{13}C_{13} = a_{11}(a_{22}a_{33} - a_{23}a_{32}) - a_{12}(a_{21}a_{33} - a_{23}a_{31}) + a_{13}(a_{21}a_{32} - a_{22}a_{31})$.
The same value is obtained by using the elements of the second or third rows (or first, second, and third columns).

Example 4.1.3. (Determinant of a triangular matrix)
The determinant of any triangular matrix equals the products of all the entries on the main diagonal. For example,

$$\begin{vmatrix} -1 & 0 & 0 \\ 6 & 3 & 0 \\ -1 & 2 & 5 \end{vmatrix} = -1 \begin{vmatrix} 3 & 0 \\ 2 & 5 \end{vmatrix} = -1 \cdot 3 \cdot 5 = -15.$$

Theory on determinants

Theorem 4.1. *We have* $\det(A) = \det(A^T)$.

Theorem 4.2. *The value of the determinant remains unchanged by multiplying the elements of any row (or column) by a given number and adding the corresponding elements of any other row (or column).*

Theorem 4.3. *If all elements of any row (or column) are zero except for one element, then the value of the determinant is equal to the product of that element by its cofactor.*

Theorem 4.4. *If $v_1, v_2, ..., v_n$ represent row vectors (or columns) of a square matrix A of order n, then $\det(A) = 0$ if and only if there exist constants (scalars) $\lambda_1, \lambda_2, ..., \lambda_n$ not all zero such that*

$$\lambda_1 v_1 + \lambda_2 v_2 + ... + \lambda_n v_n = 0, \tag{4.3}$$

where zero is the null row matrix. If Eq. (4.3) is satisfied, the vectors are linearly dependent. Otherwise, they are linearly independent. A matrix A such that $\det A = 0$ is called a singular matrix . If $\det A \neq 0$, then A is nonsingular.

In practice, a determinant of order n is evaluated by using *Theorem 2* successfully to replace all but one of the elements in a row (or column) and then using *Theorem 3* to obtain a new determinant of order $n - 1$. This procedure is continued in this manner, arriving ultimately at determinants of order 2 or 3 which are easily evaluated.

4.2 Inverse of a matrix

If A is a nonsingular square matrix of order n (where $\det A \neq 0$), then there exists a unique inverse matrix A^{-1} in the following form:

$$A^{-1} = \frac{(C_{jk})^T}{\det(A)}, \tag{4.4}$$

where (C_{jk}) is the matrix of cofactors C_{jk} and $(C_{jk})^T = C_{kj}$ is its transpose.

Example 4.2.1. Find the inverse of $A = \begin{bmatrix} -1 & -2 & 4 \\ 2 & -5 & 2 \\ 3 & -4 & -6 \end{bmatrix}$.

Solution. Using Eq. (4.2), $\det A = -1(38) - 2(18) + 4(7) = -46$. Moreover,

$$C_{11} = \begin{vmatrix} -5 & 2 \\ -4 & -6 \end{vmatrix} = 38, \quad C_{21} = -\begin{vmatrix} -2 & 4 \\ -4 & -6 \end{vmatrix} = -28, \quad C_{31} = \begin{vmatrix} -2 & 4 \\ -5 & 2 \end{vmatrix} = 16,$$

$$C_{12} = -\begin{vmatrix} 2 & 2 \\ 3 & -6 \end{vmatrix} = 18, \quad C_{22} = \begin{vmatrix} -1 & 4 \\ 3 & -6 \end{vmatrix} = -6, \quad C_{32} = -\begin{vmatrix} -1 & 4 \\ 2 & 2 \end{vmatrix} = 10,$$

$$C_{13} = \begin{vmatrix} 2 & -5 \\ 3 & -4 \end{vmatrix} = 7, \quad C_{23} = -\begin{vmatrix} -1 & -2 \\ 3 & -4 \end{vmatrix} = -10, \quad C_{33} = \begin{vmatrix} -1 & -2 \\ 2 & -5 \end{vmatrix} = 9.$$

Using Eq. (4.4), the inverse matrix is therefore

$$A^{-1} = \frac{1}{-46} \begin{bmatrix} 38 & -28 & 16 \\ 18 & -6 & 10 \\ 7 & -10 & 9 \end{bmatrix} = \begin{bmatrix} -\dfrac{19}{23} & \dfrac{14}{23} & -\dfrac{8}{23} \\[2mm] -\dfrac{9}{23} & \dfrac{3}{23} & -\dfrac{5}{23} \\[2mm] -\dfrac{7}{46} & \dfrac{5}{23} & -\dfrac{9}{46} \end{bmatrix}. \qquad \text{[answer]}$$

Check the following inverse matrix:

$$AA^{-1} = \begin{bmatrix} -1 & -2 & 4 \\ 2 & -5 & 2 \\ 3 & -4 & -6 \end{bmatrix} \begin{bmatrix} -\dfrac{19}{23} & \dfrac{14}{23} & -\dfrac{8}{23} \\[2mm] -\dfrac{9}{23} & \dfrac{3}{23} & -\dfrac{5}{23} \\[2mm] -\dfrac{7}{46} & \dfrac{5}{23} & -\dfrac{9}{46} \end{bmatrix} = \begin{bmatrix} 1 & 0 & 0 \\ 0 & 1 & 0 \\ 0 & 0 & 1 \end{bmatrix} = I = A^{-1}A.$$

4.3 **Linear systems of equations**

A set of equations having the form

$$\begin{array}{ccccccc}
a_{11}x_1 & + & a_{12}x_2 & + & \cdots & + & a_{1n}x_n = b_1, \\
a_{21}x_1 & + & a_{22}x_2 & + & \cdots & + & a_{2n}x_n = b_2, \\
\vdots & & \vdots & & \cdots & & \vdots \\
a_{m1}x_1 & + & a_{m2}x_2 & + & \cdots & + & a_{mn}x_n = b_m
\end{array} \qquad (4.5)$$

is called a system of m linear equations in the n unknowns $x_1, x_2..., x_n$. If $b_1, b_2..., b_m$ are all zero, the system is called <u>homogeneous</u>. If they are not all zero, it is called nonhomogeneous. Any set of numbers $x_1, x_2..., x_n$, which satisfies Eq. (4.5), is called a <u>solution</u> of the system.

In matrix form, Eq. (4.5) can be written as

$$\begin{bmatrix} a_{11} & a_{12} & \cdots & a_{1n} \\ a_{21} & a_{22} & \cdots & a_{2n} \\ \vdots & \vdots & \cdots & \vdots \\ a_{m1} & a_{m2} & \cdots & a_{mn} \end{bmatrix} \begin{bmatrix} x_1 \\ x_2 \\ \vdots \\ x_n \end{bmatrix} = \begin{bmatrix} b_1 \\ b_2 \\ \vdots \\ b_m \end{bmatrix}, \qquad (4.6)$$

or in a vector equation,

$$\boxed{Ax = b.}$$ (4.7)

Solution of systems of n equations in n unknowns

(i) *Inverse method.*

If $m = n$ and A is a nonsingular matrix so that A^{-1} exists, Eq. (4.7) can be solved by

$$\boxed{x = A^{-1}b,}$$ (4.8)

and the system has a unique solution.

Example 4.3.1. Solve

$$
\begin{aligned}
-x_1 &- 2x_2 + 4x_3 = -3, \\
2x_1 &- 5x_2 + 2x_3 = 7, \\
3x_1 &- 4x_2 - 6x_3 = 5.
\end{aligned}
$$

Solution. We have $\begin{bmatrix} -1 & -2 & 4 \\ 2 & -5 & 2 \\ 3 & -4 & -6 \end{bmatrix} \begin{bmatrix} x_1 \\ x_2 \\ x_n \end{bmatrix} = \begin{bmatrix} -3 \\ 7 \\ 5 \end{bmatrix}$.

From Section 4.2, $A^{-1} = \begin{bmatrix} -\dfrac{19}{23} & \dfrac{14}{23} & -\dfrac{8}{23} \\[2mm] -\dfrac{9}{23} & \dfrac{3}{23} & -\dfrac{5}{23} \\[2mm] -\dfrac{7}{46} & \dfrac{5}{23} & -\dfrac{9}{46} \end{bmatrix}$. Therefore, Eq. (4.8) gives

$$
\begin{bmatrix} x_1 \\ x_2 \\ x_3 \end{bmatrix} = \begin{bmatrix} -\dfrac{19}{23} & \dfrac{14}{23} & -\dfrac{8}{23} \\[2mm] -\dfrac{9}{23} & \dfrac{3}{23} & -\dfrac{5}{23} \\[2mm] -\dfrac{7}{46} & \dfrac{5}{23} & -\dfrac{9}{46} \end{bmatrix} \begin{bmatrix} -3 \\ 7 \\ 5 \end{bmatrix} = \begin{bmatrix} 5 \\ 1 \\ 1 \end{bmatrix}.
$$

The solution is $x_1 = 5$, $x_2 = 1$, and $x_3 = 1$.   [answer]

(ii) *Cramer's rule.*

The unknowns $x_1, x_2, ..., x_n$ can be determined from <u>Cramer's rule</u>:

$$x_1 = \frac{D_1}{D}, x_2 = \frac{D_2}{D}, ..., x_n = \frac{D_n}{D},$$ (4.9)

where $D = \det(A)$ is the determinant of the system (given by Section 4.1) and D_k, $k = 1, 2, ..., n$, is the determinant obtained from D by removing the kth row and replacing it with the vector b.

The following four cases can arise.

Case 1. $D \neq 0$, $b \neq 0$. In this case, there will be a unique solution where not all x_k will be zero.

Case 2. $D \neq 0$, $b = 0$. In this case, the only solution will be $x_1 = 0, x_2 = 0, ..., x_n = 0$ (that is, $x = 0$). This result is often called the trivial solution.

Case 3. $D = 0$, $b = 0$. In this case, there will be infinitely many solutions other than the trivial solution. At least one of the equations can be obtained from the others, that is, the equations are linearly dependent.

Case 4. $D = 0$, $b \neq 0$. In this case, infinitely many solutions will exist if and only if all of the determinants D_k in Eq. (4.9) are zero. Otherwise there will be no solution.

Example 4.3.2. Solve the system

$$\begin{aligned} -x_1 & - 2x_2 + & 4x_3 &= -3, \\ 2x_1 & - 5x_2 + & 2x_3 &= 7, \\ 3x_1 & - 4x_2 - & 6x_3 &= 5. \end{aligned}$$

Solution. By Cramer's rule,

$$x_1 = \frac{\begin{vmatrix} -3 & -2 & 4 \\ 7 & -5 & 2 \\ 5 & -4 & -6 \end{vmatrix}}{D}, \qquad x_2 = \frac{\begin{vmatrix} -1 & -3 & 4 \\ 2 & 7 & 2 \\ 3 & 5 & -6 \end{vmatrix}}{D}, \qquad x_3 = \frac{\begin{vmatrix} -1 & -2 & -3 \\ 2 & -5 & 7 \\ 3 & -4 & 5 \end{vmatrix}}{D},$$

where the determinant of the coefficients is $\begin{vmatrix} -1 & -2 & 4 \\ 2 & -5 & 2 \\ 3 & -4 & -6 \end{vmatrix} = -46$ (see Example 4.1.2). Since $D \neq 0$ and $b \neq 0$, this indicates that Case 1 applies with a unique solution. Consider

$$\begin{vmatrix} -3 & -2 & 4 \\ 7 & -5 & 2 \\ 5 & -4 & -6 \end{vmatrix} = (-3)\begin{vmatrix} -5 & 2 \\ -4 & -6 \end{vmatrix} - (-2)\begin{vmatrix} 7 & 2 \\ 5 & -6 \end{vmatrix} + 4\begin{vmatrix} 7 & -5 \\ 5 & -4 \end{vmatrix}$$

$$= (-3)(38) + 2(-52) + 4(-3) = -230.$$

Therefore, $x_1 = D_1/D = (-230)/(-46) = 5$. Evaluation of the other determinants yields $x_2 = 1$ and $x_3 = 1$. Hence the final solution is $x_1 = 5$, $x_2 = 1$, and $x_3 = 1$ (in agreement with Example 4.3.1).   [answer]

(i) *Gauss elimination*

For system Eq. (4.5), described by the vector Eq. (4.7), $A = [a_{jk}]$ is the $m \times n$ matrix

$$A = \begin{bmatrix} a_{11} & a_{12} & \cdots & a_{1n} \\ a_{21} & a_{22} & \cdots & a_{2n} \\ . & . & \cdots & . \\ a_{m1} & a_{m2} & \cdots & a_{mn} \end{bmatrix} \text{ and } x = \begin{bmatrix} x_1 \\ x_2 \\ \vdots \\ x_n \end{bmatrix}, \ b = \begin{bmatrix} b_1 \\ b_2 \\ \vdots \\ b_m \end{bmatrix}.$$

The augmented matrix of the system in Eq. (4.5) is defined by

$$\tilde{A} = \begin{bmatrix} a_{11} & \cdots & a_{1n} & b_1 \\ \cdots & \cdots & \cdots & . \\ \cdots & \cdots & \cdots & . \\ a_{m1} & \cdots & a_{mn} & b_m \end{bmatrix}.$$

In the Gauss elimination method, the augmented matrix is reduced to a "triangular form," from which the values of the unknowns are obtained by "back substitution."

Example 4.3.3. Solve the system in Example 4.3.1 by Gauss elimination.

Solution. Consider the system and its augmented matrix. The equations are given by

$$-x_1 \quad - 2x_2 \quad + 4x_3 \quad = -3,$$

$$2x_1 \quad - 5x_2 \quad + 2x_3 \quad = 7.$$

$$3x_1 \quad - 4x_2 \quad - 6x_3 \quad = 5.$$

The corresponding augmented matrix is

$$\tilde{A} = \begin{bmatrix} -1 & -2 & 4 & -3 \\ 2 & -5 & 2 & 7 \\ 3 & -4 & -6 & 5 \end{bmatrix}.$$

The three-step process is as follows:

Step 1. Elimination of x_1 from the second and third equations. Call the first equation above the pivot equation and its x_1-term the pivot element (shaded as a circle) in this step. Use this equation to eliminate x_1 in the other equations. We have

$$
\begin{array}{rrrl}
-x_1 & -2x_2 & +4x_3 & = -3, \\
 & -9x_2 & +10x_3 & = 1, \\
 & -10x_2 & +6x_3 & = -4,
\end{array}
\qquad
\left|\begin{array}{rrrr}
-1 & -2 & 4 & -3, \\
0 & -9 & 10 & 1, \\
0 & -10 & 6 & -4,
\end{array}\right|
\begin{array}{l}
\\
\text{Row } 2 + 2 \text{ Row } 1, \\
\text{Row } 3 + 3 \text{ Row } 1.
\end{array}
$$

Step 2. <u>Elimination of x_2</u> from the third equation. The first equation, which has just served as the pivot equation, remains untouched. The next step is to take the second (new) equation as the pivot equation.

$$
\begin{array}{rrrl}
-x_1 & -2x_2 & + 4x_3 & = -3, \\
\\
& -9x_2 & + 10x_3 & = 1, \\
\\
& -10x_2 & + 6x_3 & = -4.
\end{array}
$$

The elimination of x_2 gives

$$
\begin{array}{rrrl}
-x_1 & -2x_2 & +4x_3 & = -3, \\
 & -9x_2 & +10x_3 & = 1, \\
 & & -46x_3 & = -46,
\end{array}
\qquad
\left|\begin{array}{rrrr}
-1 & -2 & 4 & -3, \\
0 & -9 & 10 & 1, \\
0 & 0 & -46 & -46,
\end{array}\right|
\begin{array}{l}
\\
\\
9 \text{ Row } 3 - 10 \text{ Row } 2.
\end{array}
$$

Step 3. <u>Back substitution</u> to determine x_3, x_2, x_1. Working backward from the last to the first equation of this "triangular" system gives

$$
\begin{array}{rrl}
 & -46x_3 & = -46, \\
-9x_2 & +10x_3 & = 1, \\
-x_1 \quad -2x_2 & +4x_3 & = -3,
\end{array}
\qquad
\begin{array}{l}
x_3 = 1, \\
x_2 = \tfrac{1}{9}[10(1) - 1] = 1, \\
x_1 = 3 - 2(1) + 4(1) = 5.
\end{array}
$$

The final answer is $x_1 = 5$, $x_2 = 1$, and $x_3 = 1$ (in agreement with Examples 4.3.1 and 4.3.2).   [answer]

Determination of the inverse using *Gauss–Jordan elimination*

Gauss–Jordan elimination can be used for practically determining the inverse A^{-1} of a nonsingular $n \times n$ matrix A. The method is as follows.

(i) Using A, n systems are formed such that $Ax_{(1)} = e_{(1)}, ..., Ax_{(n)} = e_{(n)}$ has the jth component 1 and the other components 0. Thus, introducing the $n \times n$ matrices $X = [x_{(1)} ... x_{(n)}]$ and $I = [e_{(1)} ... e_{(n)}]$, one can combine the n systems into a single matrix equation $AX = I$ and the augmented matrices $[Ae_{(1)}, ..., Ae_{(n)}]$ into a single augmented matrix $\tilde{A} = [AI]$. Now $AX = I$ implies $X = A^{-1}I = A^{-1}$. To solve $AX = I$ one can apply Gauss elimination to $\tilde{A} = [AI]$ to get $[UH]$, where U is upper triangular.

(ii) The Gauss–Jordan elimination now operates on $[UH]$ and, by eliminating the entries in U above the main diagonal, reduces it to $[IK]$, which is the augmented matrix of $IX = A^{-1}$. Hence, the matrix $K = A^{-1}$ can be immediately read off.

Example 4.3.4. Find the inverse of $A = \begin{bmatrix} -1 & -2 & 4 \\ 2 & -5 & 2 \\ 3 & -4 & -6 \end{bmatrix}$ by Gauss–Jordan elimination.

The inverse of this matrix has been found in Section 4.2.

Solution.

(i) Apply the Gauss elimination to the augmented matrix $\tilde{A} = [AI]$. Then we have

$$[AI] = \left[\begin{array}{ccc|ccc} -1 & -2 & 4 & 1 & 0 & 0 \\ 2 & -5 & 2 & 0 & 1 & 0 \\ 3 & -4 & -6 & 0 & 0 & 1 \end{array}\right],$$

$$\left[\begin{array}{ccc|ccc} -1 & -2 & 4 & 1 & 0 & 0 \\ 0 & -9 & 10 & 2 & 1 & 0 \\ 0 & -10 & 6 & 3 & 0 & 1 \end{array}\right] \quad \begin{array}{l} \text{Row } 2 + 2 \cdot \text{Row } 1, \\ \text{Row } 3 + 3 \cdot \text{Row } 1, \end{array}$$

$$\left[\begin{array}{ccc|ccc} -1 & -2 & 4 & 1 & 0 & 0 \\ 0 & -9 & 10 & 2 & 1 & 0 \\ 0 & 0 & -46 & 7 & -10 & 9 \end{array}\right] \quad 9 \cdot \text{Row } 3 - 10 \cdot \text{Row } 2.$$

This augmented matrix $[UH]$ as produced by Gauss elimination agrees with Example 4.3.3.

(ii) Using additional Gauss–Jordan steps, reduce U to I.

$$\left[\begin{array}{ccc|ccc} 1 & 2 & -4 & -1 & 0 & 0 \\ 0 & 1 & \dfrac{-10}{9} & \dfrac{-2}{9} & \dfrac{-1}{9} & 0 \\ 0 & 0 & 1 & \dfrac{-7}{46} & \dfrac{5}{23} & \dfrac{-9}{46} \end{array}\right] \quad \begin{array}{l} -\text{Row } 1, \\[2ex] \left(\dfrac{-1}{9}\right) \cdot \text{Row } 2, \\[2ex] \left(\dfrac{-1}{46}\right) \cdot \text{Row } 3, \end{array}$$

$$
\begin{bmatrix}
1 & 2 & 0 & \dfrac{-37}{23} & \dfrac{20}{23} & \dfrac{-18}{23} \\[2mm]
0 & 1 & 0 & \dfrac{-9}{23} & \dfrac{3}{23} & \dfrac{-5}{23} \\[2mm]
0 & 0 & 1 & \dfrac{-7}{46} & \dfrac{5}{23} & \dfrac{-9}{46}
\end{bmatrix}
\qquad
\begin{aligned}
&\text{Row } 1 + 4 \cdot \text{Row } 3, \\[4mm]
&\text{Row } 2 + \left(\dfrac{10}{9}\right) \cdot \text{Row } 3, \\[4mm]
&
\end{aligned}
$$

$$
\begin{bmatrix}
1 & 0 & 0 & \dfrac{-19}{23} & \dfrac{14}{23} & \dfrac{-8}{23} \\[2mm]
0 & 1 & 0 & \dfrac{-9}{23} & \dfrac{3}{23} & \dfrac{-5}{23} \\[2mm]
0 & 0 & 1 & \dfrac{-7}{46} & \dfrac{5}{23} & \dfrac{-9}{46}
\end{bmatrix}
\qquad
\text{Row } 1 - 2 \cdot \text{Row } 2.
$$

The last three columns constitute A^{-1} as per Example 4.2.1.

(iii) Check the following inverse matrix:

$$
AA^{-1} =
\begin{bmatrix}
-1 & -2 & 4 \\
2 & -5 & 2 \\
3 & -4 & -6
\end{bmatrix}
\begin{bmatrix}
-\dfrac{19}{23} & \dfrac{14}{23} & -\dfrac{8}{23} \\[2mm]
-\dfrac{9}{23} & \dfrac{3}{23} & -\dfrac{5}{23} \\[2mm]
-\dfrac{7}{46} & \dfrac{5}{23} & -\dfrac{9}{46}
\end{bmatrix}
=
\begin{bmatrix}
1 & 0 & 0 \\
0 & 1 & 0 \\
0 & 0 & 1
\end{bmatrix}
= I = A^{-1} A .
$$

4.4 **Vector analysis**

This section is based on the work from Towner (2020) and Boas (2006).

4.4.1 **Vectors and fields**

Vectors

Consider the vectors $\overrightarrow{A} = (A_x, A_y, A_z)$ and $\overrightarrow{B} = (B_x, B_y, B_z)$ with the given Cartesian components as shown. One can then define the following quantities:

(a) *Scalar product:* $\overrightarrow{A} \cdot \overrightarrow{B} = AB\cos\theta = (A_x B_x + A_y B_y + A_z B_z)$, where the magnitudes of the vectors are given by $A = \sqrt{A_x^2 + A_y^2 + A_z^2} \equiv |A|$ and $B = \sqrt{B_x^2 + B_y^2 + B_z^2} \equiv |B|$ and θ is the angle between $\overrightarrow{A}$ and $\overrightarrow{B}$.

(b) *Vector product:* $\vec{C} = \vec{A} \times \vec{B} = \begin{vmatrix} \hat{i} & \hat{j} & \hat{k} \\ A_x & A_y & A_z \\ B_x & B_y & B_z \end{vmatrix} = (A_y B_z - A_z B_y)\hat{i} + (A_z B_x - A_x B_z)\hat{j} + (A_x B_y - A_y B_x)\hat{k}$, where the magnitude $|C| = AB \sin\theta$ and $\hat{i}, \hat{j}, \hat{k}$ are the unit vectors along the x, y, and z axes, respectively.

The unit vectors have the orthogonality property: $\hat{i} \cdot \hat{j} = \hat{j} \cdot \hat{k} = \hat{k} \cdot \hat{i} = 0$. The cross-products of the unit vectors are $\hat{i} \times \hat{j} = \begin{vmatrix} \hat{i} & \hat{j} & \hat{k} \\ 1 & 0 & 0 \\ 0 & 1 & 0 \end{vmatrix} = \hat{k}$. Similarly, $\hat{j} \times \hat{k} = \hat{i}$ and $\hat{k} \times \hat{i} = \hat{j}$. An orthogonal coordinate system is any set of axes for which the unit vectors satisfy the orthogonality property.

Triple products for both scalars and vectors are defined as follows:

(a) *Scalar triple product:* $\vec{A} \cdot \vec{B} \times \vec{C} = \begin{vmatrix} A_x & A_y & A_z \\ B_x & B_y & B_z \\ C_x & C_y & C_z \end{vmatrix}$.

(b) *Vector triple product:* $\vec{A} \times (\vec{B} \times \vec{C}) = (\vec{A} \cdot \vec{C})\vec{B} - (\vec{A} \cdot \vec{B})\vec{C}$.

Given the vector $\vec{A} = A_x(t)\hat{i} + A_y(t)\hat{j} + A_z(t)\hat{k}$, where only the components are functions of time t, the time derivative of $\vec{A}$ is

$$\frac{d\vec{A}}{dt} = \frac{dA_x}{dt}\hat{i} + \frac{dA_y}{dt}\hat{j} + \frac{dA_z}{dt}\hat{k}.$$

The normal rules of differentiation apply:

$$\frac{d}{dt}\left(a\vec{A}\right) = \frac{da}{dt}\vec{A} + a\frac{d\vec{A}}{dt},$$

$$\frac{d}{dt}\left(\vec{A} \cdot \vec{B}\right) = \frac{d\vec{A}}{dt} \cdot \vec{B} + \vec{A} \cdot \frac{d\vec{B}}{dt},$$

$$\frac{d}{dt}\left(\vec{A} \times \vec{B}\right) = \frac{d\vec{A}}{dt} \times \vec{B} + \vec{A} \times \frac{d\vec{B}}{dt}.$$

For the last relation, the order must be respected since $\vec{A} \times \vec{B} = -\vec{B} \times \vec{A}$.

Fields

A field is defined as a physical quantity, which has different values at different points in space, such as: (i) the temperature variation in a room, (ii) the electric field around a point charge, or (iii) the gravitational force experienced at or near the earth. The word "field" encompasses both the physical quantity and the region of interest. A scalar field results when the physical quantity is a scalar (such as temperature

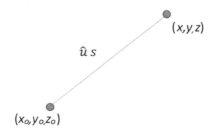

FIGURE 4.1

Schematic for the derivation of the directional derivative.

$T(x, y, z)$). A <u>vector field</u> results when the physical quantity is a vector (such as the electric field $\vec{E}(x, y, z)$ or the gravitational force $\vec{F}(x, y, z)$).

A <u>directional derivative</u> is a change in a physical quantity depending on which direction one moves. For instance, consider the scalar field $\phi(x, y, z)$. One wishes to find $(d\phi/ds)$ at a specific point (x_o, y_o, z_o) and in a given direction. Moreover, let $\hat{u}$ be a unit vector in the given direction, where $\hat{u} = a\hat{i} + b\hat{j} + c\hat{k}$, in which $|\hat{u}| = \sqrt{a^2 + b^2 + c^2} = 1$. Starting from the point (x_o, y_o, z_o) in Fig. 4.1 and going the distance s to reach the point (x, y, z), one has $(x, y, z) - (x_o, y_o, z_o) = \hat{u}s = (a\hat{i} + b\hat{j} + c\hat{k})s$. The respective components are $(x - x_o) = as$, $(y - y_o) = bs$, and $(z - z_o) = cs$. The directional derivative is evaluated from

$$\frac{d\phi}{ds} = \frac{\partial\phi}{\partial x}\frac{dx}{ds} + \frac{\partial\phi}{\partial y}\frac{dy}{ds} + \frac{\partial\phi}{\partial z}\frac{dz}{ds} = \frac{\partial\phi}{\partial x}a + \frac{\partial\phi}{\partial y}b + \frac{\partial\phi}{\partial z}c.$$

This result is the dot product of the vector $\hat{u}$ with the vector $\left(\frac{\partial\phi}{\partial x}\right)\hat{i} + \left(\frac{\partial\phi}{\partial y}\right)\hat{j} + \left(\frac{\partial\phi}{\partial z}\right)\hat{k}$:

$$\left(\frac{d\phi}{ds}\right) = \nabla\phi \cdot \hat{u}. \tag{4.10}$$

The quantity $\nabla\phi$ is called the gradient of ϕ (or grad ϕ),

$$\nabla\phi = \left(\frac{\partial\phi}{\partial x}\right)\hat{i} + \left(\frac{\partial\phi}{\partial y}\right)\hat{j} + \left(\frac{\partial\phi}{\partial z}\right)\hat{k}, \tag{4.11}$$

and ∇ can be thought of as a <u>vector operator</u>:

$$\nabla = \left(\frac{\partial}{\partial x}\right)\hat{i} + \left(\frac{\partial}{\partial y}\right)\hat{j} + \left(\frac{\partial}{\partial z}\right)\hat{k}. \tag{4.12}$$

The gradient operator can be defined for different coordinate systems (see Spiegel (1973)).

An equipotential surface is one in which the scalar field is constant on this surface, that is, $\overline{\phi(x, y, x)} = \text{constant}$. Using Eq. (4.10), we therefore have

$$\left(\frac{d\phi}{ds}\right) = \nabla\phi \cdot \hat{u} = 0.$$

Since the vector products of $\nabla\phi$ and $\hat{u}$ are zero, these vectors are orthogonal. Since $\hat{u}$ is a tangent to the equipotential surface, $\nabla\phi$ is perpendicular (or normal) to this surface.

Consider the vector field $\overrightarrow{V}(x, y, z)$ with components V_x, V_y, and V_z, each of which is a function of position:

$$\overrightarrow{V} = V_x(x, y, z)\hat{i} + V_y(x, y, z)\hat{j} + V_k(x, y, z)\hat{k}.$$

A scalar and vector product can be subsequently formed, respectively, with the vector operator ∇:

$$\nabla \cdot \overrightarrow{V} = \text{div } \overrightarrow{V} = \left(\frac{\partial V}{\partial x}\right) + \left(\frac{\partial V}{\partial y}\right) + \left(\frac{\partial V}{\partial z}\right)$$

and

$$\nabla \times \overrightarrow{V} = \text{curl } \overrightarrow{V} = \begin{vmatrix} \hat{i} & \hat{j} & \hat{k} \\ \left(\frac{\partial}{\partial x}\right) & \left(\frac{\partial}{\partial y}\right) & \left(\frac{\partial}{\partial z}\right) \\ V_x & V_y & V_z \end{vmatrix}.$$

Moreover, the scalar product ∇ with the vector $\nabla\phi$ yields the so-called <u>Laplacian</u> operator:

$$\nabla^2\phi \equiv \nabla \cdot \nabla\phi = \text{div grad } \phi = \frac{\partial}{\partial x}\left(\frac{\partial\phi}{\partial x}\right) + \frac{\partial}{\partial y}\left(\frac{\partial\phi}{\partial y}\right) + \frac{\partial}{\partial z}\left(\frac{\partial\phi}{\partial z}\right)$$

$$= \frac{\partial\phi^2}{\partial x^2} + \frac{\partial\phi^2}{\partial y^2} + \frac{\partial\phi^2}{\partial z^2}.$$

Example 4.4.1. Prove the vector identity $\nabla \times \nabla\phi = 0$.

Solution. We have

$$\nabla \times \nabla\phi = \begin{vmatrix} \hat{i} & \hat{j} & \hat{k} \\ \left(\frac{\partial}{\partial x}\right) & \left(\frac{\partial}{\partial y}\right) & \left(\frac{\partial}{\partial z}\right) \\ \left(\frac{\partial\phi}{\partial x}\right) & \left(\frac{\partial\phi}{\partial y}\right) & \left(\frac{\partial\phi}{\partial z}\right) \end{vmatrix} = \left(\frac{\partial}{\partial y}\left(\frac{\partial\phi}{\partial z}\right) - \frac{\partial}{\partial z}\left(\frac{\partial\phi}{\partial y}\right)\right)\hat{i} + \dots$$

$$= \left(\frac{\partial^2\phi}{\partial y\partial z} - \frac{\partial^2\phi}{\partial z\partial y}\right)\hat{i} + \dots = 0. \qquad \text{[answer]}$$

Example 4.4.2. The work done dW in displacing a particle an infinitesimal distance $\overrightarrow{dr}$ by a force $\overrightarrow{F}$ is $dW = \int \overrightarrow{F} \cdot \overrightarrow{dr}$. If the particle moves a macroscopic distance from point A to B along a path subject to the force $\overrightarrow{F}$, the total work done W is $W = \int_A^B \overrightarrow{F} \cdot \overrightarrow{dr}$. Moreover, if $\nabla \times \overrightarrow{F} = 0$, then the force $\overrightarrow{F}$ is termed a <u>conservative force</u>.

(a) Show that if the force is defined by a scalar potential field such that $\overrightarrow{F} = -\nabla\phi$, then the force is conservative.
(b) Show that for a conservative force, the line integral is <u>independent</u> of the path.

Solution.

(a) Substituting in $\overrightarrow{F} = -\nabla\phi$ gives $\nabla \times \overrightarrow{F} = -\nabla \times \nabla\phi = 0$ from the vector identity in Example 4.4.1.   [answer]
(b) We have $W = \int_A^B \overrightarrow{F} \cdot \overrightarrow{dr} = -\int_A^B \nabla\phi \cdot \overrightarrow{dr} = -\int_A^B d\phi = -(\phi(B) - \phi(A)) = (\phi(B) - \phi(A))$, as follows from the definition of $\nabla\phi$ in Eq. (4.11) and $\overrightarrow{dr} = \hat{i}\, dx + \hat{j}\, dy + \hat{k}\, dz$.   [answer]

Choosing A as a reference point (say, at the origin $A = (0, 0, 0)$) and B as a general point $B(x, y, z)$, one obtains $\phi(x, y, z) - \phi(0, 0, 0) = -\int_A^B \overrightarrow{F} \cdot \overrightarrow{dr}$. Hence, with a conservative force, one can find the scalar potential field (up to an arbitrary constant) by simply evaluating the line integral over any path that makes the integration easy.

4.4.2 Integral theorems

Integral theorems are of importance in electrical engineering, physics, and fluid flow problems.

Green's theorem

Consider two well-behaved functions, $P(x, y)$ and $Q(x, y)$, which are continuous and differentiable, as defined on the xy plane. The <u>Green's theorem</u> in two dimensions states

$$\iint_A \left(\frac{\partial Q}{\partial x} - \frac{\partial P}{\partial y} \right) dx\, dy = \oint_C (P\, dx + Q\, dy), \qquad (4.13)$$

where A is the area bounded by the contour C. As shown in the following example, the symbol $\oint$ indicates a line integral around a closed curve (that is, ending at the starting point), where the direction of integration is taken to be counterclockwise.

Example 4.4.3. Prove the Green's theorem given in Eq. (4.13) for two dimensions (see Fig. 4.2).

Solution.

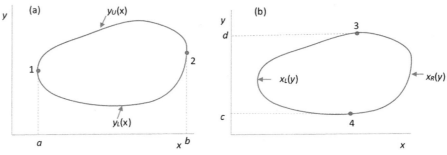

FIGURE 4.2

Schematic for the derivation of the Green's theorem in two dimensions with integration (a) over the two y paths and (b) over the two x paths.

Consider the area integral of $-\dfrac{\partial P}{\partial y}$, where the integration is carried out over y first and then over x, as in Fig. 4.2(a), yielding

$$-\iint_A \frac{\partial P}{\partial y}\,dx\,dy = -\int_a^b dx \int_{y_L}^{y_U} \frac{\partial P}{\partial y}\,dy = -\int_a^b dx \int_{y_L}^{y_U} dP$$

$$= -\int_a^b [P(x,y_U) - P(x,y_L)]\,dx$$

$$= \underbrace{\int_a^b P(x,y_L)\,dx}_{\text{line integral from 1 to 2 by lower path}} + \underbrace{\int_b^a P(x,y_U)\,dx}_{\text{line integral from 2 to 1 by upper path}}$$

$$= \oint_C P\,dx.$$

Similarly, now considering the area integral of $\dfrac{\partial Q}{\partial x}$, with integration first over x and then over y, as in Fig. 4.2(b), gives

$$\iint_A \frac{\partial Q}{\partial x}\,dx\,dy = \int_c^d dy \int_{x_L}^{x_R} \frac{\partial Q}{\partial x}\,dx = \int_c^d dy \int_{x_L}^{x_R} dQ$$

$$= \int_c^d [Q(x_R,y) - Q(x_L,y)]\,dy$$

$$= \underbrace{\int_d^c Q(x_L,y)\,dy}_{\text{line integral from 3 to 4 by left path}} + \underbrace{\int_c^d Q(x_R,y)\,dy}_{\text{line integral from 4 to 3 by right path}}$$

$$= \oint_C Q\,dy.$$

Hence, adding these two results yields Eq. (4.13).   [answer]

Using the Green's theorem one can either evaluate a line integral around a closed path or a double integral over the area enclosed, whichever is easier to perform. The divergence theorem and Stoke's theorem follow from the Green's theorem. Here one considers $P(x, y)$ and $Q(x, y)$ as two components of a vector field $\vec{V}$ defined in a plane with $\vec{V} = V_x(x, y)\hat{i} + V_y(x, y)\hat{j}$ and $V_z = 0$. These theorems can be easily generalized to three dimensions. The derivation of these theorems are given in the following sections.

Divergence theorem

Choosing $Q = V_x$ and $P = -V_y$ in Eq. (4.13) yields

$$\iint_A \left(\frac{\partial V_x}{\partial x} + \frac{\partial V_y}{\partial y} \right) dx\, dy = \oint_C (-V_y\, dx + V_x\, dy).$$

Using the definition for the divergence of $\vec{V}$ in Section 4.4.1,

$$\iint_A \text{div}\, \vec{V}\, dx\, dy = \oint_C (V_x\hat{i} + V_y\hat{j}) \cdot (\hat{i}\, dy - \hat{j}\, dx). \tag{4.14}$$

From Fig. 4.3, let $\vec{dr}$ be an infinitesimal element of the bounding contour C which encloses the area A. Here the tangent $\vec{dr} = \hat{i}\, dx + \hat{j}\, dy$. If $\hat{n}$ is the outward normal perpendicular to $\vec{dr}$, then $\hat{n} = \hat{i}\, dy - \hat{j}\, dx$ from Fig. 4.3. Since $\hat{n}$ is a unit vector, it is normalized, $\hat{n} = \dfrac{(\hat{i}\, dy - \hat{j}\, dx)}{ds}$, where $ds = \sqrt{dx^2 + dy^2}$. Hence,

$$\hat{n}\, ds = (\hat{i}\, dy - \hat{j}\, dx). \tag{4.15}$$

Using the definition of the div $\vec{V}$ from Section 4.4.1 and substituting Eq. (4.15) into Eq. (4.14) yields the divergence theorem in two dimensions:

$$\iint_A \nabla \cdot \vec{V}\, dx\, dy = \oint_C \vec{V} \cdot \hat{n}\, ds.$$

This result can be generalized to three dimensions for a volume V contained within a bounding surface S, where $d\sigma$ is an area element on the bounding surface and dV is a volume element (that is, $dV = dx\, dy\, dz$), such that

$$\boxed{\iiint_V \nabla \cdot \vec{V}\, dV = \oiint_S \vec{V} \cdot \hat{n}\, d\sigma.} \tag{4.16}$$

The divergence theorem is used in the analysis of Example 5.1.1, Section 7.2.2.

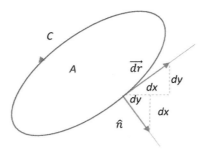

FIGURE 4.3

Schematic of the tangent and outward normal for the curve C.

Stoke's theorem

Alternatively, choosing $Q = V_y$ and $P = V_x$ in Eq. (4.13) gives

$$\iint_A \left(\frac{\partial V_y}{\partial x} - \frac{\partial V_x}{\partial y} \right) dx\, dy = \oint_C (V_x\, dx + V_y\, dy).$$

Using the definition for the curl of $\vec{V}$ in Section 4.4.1,

$$\iint_A \text{curl}\,\vec{V} \cdot \hat{k}\, dx\, dy = \oint_C (V_x \hat{i} + V_y\, \vec{j}) \cdot (\hat{i}\, dx + \hat{j}\, dy). \qquad (4.17)$$

Thus, substituting the definition of curl $\vec{V}$ from Section 4.4.1 gives <u>Stoke's theorem</u> in two dimensions:

$$\iint_A (\nabla \times \vec{V}) \cdot \hat{k}\, dx\, dy = \oint_C \vec{V} \cdot \vec{dr},$$

where $\vec{dr}$ is the tangent vector as previously shown over the bounding contour. This result can also be similarly generalized to three dimensions, where the area A instead of being confined to the xy plane is an open surface σ and the contour C is the curve bounding the surface, as shown in Fig. 4.4. Thus, from Fig. 4.4, if $\hat{n}$ is the unit vector normal to the surface, the final result is

$$\boxed{\iint_\sigma (\nabla \times \vec{V}) \cdot \hat{n}\, d\sigma = \oint_C \vec{V} \cdot \vec{dr}.} \qquad (4.18)$$

Example 4.4.4. Stoke's theorem is important in magnetism. Using Ampere's Law, $\oint_C \vec{H} \cdot \vec{dr} = I$, derive Maxwell's equation, $\nabla \times \vec{H} = \vec{J}$, using Stoke's theorem. Here $\vec{H} = \vec{B}/\mu_o$, $\vec{B}$ is the magnetic field, μ_o is the permeability, and I is the total current crossing any surface bounded by C. The integral C is taken around a closed contour that bounds an open surface S and J is the current crossing a perpendicular area.

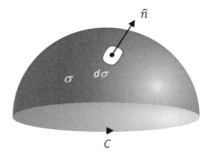

FIGURE 4.4

Schematic showing an outward normal on an open surface σ with a contour C.

Solution.

The quantity $\vec{J} \cdot \hat{n} d\sigma$ is the amount of current that crosses a surface element $d\sigma$. Therefore, the total current over a surface S bounded by a contour C is $\iint_S \vec{J} \cdot \hat{n} d\sigma$. Hence, by Ampere's Law,

$$\oint_C \vec{H} \cdot \vec{dr} = \iint_S \vec{J} \cdot \hat{n} d\sigma.$$

Using Stoke's theorem in Eq. (4.18),

$$\oint_C \vec{H} \cdot \vec{dr} = \iint_\sigma (\nabla \times \vec{H}) \cdot \hat{n} d\sigma = \iint_S \vec{J} \cdot \hat{n} d\sigma.$$

Since this result is valid for any surface S bounded by C, the result is as follows:

$$\nabla \times \vec{H} = \vec{J}. \qquad \text{[answer]}$$

Problems

4.1   Given the following system of equations:

$$3y + 2x = z + 1,$$
$$3x + 2z = 8 - 5y,$$
$$3z - 1 = x - 2y,$$

solve for x, y, and z using the following methods:

(**a**)   Inverse method;

(**b**)   Cramer's rule;

(**c**)   Gauss elimination;

(**d**)   Maple software package.

4.2 Find the directional derivative of $\phi = x^2 y + xz$ at $(1, 2, -1)$ in the direction $\overrightarrow{A} = 2\hat{i} - 2\hat{j} + \hat{k}$. Hint: Determine the vector $\hat{u} = \dfrac{\overrightarrow{A}}{|A|}$, find $\nabla\phi$ at $(1, 2, -1)$, and then calculate $\left(\dfrac{d\phi}{ds}\right)$ from Eq. (4.10).

4.3 What is the work done if $\overrightarrow{F} = xy\hat{i} - y^2\hat{j}$ and $\overrightarrow{dr} = \hat{i}dx + \hat{j}dy$ assuming the path is:

(a) a straight line where $y = \frac{1}{2}x$ from $A = (0, 0)$ to $B = (2, 1)$?

(b) parabolic where $y = \frac{1}{4}x^2$ from $A = (0, 0)$ to $B = (2, 1)$?

(c) Comment if the evaluated work is the same or different in (a) and (b).

4.4 Let the force $\overrightarrow{F} = (2xy - z^3)\hat{i} + x^2\hat{j} - (3xz^2 + 1)\hat{k}$.

(a) Show that $\overrightarrow{F}$ is conservative.

(b) Find $\phi(x, y, z) - \phi(0, 0, 0)$. Note that from (a), since the force is conservative, any path can be taken to evaluate the line integral. Therefore, use simple straight line paths along the same directions as the axes starting from the origin $(0, 0, 0)$ to $(x, 0, 0)$, then from $(x, 0, 0)$ to $(x, y, 0)$, and finally up to the final endpoint from $(x, y, 0)$ to (x, y, z).

Partial differential equations

Partial differential equations involve functions of several independent variables that arise routinely in nature. They describe physical laws in many areas, including fluid mechanics, heat and mass transfer, propagation of sound, elasticity, electrostatics and electrodynamics (for example, Maxwell's equations), and relativity, to name a few.

The derivation of important partial differential equations that arise in engineering is presented. Important analytical methods, including separation of variables and transform techniques, as employed for solution of various types of engineering problems that arise in heat transfer, fluid flow, and diffusion theory, are discussed.

Definitions

A partial differential equation is an equation containing an unknown function of two or more variables and its partial derivatives with respect to these variables.

The order of a partial differential equations is that of the highest-order derivatives. For example, $\dfrac{\partial^2 u}{\partial x \partial y} = 2x - y$ is a partial differential equation of order 2.

A solution of a partial differential equation is any function that satisfies the equation identically.

A general solution is a solution that contains a number of arbitrary independent functions equal to the order of the equation. A particular solution is one that is obtained from the general solution by a particular choice of arbitrary functions. For example, the general solution of the partial differential equation $\dfrac{\partial^2 u}{\partial x \partial y} = 2x - y$ is $u = x^2 y - \frac{1}{2}xy^2 + F(x) + G(y)$. With $F(x) = 2\sin x$ and $G(y) = 3y^4 - 5$, the particular solution is $u = x^2 y - \frac{1}{2}xy^2 + 2\sin x + 3y^4 - 5$.

A singular solution is one that cannot be obtained from the general solution by a particular choice of arbitrary functions.

A boundary value problem involving a partial differential equation seeks all solutions of a partial differential equation which satisfy conditions called boundary conditions.

When time is one of the variables, the solution u (or $\dfrac{\partial u}{\partial t}$ or both) must satisfy initial conditions at $t = 0$.

Advanced Mathematics for Engineering Students. https://doi.org/10.1016/B978-0-12-823681-9.00013-7

Linear partial differential equations

The general linear partial differential equation of order 2 in two independent variables has the form

$$A\frac{\partial^2 u}{\partial x^2} + B\frac{\partial^2 u}{\partial x \partial y} + C\frac{\partial^2 u}{\partial y^2} + D\frac{\partial u}{\partial x} + E\frac{\partial u}{\partial y} + Fu = G, \tag{5.1}$$

where A, B, C, D, E, F, and G may depend on x and y (but not u). If a second-order equation with independent variables x and y does not have the form of Eq. (5.1), it is nonlinear. If $G = 0$, the equation is homogeneous; otherwise it is nonhomogeneous. Generalizations to higher-order equations are easily made.

Because of the nature of the solutions of Eq. (5.1), the equation is

(i) elliptic if $B^2 - 4AC < 0$,
(ii) hyperbolic if $B^2 - 4AC > 0$,
(iii) parabolic if $B^2 - 4AC = 0$.

5.1 Derivation of important partial differential equations

Heat conduction and diffusion equation

The heat conduction equation is given by

$$\boxed{\frac{\partial u}{\partial t} = \kappa \nabla^2 u.} \tag{5.2}$$

Here $u(x, y, z, t)$ is the temperature of a solid at position (x, y, z) at time t. The constant, κ, called the thermal diffusivity, equals $K/(\sigma\rho)$, where the thermal conductivity K, the specific heat σ, and the density ρ are assumed constant.

The one-dimensional heat conduction equation reduces to $\dfrac{\partial u}{\partial t} = \kappa \dfrac{\partial^2 u}{\partial x^2}$.

Example 5.1.1. (Derivation of the heat conduction equation)

Solution.
Let V be an arbitrary volume lying within a solid and S denote its surface. Using the Fourier law for the heat flux,

$$\vec{J} = -K\vec{\nabla}u,$$

the total quantity of heat leaving S per unit time is

$$\iint_S \left(-K\vec{\nabla}u\right) \cdot \hat{n}\, dS,$$

where $\hat{n}$ is the normal to the surface S. Thus, the quantity of heat entering S per unit time using the divergence theorem in Eq. (4.16) is

$$\iint_S \left(K \vec{\nabla} u \right) \cdot \hat{n} \, dS = \iiint_V \vec{\nabla} \cdot \left(K \vec{\nabla} u \right) dV. \tag{5.3}$$

The heat contained in a volume V is given by

$$\iiint_V \sigma \rho u \, dV.$$

Then the rate of increase of heat is

$$\frac{\partial}{\partial t} \iiint_V \sigma \rho u \, dV = \iiint_V \sigma \rho \frac{\partial u}{\partial t} \, dV. \tag{5.4}$$

Equating Eq. (5.3) and Eq. (5.4),

$$\iiint_V \sigma \rho \frac{\partial u}{\partial t} \, dV = \iiint_V \vec{\nabla} \cdot \left(K \vec{\nabla} u \right) dV$$

or

$$\iiint_V \left[\sigma \rho \frac{\partial u}{\partial t} - \vec{\nabla} \cdot \left(K \vec{\nabla} u \right) \right] dV = 0.$$

Since V is arbitrary, the integrand (assumed continuous) must be identically zero so that

$$\sigma \rho \frac{\partial u}{\partial t} = \vec{\nabla} \cdot \left(K \vec{\nabla} u \right)$$

or, if K, σ, and ρ are constants,

$$\frac{\partial u}{\partial t} = \frac{K}{\sigma \rho} \vec{\nabla} \cdot \vec{\nabla} u = \kappa \nabla^2 u.$$

Moreover, for steady-state heat flow (such that $\frac{\partial u}{\partial t} = 0$), the equation reduces to Laplace's equation

$$\nabla^2 u = 0.$$

Example 5.1.2. (Derivation of the diffusion equation)

Solution.
Consider the volume element in Fig. 5.1 over which the mass balance is made, where F_x is the rate of transfer in the x direction through a unit area.

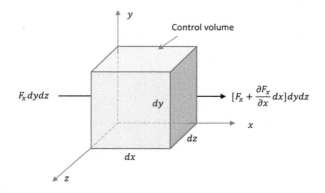

FIGURE 5.1

Volume element for derivation of the diffusion equation.

In the x direction, the net mass flow across the opposing faces is

$$Input - Output = \left[F_x - \left(F_x + \frac{\partial F_x}{\partial x} dx \right) \right] dy\, dz, \quad (5.5)$$

and similar expressions hold for the other two directions. In diffusion theory, the rate of transfer of a diffusing substance through a unit area is given by Fick's first law of diffusion:

$$F = -D\vec{\nabla}C \quad \left(\text{for example, } F_x = -D\frac{\partial C}{\partial x} \right), \quad (5.6)$$

where C is the concentration of the diffusing substance and D is a diffusion coefficient ($cm^2\ s^{-1}$). The accumulation rate in the volume element is given by

$$\text{Accumulation rate} = \frac{\partial C}{\partial t} dx\, dy\, dz. \quad (5.7)$$

Equating Eq. (5.5) and Eq. (5.7) for all directions and dividing by $dx\, dy\, dz$ yields

$$\frac{\partial C}{\partial t} = -\left(\frac{\partial F_x}{\partial x} + \frac{\partial F_y}{\partial y} + \frac{\partial F_z}{\partial z} \right). \quad (5.8)$$

Using Eq. (5.6), one obtains

$$\boxed{\frac{\partial C}{\partial t} = D\nabla^2 C} \quad (5.9)$$

if D is constant. This equation is known as the diffusion equation (Fick's second law) and has the same form as the heat conduction equation in Eq. (5.2).

Vibrating string equation

The one-dimensional wave equation is given by

$$\frac{\partial^2 u}{\partial t^2} = c^2 \frac{\partial^2 u}{\partial x^2}. \tag{5.10}$$

This equation is applicable to the small transverse vibrations of a taut, flexible string (for example, a violin string), initially located on the x axis and set into motion (see Fig. 5.2).

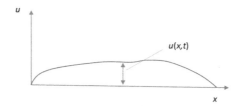

FIGURE 5.2

Displacement of a string.

The function $u(x, t)$ is the displacement at any point x of the string at time t. The constant $c^2 = T/\rho$, where T is the (constant) tension of the string and ρ is the (constant) mass per unit length of the string. It is assumed that no external forces act on the string but that it vibrates only due to its elasticity.

The equation can be generalized to higher dimensions, for example vibrations of a membrane (drum head) in two dimensions:

$$\frac{\partial^2 u}{\partial t^2} = c^2 \left(\frac{\partial^2 u}{\partial x^2} + \frac{\partial^2 u}{\partial y^2} \right). \tag{5.11}$$

Laplace's equation

Laplace's equation is given by

$$\nabla^2 u = 0. \tag{5.12}$$

This equation occurs in many fields. In the theory of heat conduction, u is the steady-state temperature. In the theory of gravitation or electricity, u represents the gravitational or electric potential, respectively. Hence, this equation is called the potential equation.

Poisson's equation

Possion's equation is an extension of Laplace's equation. For example, Newton's law of gravity is described by $\nabla^2 \phi = 4\pi G \rho_0$, where ϕ is the gravitational potential, G is the gravitational constant, and ρ_0 is the density of the matter field.

5.2 Analytical methods of solution

There are many methods by which boundary (and initial) value problems involving linear partial differential equations can be solved. The following are among the most important ones.

5.2.1 General solutions

The general solution is first found and then the particular solution which satisfies the boundary (and initial) conditions. The following theorems are of fundamental importance.

Superposition principle

If $u_1, u_2, ..., u_n$ are solutions of a linear homogeneous partial differential equation, then $c_1 u_1 + c_2 u_2 + ... + c_n u_n$ is also a solution to the general solution of the homogeneous equation, where $c_1, c_2, ..., c_n$ are arbitrary constants. Sometimes, general solutions can be found by using methods of ordinary differential equations.

General solution of linear nonhomogeneous partial differential equations

A general solution of a linear nonhomogeneous partial differential equation can be obtained by adding a particular solution of the nonhomogeneous equation to the general solution of the homogeneous equation.

Example 5.2.1. Solve $\dfrac{\partial^2 z}{\partial x \partial y} = x^2 y$ with boundary conditions $z(x, 0) = x^2$, $z(1, y) = \cos y$.

Solution.

Given $\dfrac{\partial}{\partial x}\left(\dfrac{\partial z}{\partial y}\right) = x^2 y$ and integrating with respect to x,

$$\frac{\partial z}{\partial y} = \frac{1}{3}x^3 y + F(y).$$

Integrating with respect to y yields

$$z = \frac{1}{6}x^3 y^2 + \int F(y)\,dy + G(x).$$

Hence, the general solution is

$$z(x, y) = \frac{1}{6}x^3 y^2 + H(y) + G(x),$$

where $H(y)$ and $G(x)$ are arbitrary (essential) functions. Using the boundary conditions $z(x, 0) = x^2 = H(0) + G(x)$ implies $G(x) = x^2 - H(0)$. Therefore, $z = \frac{1}{6}x^3 y^2 + H(y) + x^2 - H(0)$.

Since $z(1, y) = \cos y$, we have $\cos y = \frac{1}{6}y^2 + H(y) + 1 - H(0)$. Hence, $H(y) = \cos y - \frac{1}{6}y^2 - 1 + H(0)$.

The particular solution is finally $z = \frac{1}{6}x^3y^2 + \cos y - \frac{1}{6}y^2 + x^2 - 1$.   [answer]

If $A, B, ..., F$ are constants and $G = 0$ in the general linear partial differential equation of Eq. (5.1), the general solution of the <u>homogeneous</u> equation can be found by assuming

$$u = e^{ax+by},$$

where a and b are constants to be determined.

Example 5.2.2. Find solutions of $\dfrac{\partial^2 u}{\partial x^2} + 3\dfrac{\partial^2 u}{\partial x \partial y} + 2\dfrac{\partial^2 u}{\partial y^2} = 0$.

Solution.
Assume $u = e^{ax+by}$. Therefore substituting this expression into the partial differential equation gives

$$\left(a^2 + 3ab + 2b^2\right)e^{ax+by} = 0 \quad \text{or} \quad \left(a^2 + 3ab + 2b^2\right) = 0.$$

Thus, $(a + b)(a + 2b) = 0$, which implies $a = -b$ and $a = -2b$. If $a = -b$, then $e^{-bx+by} = e^{b(y-x)}$ is a solution for any value of b. Likewise, if $a = -2b$, then $e^{b(y-2x)}$ is also a solution for any value of b. Since the equation is linear and homogeneous, sums of the solutions are solutions. For example, $3e^{2(y-x)} - 2e^{3(y-x)} + 5e^{\pi(y-x)}$ is a solution (among others). In fact, $F(y - x)$, where F is arbitrary is a solution. Similarly, $G(y - 2x)$ is a solution where G is arbitrary.

The general solution is $u = F(y - x) + G(y - 2x)$.   [answer]

5.2.2 Separation of variables

In the separation of variables method, a solution can be expressed as a product of unknown functions, each of which depends on only one of the independent variables. The resulting equation must be separable, so that one side depends only on one variable while the other side depends on the remaining variables so that each side must be equal to a constant. By repetition of this method, the unknown functions can be determined. Superposition of these solutions lead to the actual solution. This method often makes use of Fourier series, Fourier integrals, Bessel series, and Legendre series.

Example 5.2.3. Vibrating string (wave equation) [solution by Fourier series].

A string of length L is stretched between points $(0, 0)$ and $(L, 0)$ on the x axis (see Fig. 5.3). At time $t = 0$, it has a shape given by $f(x)$, $0 < x < L$, and it is released with an initial velocity $g(x)$. Find the displacement of the string at any time later.

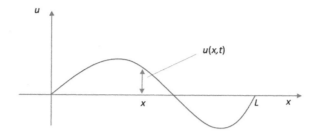

FIGURE 5.3

Vibration of a string of length L.

The equation of the vibrating string is

$$\frac{\partial^2 u}{\partial t^2} = c^2 \frac{\partial^2 u}{\partial x^2}, \quad 0 < x < L, \; t > 0, \tag{5.13}$$

where $u(x,t)$ is the displacement from the x axis at time t. Since the ends of the string are fixed at $x = 0$ and $x = L$,

$$u(0,t) = u(L,t) = 0, \quad t > 0. \tag{5.14}$$

The initial shape of the string is given by $f(x)$:

$$u(x,0) = f(x), \quad 0 < x < L, \tag{5.15}$$

and the initial velocity is given by $g(x)$:

$$\left.\frac{\partial u}{\partial t}\right|_{t=0} = u_t(x,0) = g(x), \quad 0 < x < L. \tag{5.16}$$

Solution.

To solve this boundary value problem, let $u = X(x)T(t)$. Substituting this relation into the partial differential equation in Eq. (5.13) gives

$$XT'' = c^2 X'' T \quad \text{or} \quad \frac{T''}{c^2 T} = \frac{X''}{X}.$$

Letting the separation constant equal $-\lambda^2$, one has

$$T'' + \lambda^2 c^2 T = 0, \quad X'' + \lambda^2 X = 0.$$

Note that choosing the separation constant equal to λ^2 or zero leads to a trivial solution $X = 0$ following application of the boundary conditions. Thus,

$$T = A_1 \sin \lambda ct + B_1 \cos \lambda ct, \quad X = A_2 \sin \lambda x + B_2 \cos \lambda x.$$

A solution is given by

$$u(x,t) = XT = (A_2 \sin \lambda x + B_2 \cos \lambda x)(A_1 \sin \lambda ct + B_1 \cos \lambda ct).$$

The boundary condition $u(0,t) = 0$ implies $B_2 = 0$. Therefore,

$$u(x,t) = \sin \lambda x (A_1 \sin \lambda ct + B_1 \cos \lambda ct).$$

From the boundary condition $u(L,t) = 0$, one therefore has $\sin \lambda L (A_1 \sin \lambda ct + B_1 \cos \lambda ct) = 0$ so that $\sin \lambda L = 0$, $\lambda L = n\pi$ or $\lambda = \dfrac{n\pi}{L}$ (n an integer) since the second factor must not be equal to zero.

Thus, there are infinitely many solutions of Eq. (5.13) satisfying the boundary conditions in Eq. (5.14):

$$u_n(x,t) = (A_n \sin \lambda_n t + B_n \cos \lambda_n t) \sin \frac{n\pi x}{L},$$

where $\lambda_n = \dfrac{cn\pi}{L}$ ($n = 1, 2...$). Since Eq. (5.13) is linear and homogeneous, it follows from the superposition principle that a solution to Eq. (5.13) is the infinite series

$$u(x,t) = \sum_{n=1}^{\infty} u_n(x,t) = \sum_{n=1}^{\infty} (A_n \sin \lambda_n t + B_n \cos \lambda_n t) \sin \frac{n\pi x}{L}. \qquad (5.17)$$

Here to satisfy the initial conditions in Eq. (5.15) and Eq. (5.16), it is necessary to superimpose solutions. Note that u_n are called the eigenfunctions and λ_n the eigenvalues of the vibrating string. Using the initial condition in Eq. (5.15),

$$u(x,0) = \sum_{n=1}^{\infty} B_n \sin \frac{n\pi x}{L} = f(x),$$ and from the theory of the <u>Fourier sine series</u> in Section 2.5.6,

$$B_n = \frac{2}{L} \int_0^L f(x) \sin \frac{n\pi x}{L} \, dx \qquad (n = 1, 2, ...). \qquad (5.18)$$

Using the initial condition in Eq. (5.16),

$$\frac{\partial u}{\partial t}\bigg|_{t=0} = \left[\sum_{n=1}^{\infty} (A_n \lambda_n \cos \lambda_n t - B_n \lambda_n \sin \lambda_n t) \sin \frac{n\pi x}{L} \right]_{t=0}$$

$$= \sum_{n=1}^{\infty} A_n \lambda_n \sin \frac{n\pi x}{L} = g(x),$$

and using the theory of the Fourier sine series,

$$A_n = \frac{2}{cn\pi} \int_0^L g(x) \sin \frac{n\pi x}{L} \, dx \qquad (n = 1, 2, ...). \qquad (5.19)$$

Thus, Eq. (5.17) to Eq. (5.19) constitute a solution of Eq. (5.13) subject to conditions Eq. (5.14) to Eq. (5.16). [answer]

Example 5.2.4. Heat conduction equation [solution by the Fourier series].

Find the temperature in a bar in which the ends of the bar are kept at zero temperature and the initial temperature in the bar is $f(x)$ (see Fig. 5.4).

0   x=L

FIGURE 5.4

Heat conduction in a bar of length L.

The heat conduction equation for a temperature $u(x, t)$ is

$$\frac{\partial u}{\partial t} = \kappa \frac{\partial^2 u}{\partial x^2}, \quad 0 < x < L, \ t > 0, \tag{5.20}$$

which is subject to boundary conditions

$$u(0, t) = u(L, t) = 0, \ t > 0 \tag{5.21}$$

and initial condition

$$u(x, 0) = f(x), \ 0 < x < L. \tag{5.22}$$

Solution.
To solve this boundary value problem, let $u = X(x)T(t)$. Substituting this relation into the partial differential equation in Eq. (5.20) gives

$$XT' = \kappa X''T \quad \text{or} \quad \frac{X''}{X} = \frac{T'}{\kappa T}.$$

Each side must be equal to a constant $-\lambda^2$ (note that choosing the separation constant equal to λ^2, the resulting solution does not satisfy the boundedness condition for real values of λ) so that

$$X'' + \lambda^2 X = 0, \quad T' + \kappa \lambda^2 T = 0,$$

with solutions

$$X = A_1 \cos \lambda x + B_1 \sin \lambda x, \quad T = c_1 e^{-\kappa \lambda^2 t}.$$

A solution of the partial differential equation is thus given by

$$u(x, t) = XT = (A \cos \lambda x + B \sin \lambda x) e^{-\kappa \lambda^2 t}.$$

The boundary condition $u(0, t) = A e^{-\kappa \lambda^2 t} = 0$ implies $A = 0$. Therefore,

$$u(x, t) = B \sin \lambda x \, e^{-\kappa \lambda^2 t}.$$

Using the second boundary condition, $u(L, t) = B \sin \lambda L \, e^{-\kappa \lambda^2 t} = 0$. If $B = 0$, the solution is identically zero, so one must choose $\sin \lambda L = 0$ or $\lambda L = n\pi$, that is, $\lambda = \dfrac{n\pi}{L}$ $(n = 1, 2, ...)$. Hence, the eigenfunctions

$$u_n(x, t) = B_n \sin \frac{n\pi x}{L} e^{-\kappa \lambda_n^2 t} \quad (n = 1, 2...)$$

are solutions of the heat conduction equation, with corresponding eigenvalues $\lambda_n = \dfrac{n\pi}{L}$.

To satisfy the initial condition, it is necessary to superimpose an infinite number of solutions:

$$\boxed{u(x, t) = \sum_{n=1}^{\infty} u_n(x, t) = \sum_{n=1}^{\infty} B_n \sin \frac{n\pi x}{L} e^{-\kappa \lambda_n^2 t}} \qquad \left(\lambda_n = \frac{n\pi}{L} \right). \qquad (5.23)$$

From Eq. (5.23) and the initial condition,

$$u(x, 0) = \sum_{n=1}^{\infty} B_n \sin \frac{n\pi x}{L} = f(x).$$

The B_n coefficients are obtained from the Fourier sine series,

$$\boxed{B_n = \frac{2}{L} \int_0^L f(x) \sin \frac{n\pi x}{L} \, dx} \qquad (n = 1, 2, ...). \qquad (5.24)$$

Thus, Eq. (5.23) and Eq. (5.24) constitute a solution of the problem.   [answer]

Example 5.2.5. Heat conduction equation [sinusoidal initial temperature].

Find the temperature $u(x, t)$ in a lateral insulated copper bar 80 cm long if the initial temperature is $100 \sin \left(\dfrac{\pi x}{80} \right)$ °C and the ends are kept at 0°C. Physical data for copper include: density of $8.92 \ \dfrac{g}{cm^3}$, specific heat of $0.092 \ \dfrac{cal}{g°C}$, and thermal conductivity of $0.95 \ \dfrac{cal}{cm \cdot s°C}$.

Solution.
The initial condition gives $u(x, 0) = \displaystyle\sum_{n=1}^{\infty} B_n \sin \frac{n\pi x}{80} = f(x) = 100 \sin \left(\frac{\pi x}{80} \right)$ so

that $B_1 = 100$, $B_2 = B_3 = ... = 0$. In Eq. (5.23), $\lambda_1^2 = \dfrac{\pi^2}{L^2}$ and $\kappa = \dfrac{K}{\sigma \rho} = 0.95/$

$(0.092 \times 8.92) = 1.158 \; \dfrac{\text{cm}^2}{\text{s}}$. Hence, $\kappa\lambda_1^2 = 1.158(\pi)^2/80^2 = 0.001785$. Thus, the solution is $u(x,t) = 100 \sin \dfrac{\pi x}{80} e^{-0.001785t}$, where t is in seconds.   [answer]

Example 5.2.6. Laplace's equation [two-dimensional heat flow in steady state].

If the heat flow is in steady state such that $\dfrac{\partial u}{\partial t} = 0$, the heat conduction equation reduces to the Laplace equation (in two dimensions):

$$\nabla^2 u = \frac{\partial^2 u}{\partial x^2} + \frac{\partial^2 u}{\partial y^2} = 0. \tag{5.25}$$

This equation must be solved for a given boundary condition on the boundary curve C for some region of the xy plane. This boundary value problem is called a:

(i) Dirichlet problem if u is prescribed on C;

(ii) Neumann problem if $u_n = \dfrac{\partial u}{\partial x}$ (that is, the normal derivative) is prescribed on C;

(iii) Mixed problem if u is prescribed on a portion of C and u_n on the rest of C.

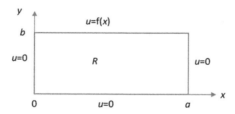

FIGURE 5.5

Dirichlet problem for heat flow in a plate.

For example, consider the Dirichlet problem in Fig. 5.5 for the rectangle R. The heat conduction equation in Eq. (5.25) is subject to the boundary conditions

$$u(0, y) = u(a, y) = 0 \tag{5.26}$$

and

$$u(x, 0) = 0, \qquad u(x, b) = f(x). \tag{5.27}$$

Solution.

Using the separation of variables, let $u = X(x)Y(y)$. Substituting this relation into Eq. (5.25) yields

$$X''Y + XY'' = 0 \quad \text{or} \quad \frac{X''}{X} = -\frac{Y''}{Y} = -\lambda^2.$$

Thus, $X'' + \lambda^2 X = 0$, with $X(0) = 0$, $X(a) = 0$ (see Eq. (5.26)). This ordinary differential equation gives $\lambda = \left(\dfrac{n\pi}{a}\right)$ and the corresponding nonzero solutions

$$X(x) = X_n(x) = \sin \frac{n\pi}{a} x \qquad (n = 1, 2...).$$

Similarly, the equation for Y becomes

$$Y'' - \lambda_n^2 Y = 0, \qquad \lambda_n^2 = \left(\frac{n\pi}{a}\right)^2,$$

with the solution

$$Y(y) = Y_n(y) = A_n e^{\lambda_n y} + B_n e^{-\lambda_n y}.$$

The boundary condition in Eq. (5.27) gives $Y(0) = A_n + B_n = 0$, which implies $B_n = -A_n$. Therefore, the solution can be written as

$$Y(y) = A_n^* \sinh \frac{n\pi y}{a} \qquad (A_n^* = 2A_n).$$

A solution of the partial differential equation is thus given by

$$u(x, y) = A_n^* \sin \frac{n\pi x}{a} \sinh \frac{n\pi y}{a}.$$

For the boundary condition $u(x, b) = f(x)$, one can consider the infinite series

$$u(x, y) = \sum_{n=1}^{\infty} u_n(x, y).$$

Therefore, $u(x, b) = \displaystyle\sum_{n=1}^{\infty} \left(A_n^* \sinh \frac{n\pi b}{a} \right) \sin \frac{n\pi x}{a} = f(x)$. Again, this is a Fourier sine series (with Fourier coefficients $b_n = A_n^* \sinh \dfrac{n\pi b}{a}$). Therefore, $b_n = A_n^* \sinh \dfrac{n\pi b}{a} = \dfrac{2}{a} \displaystyle\int_0^a f(x) \sin \frac{n\pi x}{a} dx$.

The solution of the problem is

$$\boxed{u(x, y) = \sum_{n=1}^{\infty} A_n^* \sin \frac{n\pi x}{a} \sinh \frac{n\pi y}{a},} \qquad (5.28)$$

where

$$\boxed{A_n^* = \frac{2}{a \sinh \dfrac{n\pi b}{a}} \int_0^a f(x) \sin \frac{n\pi x}{a} dx.} \qquad \text{[answer]} \qquad (5.29)$$

Example 5.2.7. Cylindrical heat conduction equation [solution by the Fourier–Bessel series].

An infinitely long cylinder of unit radius has a constant initial temperature u_o. At time $t = 0$ a temperature of $0°C$ is applied to the surface and is maintained. Find the temperature at any point of the cylinder at any later time t (see Fig. 5.6).

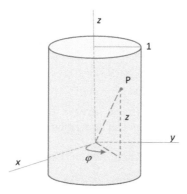

FIGURE 5.6

Heat conduction in an infinitely long cylinder.

For an infinitely long cylinder, the heat conduction equation is independent of z and φ and reduces to

$$\frac{\partial u}{\partial t} = \kappa \left(\frac{\partial^2 u}{\partial r^2} + \frac{1}{r} \frac{\partial u}{\partial r} \right), \quad 0 < r < 1, \ t > 0, \tag{5.30}$$

subject to the boundary conditions

$$u(1, t) = 0, \quad |u(r, t)| < M \tag{5.31}$$

and initial condition

$$u(r, 0) = u_o. \tag{5.32}$$

The second boundary condition in Eq. (5.31) is a reflection that the temperature must be bounded.

Solution.
Using separation of variables, $u = R(r)T(t)$, one obtains

$$RT' = \kappa \left[TR'' + \frac{T}{r} R' \right] \quad \text{or} \quad \frac{T'}{\kappa T} = \left[\frac{R''}{R} + \frac{1}{r} \frac{R'}{R} \right] = -\lambda^2.$$

Thus,

$$r^2 R'' + r R' + r^2 \lambda^2 R = 0, \quad T' + \kappa \lambda^2 T = 0,$$

with solutions as shown in Section 2.5.3 for the Bessel equation:

$$R = A_1 J_0(\lambda r) + B_1 Y_0(\lambda r), \quad T = c_1 e^{-\kappa \lambda^2 t}.$$

A solution of the partial differential equation is thus given by

$$u(r, t) = (A J_0(\lambda r) + B Y_0(\lambda r)) e^{-\kappa \lambda^2 t}.$$

Since $Y_0(\lambda r)$ is unbounded as r approaches 0 (see Fig. 2.4) and $|u(r, t)|$ must be bounded in Eq. (5.31), this condition implies $B = 0$. Therefore,

$$u(r, t) = A J_0(\lambda r) e^{-\kappa \lambda^2 t}.$$

Using the boundary condition $u(1, t) = A J_0(\lambda) e^{-\kappa \lambda^2 t} = 0$, we obtain $J_0(\lambda) = 0$, which has infinitely many zeros for $\lambda = \lambda_n (n = 1, 2, ...)$. Hence, the eigenfunctions are

$$u_n(r, t) = A_n J_0(\lambda_n r) e^{-\kappa \lambda_n^2 t} \quad (n = 1, 2, ...),$$

with corresponding eigenvalues λ_n. To satisfy the initial condition in Eq. (5.32), it is necessary to superimpose an infinite number of solutions:

$$u(r, t) = \sum_{n=1}^{\infty} u_n(r, t) = \sum_{n=1}^{\infty} A_n J_0(\lambda_n r) e^{-\kappa \lambda_n^2 t}.$$

From the initial condition in Eq. (5.32):

$$u(r, 0) = \sum_{n=1}^{\infty} A_n J_0(\lambda_n r) = u_o.$$

Hence, this is a Fourier–Bessel series (see Example 2.5.12), and the coefficients A_n are given by

$$A_n = \frac{2}{J_1{}^2(\lambda_n)} \int_0^1 r f(r) J_0(\lambda_n r) \, dr$$

$$= \frac{2u_o}{J_1{}^2(\lambda_n)\lambda_n^2} \int_0^{\lambda_n} \xi J_0(\xi) \, d\xi.$$

The second equality arises with $f(r) = u_o$. Using the recursive formula for Bessel functions (see Section 2.5.3), $\dfrac{d}{dx}[x^\nu J_\nu(x)] = x^\nu J_{\nu-1}(x)$, or equivalently (with $\nu = 1$) the relation $\int x J_0(x) \, dx = x J_1(x)$; therefore, A_n becomes

$$A_n = \frac{2u_o}{J_1{}^2(\lambda_n)\lambda_n^2} \cdot \lambda_n J_1(\lambda_n) = \frac{2u_o}{\lambda_n J_1(\lambda_n)}.$$

The solution of the problem is therefore

$$u(r, t) = 2u_o \sum_{n=1}^{\infty} \frac{J_0(\lambda_n r)}{\lambda_n J_1(\lambda_n)} e^{-\kappa \lambda_n^2 t}.$$ [answer]

The above solution is identical to that given in Example 5.2.12 using Laplace transforms and the complex inversion formula.

Example 5.2.8. Heat conduction equation [solution by Fourier integrals].

Consider the problem of Example 5.2.4 for heat conduction in a bar, but in this case the bar is semiinfinite in length. In this analysis, the Fourier series can be replaced by Fourier integrals (see Section 3.2.1).

A semiinfinite bar ($x \geq 0$) whose surface is insulated has an initial temperature $f(x)$. A temperature of zero is suddenly applied to the end $x = 0$ and maintained. Find the temperature $u(x, t)$ at any point x at time t. The heat conduction equation is

$$\frac{\partial u}{\partial t} = \kappa \frac{\partial^2 u}{\partial x^2}, \quad x > 0, \ t > 0,$$ (5.33)

with boundary conditions

$$u(0, t) = 0, \qquad |u(x, t)| < M.$$ (5.34)

The initial condition is

$$u(x, 0) = f(x).$$ (5.35)

Solution.
By separation of variables (see Example 5.2.4),

$$u(x, t) = XT = (A \cos \lambda x + B \sin \lambda x) e^{-\kappa \lambda^2 t}.$$

Using the first boundary condition in Eq. (5.34), $A = 0$ so that

$$u(x, t) = B \sin \lambda x \, e^{-\kappa \lambda^2 t}.$$ (5.36)

Since there is no restriction on λ one can replace B in Eq. (5.36) by a function $B(\lambda)$ and still have a solution. Furthermore, one can integrate over λ from 0 to ∞ and still have a solution. This approach is the analogy of the superposition theorem for discrete values of λ used in connection with a Fourier series. Thus, a possible solution is

$$u(x, t) = \int_0^{\infty} B(\lambda) \sin \lambda x \, e^{-\kappa \lambda^2 t} \, d\lambda.$$ (5.37)

Using the initial condition in Eq. (5.35),

$$u(x,0) = \int_0^\infty B(\lambda) \sin \lambda x \, d\lambda = f(x). \qquad (5.38)$$

Recalling Fourier's integral theorem in Section 3.2.1,

$$f(x) = \int_0^\infty [A(\lambda) \cos \lambda x + B(\lambda) \sin \lambda x] \, d\lambda, \qquad (5.39)$$

where

$$\begin{aligned} A(\lambda) &= \frac{1}{\pi} \int_{-\infty}^\infty f(x) \cos \lambda x \, dx, \\ B(\lambda) &= \frac{1}{\pi} \int_{-\infty}^\infty f(x) \sin \lambda x \, dx. \end{aligned} \qquad (5.40)$$

Now since the function $f(x)$ in Eq. (5.38) must be odd, $A(\lambda) = 0$ and $B(\lambda)$ is

$$B(\lambda) = \frac{2}{\pi} \int_0^\infty f(x) \sin \lambda x \, dx = \frac{2}{\pi} \int_0^\infty f(v) \sin \lambda v \, dv.$$

Substituting this expression in Eq. (5.37) yields

$$u(x,t) = \frac{2}{\pi} \int_0^\infty \int_0^\infty f(v) e^{-\kappa \lambda^2 t} \sin \lambda v \sin \lambda x \, d\lambda \, dv. \qquad (5.41)$$

The equation $\sin \lambda v \sin \lambda x = \frac{1}{2}[\cos \lambda(v-x) - \cos \lambda(v+x)]$ implies

$$\begin{aligned} u(x,t) &= \frac{1}{\pi} \int_0^\infty \int_0^\infty f(v) e^{-\kappa \lambda^2 t} [\cos \lambda(v-x) - \cos \lambda(v+x)] \, d\lambda \, dv \\ &= \frac{1}{\pi} \int_0^\infty f(v) \left[\int_0^\infty e^{-\kappa \lambda^2 t} \cos \lambda(v-x) \, d\lambda \right. \\ &\qquad \left. - \int_0^\infty e^{-\kappa \lambda^2 t} \cos \lambda(v+x) \, d\lambda \right] dv. \end{aligned}$$

Using the result $\int_0^\infty e^{-\alpha \lambda^2} \cos \beta \lambda \, d\lambda = \frac{1}{2} \sqrt{\frac{\pi}{\alpha}} e^{\frac{-\beta^2}{4\alpha}}$, one finds

$$u(x,t) = \frac{1}{2\sqrt{\pi \kappa t}} \left[\int_0^\infty f(v) e^{-\frac{(v-x)^2}{4\kappa t}} \, dv - \int_0^\infty f(v) e^{-\frac{(v+x)^2}{4\kappa t}} \, dv \right].$$

Finally letting $\dfrac{(v-x)}{2\sqrt{\kappa t}} = \omega$ in the first integral and $\dfrac{(v+x)}{2\sqrt{\kappa t}} = \omega$ in the second one,

$$u(x,t) = \frac{1}{\sqrt{\pi}}\left[\int_{\frac{-x}{2\sqrt{\kappa t}}}^{\infty} e^{-\omega^2} f(2\omega\sqrt{\kappa t} + x)\,d\omega - \int_{\frac{x}{2\sqrt{\kappa t}}}^{\infty} e^{-\omega^2} f(2\omega\sqrt{\kappa t} - x)\,d\omega\right].$$

[answer]

Moreover, in the case that the initial temperature $f(x) = u_o$, which equals a constant,

$$u(x,t) = \frac{u_o}{\sqrt{\pi}}\left[\int_{\frac{-x}{2\sqrt{\kappa t}}}^{\infty} e^{-\omega^2}\,d\omega - \int_{\frac{x}{2\sqrt{\kappa t}}}^{\infty} e^{-\omega^2}\,d\omega\right] = \left(\frac{u_o}{\sqrt{\pi}}\right)\int_{\frac{-x}{2\sqrt{\kappa t}}}^{\frac{x}{2\sqrt{\kappa t}}} e^{-\omega^2}\,d\omega,$$

we have

$$u(x,t) = \left(\frac{2u_o}{\sqrt{\pi}}\right)\int_{0}^{\frac{x}{2\sqrt{\kappa t}}} e^{-\omega^2}\,d\omega = u_o\,\mathrm{erf}\left(\frac{x}{2\sqrt{\kappa t}}\right).$$

[answer]   (5.42)

Here the <u>error function</u> is defined as $\mathrm{erf}x = \dfrac{2}{\sqrt{\pi}}\displaystyle\int_{0}^{x} e^{-\omega^2}\,d\omega$, whose value is tabulated.

Example 5.2.9. Neutron diffusion and buckling [solution of a two-dimensional wave equation for a cylindrical geometry].

Consider a cylindrical nuclear reactor with height H and radius R. To determine the physical size of the reactor, one requires an evaluation of the so-called <u>geometric buckling</u> B^2 for the reactor. This quantity arises as the eigenvalue for solution of the wave equation:

$$\nabla^2\phi + B^2\phi = 0.$$

The solution of this equation describes the steady-sate spatial distribution of the neutron flux ϕ in the reactor. To design a critical reactor, one sets B^2 equal to the material buckling B^2 (that is determined from the material properties of the moderator and fuel system) (Lewis et al., 2017). The term buckling arises from the situation of a loaded column, where equivalently the curvature of the flux must be contained within the physical boundaries of the reactor.

(a) Solve the wave equation to determine the neutron flux distribution ϕ.
(b) Evaluate the geometric buckling B^2 as the lowest value of the eigenvalue. One can further optimize the physical dimensions of the reactor by optimizing the buckling equation, as shown in the Lagrange multiplier problem of Example 14.2.1.

Solution.

A cylindrical reactor is the most common geometry where the wave equation becomes

$$\left(\frac{1}{r} \frac{\partial}{\partial r} \left(r \frac{\partial}{\partial r} \right) + \frac{1}{r^2} \frac{\partial^2}{\partial \varphi^2} + \frac{\partial^2}{\partial z^2} \right) \phi_\lambda = -\lambda^2 \phi_\lambda.$$

Assuming azimuthal symmetry, the separation of variables technique yields

$$\phi_\lambda(r, z) = R(r)Z(z),$$

where

$$\frac{1}{R} \left(\frac{\partial^2 R}{\partial r^2} + \frac{1}{r} \frac{\partial R}{\partial r} \right) + \frac{1}{Z} \frac{\partial^2 Z}{\partial z^2} + \lambda^2 = 0.$$

As follows with this solution technique,

$$\frac{1}{Z} \frac{d^2 Z}{dz^2} = -\alpha^2 \text{ and } \frac{1}{R} \left(\frac{d^2 R}{dr^2} + \frac{1}{r} \frac{dR}{dr} \right) = -\beta^2,$$

where $\lambda^2 = \alpha^2 + \beta^2$. The eigenfunction solutions of these two ordinary differential equations are

$$Z_m(z) = A_1 \cos(\alpha_m z) + B_1 \sin(\alpha_m z) \text{ and } R_n(r) = C_1 J_0(\beta_n r) + D_1 Y_0(\beta_n r).$$

If the reactor is symmetric about the $z = 0$ plane for the origin of the coordinate system located at the center of the reactor, $B_1 = 0$. Furthermore, the neutron flux vanishes at the boundary $H/2$:

$$Z_m(H/2) = A_1 \cos(\alpha_m H/2) = 0, \qquad \alpha_m = m\pi/H, \text{ where } m = 1, 3, 5....$$

Since Y_0 goes to infinity at $r = 0$ from Fig. 2.4, $D_1 = 0$. The flux also vanishes at the radius $r = R$ so that

$$R_n(R) = C_1 J_0(\beta_n R) = 0, \qquad \beta_n = x_n/R,$$

where x_n are the zeros of the J_0 function in Fig. 2.4. Hence, the eigenfunctions are

$$\phi_\lambda(r, z) = A_n \cos\left(\frac{m\pi}{H} z\right) J_0\left(\frac{x_n}{R} r\right).$$

Using the lowest values of m and n, the neutron flux distribution is given by

$$\boxed{\phi(r, z) = A \cos\left(\frac{\pi}{H} z\right) J_0\left(\frac{2.405 r}{R}\right).}$$ [answer]

(b) The buckling B^2 is given by

$$B^2 = \lambda^2 = \left(\frac{\pi}{H}\right)^2 + \left(\frac{2.405}{R}\right)^2. \qquad \text{[answer]}$$

5.2.3 Fourier and Laplace transform methods

Operational methods (that is, the Fourier and Laplace transforms) can be used to solve partial differential equations. Here the transform of the partial differential equation and associated boundary conditions are first obtained with respect to one of the independent variables. The resulting equation for the transform is solved and then the required solution by taking the inverse transform.

Example 5.2.10. Heat conduction equation [solution by the Fourier sine transforms].

Consider Example 5.2.8 for heat conduction in a semiinfinite bar where

$$\frac{\partial u}{\partial t} = \kappa \frac{\partial^2 u}{\partial x^2}, \quad x > 0, \ t > 0, \tag{5.43}$$

with boundary conditions

$$u(0, t) = 0, \qquad |u(x, t)| < M, \quad t > 0, \tag{5.44}$$

and initial condition

$$u(x, 0) = f(x), \qquad 0 \le x < \infty. \tag{5.45}$$

Solution.
One can apply a Fourier sine transform with respect to x, since x varies from 0 to ∞ and then solve the ordinary differential equation.

Letting $\hat{u}_s(w, t) = \mathscr{F}_s\{u(x, t)\}$ and applying the property of the sine transform (see Section 3.2.3),

$$\mathscr{F}_s\{u''(x)\} = -w^2 \mathscr{F}_s\{u(x)\} + \sqrt{\frac{2}{\pi}} \, w \, u(0),$$

the transform of Eq. (5.43) becomes

$$\frac{\partial \hat{u}_s}{\partial t} = -\kappa w^2 \hat{u}_s + \kappa \sqrt{\frac{2}{\pi}} \, w \, u(0).$$

From the boundary condition in Eq. (5.44),

$$\frac{\partial \hat{u}_s}{\partial t} = -\kappa w^2 \hat{u}_s,$$

which has the solution

$$\hat{u}_s(w, t) = C(w)e^{-\kappa w^2 t}. \tag{5.46}$$

The transform of the initial condition in Eq. (5.45) gives

$$\hat{u}_s(w, 0) = \hat{f}(w). \tag{5.47}$$

Thus, $\hat{u}_s(w, 0) = C(w)e^0 = \hat{f}(w)$, which implies $C(w) = \hat{f}(w)$. Hence,

$$\hat{u}_s(w, t) = \hat{f}(w)e^{-\kappa w^2 t}. \tag{5.48}$$

Taking the inverse sine transform of Eq. (5.48) gives

$$u(x, t) = \sqrt{\frac{2}{\pi}} \int_0^\infty \hat{f}(w) \sin wx \, e^{-\kappa w^2 t} \, dw.$$

Inserting in the Fourier sine transform of $\hat{f}(w)$,

$$\hat{f}(w) = \sqrt{\frac{2}{\pi}} \int_0^\infty f(v) \sin wv \, dv,$$

gives

$$\begin{aligned}
u(x, t) &= \frac{2}{\pi} \int_0^\infty \left[\int_0^\infty f(v) \sin wv \, dv \right] \sin wx \, e^{-\kappa w^2 t} \, dw \\
&= \frac{2}{\pi} \int_0^\infty \int_0^\infty f(v)e^{-\kappa w^2 t} \sin wv \sin wx \, dw \, dv.
\end{aligned}$$

By changing the dummy variable of integration from w to λ yields the final solution:

$$\boxed{u(x, t) = \frac{2}{\pi} \int_0^\infty \int_0^\infty f(v)e^{-\kappa \lambda^2 t} \sin \lambda v \sin \lambda x \, d\lambda \, dv.} \qquad \text{[answer]}$$

As expected, this solution is identical to Eq. (5.41) as obtained by the method of Fourier integrals in Example 5.2.8.

Example 5.2.11. Heat conduction equation [solution by Laplace transforms].

Again consider Example 5.2.8 for the heat conduction in a semiinfinite bar, where the initial temperature distribution is constant such that $f(x) = u_o$. Thus, the problem is defined by

$$\frac{\partial u}{\partial t} = \kappa \frac{\partial^2 u}{\partial x^2}, \quad x > 0, \ t > 0, \tag{5.49}$$

with boundary conditions

$$u(0, t) = 0, \qquad |u(x, t)| < M, \quad t > 0, \tag{5.50}$$

and initial condition

$$u(x, 0) = u_o, \qquad 0 \le x < \infty. \tag{5.51}$$

Solution.
Similarly, one can use the Laplace transform method with respect to t and then solve
the ordinary differential equation.

Letting $U(x, s) = \mathscr{L}\{u(x, t)\}$ and using the property

$$\mathscr{L}\{u'(t)\} = s\mathscr{L}\{u(t)\} - u(0),$$

the transform of Eq. (5.49) becomes

$$sU(x, s) - u(x, 0) = \kappa \frac{d^2 U(x, s)}{dx^2}.$$

Using the initial condition in Eq. (5.51),

$$\kappa \frac{d^2 U}{dx^2} - sU = -u_o. \tag{5.52}$$

The homogeneous equation of Eq. (5.52) is

$$\kappa \frac{d^2 U}{dx^2} - sU = 0 \quad \text{or} \quad \frac{d^2 U}{dx^2} - \frac{U}{L^2} = 0 \quad \left(\text{where } L^2 = \frac{\kappa}{s}\right),$$

which has the solution

$$U_h(x, s) = A(s)e^{-\frac{x}{L}} + B(s)e^{\frac{x}{L}}.$$

The particular solution of Eq. (5.52) is

$$U_p(x, s) = \frac{u_o}{s}.$$

Therefore, the general solution of Eq. (5.52) is

$$U(x, s) = U_h + U_p = A(s)e^{-\frac{x}{L}} + B(s)e^{\frac{x}{L}} + \frac{u_o}{s}.$$

Since $u(x, t)$ must remain finite (see the second boundary condition in Eq. (5.50)),
$U(x, s)$ must be finite as $x \to \infty$. As such, $B = 0$ so that

$$U(x, s) = A(s)e^{-\frac{x}{L}} + \frac{u_o}{s}.$$

Taking the transform of the first boundary condition in Eq. (5.50) gives $U(0, s) = 0$
so that $U(0, s) = A + \frac{u_o}{s} = 0$. Hence $A = -\frac{u_o}{s}$. The solution of the transformed

equation is therefore

$$U(x, s) = \frac{u_o}{s}\left[1 - e^{-\frac{x}{L}}\right] = u_o\left[\frac{1 - e^{-\frac{x\sqrt{s}}{\sqrt{\kappa}}}}{s}\right].$$

Using standard Laplace transform tables (Spiegel, 1973),

$$\mathscr{L}^{-1}\left\{\frac{1 - e^{-a\sqrt{s}}}{s}\right\} = \text{erf}\left(\frac{a}{2\sqrt{t}}\right).$$

Hence, $\mathscr{L}^{-1}\{U(x, s)\} = u_o\mathscr{L}^{-1}\left\{\frac{1 - e^{-\frac{x\sqrt{s}}{\sqrt{\kappa}}}}{s}\right\} = u_o\text{erf}\left(\frac{x}{2\sqrt{\kappa t}}\right).$

The solution is $\boxed{u(x, t) = u_o\text{erf}\left(\frac{x}{2\sqrt{\kappa t}}\right).}$ [answer]

As expected, this solution is identical to Eq. (5.42) as obtained by the method of Fourier integrals in Example 5.2.8.

Example 5.2.12. Cylindrical heat conduction equation [solution by Laplace transform and inversion formula].

Solve the heat conduction equation for the temperature in the infinitely long circular cylinder for the problem given in Example 5.2.7.

Solution.
From Fig. 5.6, the heat conduction equation $\frac{\partial u}{\partial t} = \kappa\nabla^2 u$ reduces to

$$\frac{\partial u}{\partial t} = \kappa\left(\frac{\partial^2 u}{\partial r^2} + \frac{1}{r}\frac{\partial u}{\partial r}\right), \qquad 0 < r < 1, \ t > 0, \tag{5.53}$$

with boundary and initial conditions

$$u(1, t) = 0, \ |u(r, t)| < M \text{ and } u(r, 0) = u_o. \tag{5.54}$$

Replacing κt by t in Eq. (5.53) yields $\frac{\partial u}{\partial t} = \frac{\partial^2 u}{\partial r^2} + \frac{1}{r}\frac{\partial u}{\partial r}$, and taking Laplace transforms, $sU - \underbrace{u(r, 0)}_{u_o} = \frac{d^2 U}{dr^2} + \frac{1}{r}\frac{dU}{dr}$, which implies

$$\frac{d^2 U}{dr^2} + \frac{1}{r}\frac{dU}{dr} - sU = -u_o. \tag{5.55}$$

The solution of the homogeneous equation of Eq. (5.55), $\dfrac{d^2U}{dr^2} + \dfrac{1}{r}\dfrac{dU}{dr} - sU = 0$, is given by the Bessel function:

$$U_h(r, s) = c_1 J_0\left(i\sqrt{s}\,r\right) + c_2 Y_0\left(i\sqrt{s}\,r\right). \tag{5.56}$$

The particular solution is

$$U_p(r, s) = \frac{u_o}{s}. \tag{5.57}$$

Therefore, the general solution is

$$U(r, s) = U_h + U_p = c_1(s) J_0\left(i\sqrt{s}\,r\right) + c_2(s) Y_0\left(i\sqrt{s}\,r\right) + \frac{u_o}{s}. \tag{5.58}$$

Since $Y_0\left(i\sqrt{s}\,r\right)$ is unbounded as $r \longrightarrow 0$ (see Fig. 2.4), $c_2 = 0$ and Eq. (5.58) becomes $U(r, s) = c_1 J_0\left(i\sqrt{s}\,r\right) + \dfrac{u_o}{s}$. Now, from the boundary condition, $U(1, s) = 0$, which yields $c_1 J_0\left(i\sqrt{s}\right) + \dfrac{u_o}{s} = 0$ so that $c_1 = \dfrac{-u_o}{s J_0\left(i\sqrt{s}\right)}$. Therefore, $U(r, s) = \dfrac{u_o}{s} - \dfrac{u_o}{s}\dfrac{J_0\left(i\sqrt{s}\,r\right)}{J_0\left(i\sqrt{s}\right)}$.

By the inversion formula in Eq. (3.9), $u(r, t) = u_o - \dfrac{u_o}{2\pi i}\displaystyle\int_{\gamma-i\infty}^{\gamma+i\infty} \dfrac{e^{st} J_0\left(i\sqrt{s}\,r\right)}{s J_0\left(i\sqrt{s}\right)}\, ds$.

Using the residue theorem in Section 3.1.6, the function $J_0\left(i\sqrt{s}\right)$ has simple poles, where $i\sqrt{s} = \lambda_1, \lambda_2, ..., \lambda_n$ in which λ_n is the nth root of J_0. Thus, the integrand has simple poles at $s = -\lambda_n^2$, $n = 1, 2, 3, ...$, and at $s = 0$. The residues at these poles are the following:

Residue of integrand at $s = 0$:

$$\lim_{s\to 0} \frac{\cancel{s}\, e^{st} J_0\left(i\sqrt{s}\,r\right)}{\cancel{s}\, J_0\left(i\sqrt{s}\right)} = 1.$$

Residue of integrand at $s = -\lambda_n^2$:

$$\lim_{s\to -\lambda_n^2} \frac{(s + \lambda_n^2) e^{st} J_0\left(i\sqrt{s}\,r\right)}{s J_0\left(i\sqrt{s}\right)} = \lim_{s\to -\lambda_n^2}\left\{\frac{(s + \lambda_n^2)}{J_0\left(i\sqrt{s}\right)}\right\} \lim_{s\to -\lambda_n^2}\left\{\frac{e^{st} J_0\left(i\sqrt{s}\,r\right)}{s}\right\}$$

$$= \lim_{s\to -\lambda_n^2}\left\{\frac{1}{J_0'\left(i\sqrt{s}\right)\dfrac{i}{2\sqrt{s}}}\right\}\left\{\frac{e^{-\lambda_n^2 t} J_0\left(\lambda_n r\right)}{-\lambda_n^2}\right\}$$

$$= \left\{-\frac{2 e^{-\lambda_n^2 t} J_0\left(\lambda_n r\right)}{\lambda_n J_1(\lambda_n)}\right\}.$$

In the derivation of the last line, l'Hopital's rule has been used in evaluating the limit and $J_0'(u) = -J_1(u)$. Then,

$$u(r, t) = u_o - u_o \left\{ 1 - \sum_{n=1}^{\infty} \frac{2e^{-\lambda_n^2 t} J_0(\lambda_n r)}{\lambda_n J_1(\lambda_n)} \right\} = 2u_o \sum_{n=1}^{\infty} \frac{e^{-\lambda_n^2 t} J_0(\lambda_n r)}{\lambda_n J_1(\lambda_n)}.$$

Replacing t by κt yields the required solution:

$$u(r, t) = 2u_o \sum_{n=1}^{\infty} \frac{J_0(\lambda_n r)}{\lambda_n J_1(\lambda_n)} e^{-\lambda_n^2 \kappa t}. \qquad \text{[answer]}$$

As expected, this solution is identical to Example 5.2.7 as derived using a Fourier–Bessel series.

Example 5.2.13. Surface heat conduction into a flowing fluid [solution by variable transformation].

An infinitely wide flat plate is held at a constant temperature T_o. The plate is immersed into a laminar flowing fluid that has a constant density and thermal conductivity k with a bulk temperature T_1. With the origin taken at the edge of the plate (see Fig. 5.7), the velocity distribution is $V_x = \beta y$, $V_y = 0$, and $V_z = 0$, where β is a constant. Molecular transport of heat is assumed to only occur in the y direction.

(a) Determine the temperature distribution within the fluid.
(b) Determine the heat transfer coefficient between the fluid and plate.

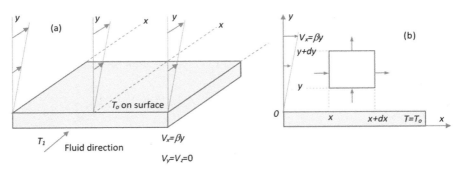

FIGURE 5.7

(a) Schematic of a fluid flowing over a flat plate. (b) Volume element of the fluid (the depth along the volume element is taken as unity).

Solution.
A heat balance over the volume element at steady state gives

Input = conduction from below and enthalpy flowing in from the left

$$= -k\,\partial x \cdot 1 \cdot \frac{\partial T}{\partial y} + C_p V_x \rho\, \partial y \cdot 1 \cdot (T - Tref),$$

Output = [conduction from below plus conduction in the element]

+ [enthalpy flowing in from the left and enthalpy in volume element]

$$= -\left[k\partial x \frac{\partial T}{\partial y} + \partial\left(k\partial x \frac{\partial T}{\partial y} \right) \right] + C_p V_x \rho \partial y (T - Tref)$$

$$+ \partial [C_p V_x \rho \partial y (T - Tref)],$$

where T is the temperature of the fluid at x and y and C_p is the specific heat. Since the accumulation at steady state is Accumulation = Input − Output = 0, we have

$$\partial\left(k\partial x \frac{\partial T}{\partial y} \right) = \partial [C_p V_x \rho \partial y (T - Tref)$$

or

$$\frac{\partial}{\partial y}\left(k \frac{\partial T}{\partial y} \right) = \frac{\partial}{\partial x}\left(C_p V_x \rho y T \right).$$

Given $V_x = \beta y$ and β, k, C_p, and ρ are constants, one obtains the partial differential equation for the following problem:

$$\frac{\partial T}{\partial x} = \frac{A}{y}\frac{\partial^2 T}{\partial y^2},$$

where $A = \dfrac{k}{\beta \rho C_p}$. This problem is subject to the following boundary conditions:

(i) leading edge of plate: $x = 0$, $y > 0$, $T = T_1$,
(ii) above the plate: $x > 0$, $y = \infty$, $T = T_1$,
(iii) on the plate: $x > 0$, $y = 0$, $T = T_0$.

With a change of variables, $\dfrac{T - T_1}{T_0 - T_1}$, the differential equation becomes

$$\frac{A}{y}\frac{\partial^2 \theta}{\partial y^2} = \frac{\partial \theta}{\partial x}.$$

The boundary conditions are therefore: (i) $x = 0$, $y > 0$, $T = T_1$, $\theta = 0$, (ii) $x > 0$, $y = \infty$, $T = T_1$, $\theta = 0$, and (iii) $x > 0$, $y = 0$, $T = T_0$, $\theta = 1$. Moreover, given these boundary conditions, one can assume a solution of the form $\theta = f\left(\dfrac{y}{x^n} \right) = f(\eta)$ with boundary conditions (i) $\theta = 0$, $x = 0$, $y > 0$, which implies $\eta = \infty$, and (ii) $\theta = 1$, $x > 0$, $y = 0$, which implies $\eta = 0$.

For this transformed equation, $\dfrac{\partial \theta}{\partial x} = \dfrac{d\theta}{d\eta}\dfrac{\partial \eta}{\partial x} = \dfrac{-ny}{x^{n+1}}\dfrac{d\theta}{d\eta} = \dfrac{-n\eta}{x}\dfrac{d\theta}{d\eta}$ and $\dfrac{\partial \theta}{\partial y} =$

$\dfrac{d\theta}{d\eta}\dfrac{\partial \eta}{\partial y} = \dfrac{1}{x^n}\dfrac{d\theta}{d\eta}$. Taking the derivative again for the latter equation, $\dfrac{\partial^2 \theta}{\partial y^2} =$

$\frac{1}{x^n}\frac{\partial}{\partial y}\frac{d\theta}{d\eta} = \frac{1}{x^n}\frac{d^2\theta}{d\eta^2}\frac{\partial\eta}{\partial y} = \frac{1}{x^{2n}}\frac{d^2\theta}{d\eta^2}$. Thus, substituting these expressions into the

differential equation yields $\frac{-n\eta}{x}\frac{d\theta}{d\eta} = \frac{A}{y}\frac{1}{x^{2n}}\frac{d^2\theta}{d\eta^2}$. Since $\eta = yx^{-n}$, this equation

becomes

$$\frac{-n\eta}{x}\frac{d\theta}{d\eta} = \frac{A}{x^{3n}}\frac{1}{\eta}\frac{d^2\theta}{d\eta^2}.$$

Setting $n = 1/3$ so that the factor x cancels on both sides of the equation gives the ordinary differential equation

$$\boxed{\frac{d^2\theta}{d\eta^2} + \frac{\eta^2}{3A}\frac{d\theta}{d\eta} = 0.}$$

Setting $\frac{d\theta}{d\eta} = v$ and separating variables gives $\int\frac{dv}{v} = -\int\frac{\eta^2\,d\eta}{3A}$. On integrating,

the solution is $v = \frac{d\theta}{d\eta} = Be^{-\eta^3/(9A)}$ or $\int_0^\theta d\theta = B\int_\infty^\eta e^{-\eta^3/(9A)}d\eta$. Integrating again

gives $\theta = B\int_\infty^\eta e^{-\eta^3/(9A)}d\eta$. Applying the boundary condition that $\theta = 1$ when $\eta =$

0, the constant B is evaluated as $B = -\dfrac{1}{\int_0^\infty e^{-\eta^3/(9A)}d\eta}$. Therefore, the final solution

is

$$\boxed{\theta = \frac{T - T_1}{T_o - T_1} = \frac{\int_\eta^\infty e^{-\eta^3/(9A)}\,d\eta}{\int_0^\infty e^{-\eta^3/(9A)}\,d\eta}.}$$

(b) The heat transfer coefficient h at the surface of the plate can be determined by equating convection from the surface to conduction through the fluid:

$$h(T_o - T_1) = k\left(\frac{\partial T}{\partial y}\right)_{y=0}. \tag{5.59}$$

From part (a), $\frac{\partial T}{\partial y} = \frac{\partial\theta}{\partial y}[T_o - T_1]$, where $\frac{\partial\theta}{\partial y} = \frac{1}{x^{1/3}}\frac{d\theta}{d\eta} = \frac{Be^{-\eta^3/(9A)}}{x^{1/3}}$. Using the

expression for B from part (a) and substituting these two derivatives into the heat balance in Eq. (5.59) gives

$$h = \frac{-k}{x^{1/3}}\frac{e^{-\eta^3/(9A)}}{\int_0^\infty e^{-\eta^3/(9A)}\,d\eta}.$$

Integrating and evaluating the heat transfer coefficient at the surface of the plate $y = \eta = 0$,

$$h = 0.43k \left(\frac{\beta \rho C_p}{kx}\right)^{1/3}.$$

Problems

5.1 Consider a solid ball of radius R, with a constant thermal diffusivity κ. Initially, the ball has a uniform temperature u_o. If this ball is dropped into an ice-water bath of temperature zero, show that the temperature distribution $u(r, t)$ in the ball can be described by the following partial differential equation when only radial heat conduction is considered: $\frac{\partial u}{\partial t} = \kappa \left[\frac{1}{r^2} \frac{\partial}{\partial r} \left(r^2 \frac{\partial u}{\partial r}\right)\right]$, with the initial condition $u(r, t) = u_o$, $0 < r < R$, $t = 0$, and boundary conditions $u(r, t)$ is finite at $r = 0$, $t > 0$, and $u(r, t) = 0$, $r = R$, $t > 0$.
Using the transformation $\xi = ur$, show that the partial differential equation becomes $\frac{\partial \xi}{\partial t} = \kappa \frac{\partial^2 \xi}{\partial r^2}$, subject to an initial condition $\xi(r, t) = u_o r$, $0 < r < R$, $t = 0$, and boundary conditions $\xi(r, t) = 0$, $r = 0$, $t > 0$, and $\xi(r, t) = 0$, $r = R$, $t > 0$. Solve this transformed problem for $\xi(r, t)$. What is the corresponding solution for $u(r, t)$?

5.2 A chemical pollutant is released into the atmosphere, where it diffuses radially with a constant diffusion coefficient D. It can be assumed that the initial concentration of the pollutant is zero, where the pollutant is a point source of strength S (atoms s^{-1}). The diffusion equation for the pollutant concentration $C(r, t)$ (atoms m^{-3}) is given by

$$\frac{\partial C}{\partial t} = D \left[\frac{1}{r^2} \frac{\partial}{\partial r} \left(r^2 \frac{\partial C}{\partial r}\right)\right].$$

The source condition can be represented as a boundary condition at $r = 0$. Using Fick's first law of diffusion, where the flux $F = -D\frac{\partial C}{\partial r}$, the release rate of the pollutant (at $r = 0$) is equal to the source strength such that $\lim_{r \to 0} (4\pi r^2 F) = S$. The other conditions follow for the given physical problem: $C = 0$ as $r \to \infty$, $t > 0$, and $C = 0$, $0 < r < \infty$, $t = 0$.

(a) As proved in Problem 5.1, with the transformation $\xi = Cr$, the diffusion equation reduces to $\frac{\partial \xi}{\partial t} = D\frac{\partial^2 \xi}{\partial r^2}$. Using this transformation, show that the initial and boundary conditions become $\xi(r, t) = 0$, $0 < r < R$, $t = 0$, and $\xi(r, t) = \frac{S}{4\pi D}$, $r = 0$, $t > 0$, and $\xi(r, t)$ is finite as $r \to \infty$, $t > 0$. Using a Laplace transform method, show that the solution, subject to the given conditions, is given by

$\xi(r,t) = \dfrac{S}{4\pi D} \text{erfc}\left(\dfrac{r}{2\sqrt{Dt}}\right)$, where the latter function is the complementary error function.

(b) What is the corresponding solution for $C(r,t)$?

(c) Briefly mention why the method of separation of variables is inappropriate for this given problem.

(d) Assuming a diffusivity $D = 10^{-2}$ m^{-2} s^{-1} and a source strength $S = 100$ atoms s^{-1}, what is the concentration at a distance of 100 m from the point source after 10^4 s?

5.3 Consider the radial diffusion of material in a sphere of radius a, with a concentration distribution $C(r,t)$ and constant diffusivity D. It is assumed that: (i) the concentration at the center of the sphere is finite for $t > 0$, (ii) the surface of the sphere is maintained at a zero concentration for $t > 0$, and (iii) initially there is a uniform concentration C_o throughout the sphere $0 < r < a$.

Defining the variables $x = \dfrac{r}{a}$, $\tau = \dfrac{Dt}{a^2}$, and $u = \dfrac{C}{C_o}\dfrac{r}{a} = \dfrac{C}{C_o}x$, the transformed problem is $\dfrac{\partial u}{\partial \tau} = \dfrac{\partial^2 u}{\partial x^2}$ with conditions $u(x,\tau) = x$, $0 < x < 1$, $\tau = 0$, $u(x,\tau) = 0$, $x = 0$, $\tau > 0$, and $u(x,\tau) = 0$, $x = 1$, $\tau > 0$.

Using the Laplace transform method with respect to the variable τ, such that $U(x,s) = \mathcal{L}\{u(x,\tau)\}$, show that the transformed solution is $U(x,s) = \dfrac{1}{s}\left[x - \dfrac{\sinh x\sqrt{s}}{\sinh\sqrt{s}}\right]$. What are the corresponding solutions for $u(x,\tau)$ and $C(r,t)$?

(a) The release fraction $F(t)$ can be defined as the total amount of material which has diffused through the surface of the sphere at time t divided by the initial amount of material in the sphere, $F(t) = \dfrac{4\pi a^2 \displaystyle\int_0^t J(t)dt}{\frac{4}{3}\pi a^3 C_o} = \dfrac{3}{aC_o}\displaystyle\int_0^t J(t)dt$. The flux of

material $J(t)$ is evaluated from Fick's law of diffusion: $J(t) = -D\dfrac{\partial C(r,t)}{\partial r}\bigg|_{r=a}$.

Hence, show that $F(\tau) = \dfrac{3a}{DC_o}\displaystyle\int_0^\tau J(\tau)d\tau = -3\displaystyle\int_0^\tau \left(\dfrac{du}{dx}\right)_{x=1} d\tau$, and defining the

Laplace transform $F(s) = \mathcal{L}\{F(\tau)\}$, show that $F(s) = -\dfrac{3}{s}\dfrac{dU(x,s)}{dx}\bigg|_{x=1}$.

(b) Using the solution for $U(x,s)$ and the latter expression for $F(s)$ from part (a), show that $F(s) = 3\left[\dfrac{\coth\sqrt{s}}{s^{3/2}} - \dfrac{1}{s^2}\right]$.

(c) An infinite series will eventually result for the release fraction $F(t)$ which is more difficult to evaluate. However, an analytical form for $F(t)$ is possible by considering a "short-time" approximation for the condition that $\tau \ll 1$. Show that this condition is equivalent to $s \gg 1$ in Laplace transform space. Determine an analytic expression for $F(t)$ by applying the condition $s \gg 1$ and taking the inverse transform of this resultant expression.

5.4   A standard model for describing the dispersion of chemical pollutants in the atmosphere is the Gaussian plume model. In this model it is assumed that the wind carries the pollutant in the x direction with an average wind speed u, while the material diffuses in the other directions. In the case of a plane source of pollutant at $x = 0$, assuming only diffusion in the y direction (with a diffusivity D), the steady-state transport equation for the concentration C is $u \dfrac{\partial C}{\partial x} = D \dfrac{\partial^2 C}{\partial y^2}$. Using the transformation $\tau = x D / u$, this equation can be written as $\dfrac{\partial C}{\partial \tau} = \dfrac{\partial^2 C}{\partial y^2}$, subject to the conditions

$$C = S\delta(y), \quad -\infty < y < \infty, \quad \tau = 0,$$

$$C \text{ is finite at } y = 0, \text{ which implies } \frac{\partial C}{\partial y} = 0, \quad y = 0, \tau > 0,$$

$$C = 0, \quad y = \pm\infty, \quad \tau > 0.$$

The condition at $\tau = 0$ accounts for the plane source S of material being released into the atmosphere (that is, at $x = \tau = 0$). Here $\delta(y)$ is a so-called distribution function which has the property that $\displaystyle\int_{-\infty}^{\infty} \delta(y - a) f(y) \, dy = f(a)$ and $\displaystyle\int_{-\infty}^{\infty} S\delta(y) \, dy = S$.

(a)   Using a Fourier cosine transform method such that $\hat{C}_c(w, \tau) = \mathscr{F}_c\{C(y, \tau)\}$, show that the transformed partial differential equation is

$$\frac{\partial \hat{C}_c(w, \tau)}{\partial \tau} = -w^2 \hat{C}_c(w, \tau) - \sqrt{\frac{2}{\pi}} \frac{\partial C}{\partial y}\bigg|_{y=0}.$$

(b)   Using the given conditions and the property of the distribution function, show that the transformed solution in part (a) is $\hat{C}(w, \tau) = \dfrac{S}{\sqrt{2\pi}} e^{-w^2 \tau}$.

(c)   Taking the inverse Fourier cosine transform of the expression in part (b), show that the final solution is given by $C(y, \tau) = \dfrac{S}{\sqrt{4\pi\tau}} e^{-y^2/(4\tau)}$. Why is the method of separation of variables inappropriate for this problem?

5.5   The model in Problem 5.4 can be generalized further to account for diffusion in both the y and z directions, with a time-dependent diffusive-convective equation for the concentration of the pollutant in the atmosphere $C(x, y, z)$ (atoms m^{-3}) and point source of pollutant S (atoms s^{-1}). If the movement of the effluent in the x direction from the wind (with a constant speed u m/s) is much greater than that of diffusion, the following isotropic diffusion equation applies: $\dfrac{\partial C}{\partial t} = D\left(\dfrac{\partial^2 C}{\partial y^2} + \dfrac{\partial^2 C}{\partial z^2} - u \dfrac{\partial C}{\partial x}\right)$, where D is the diffusion coefficient (m^2 s^{-1}). For steady-state conditions, assuming $D = 1$ m^2 s^{-1} and $u = 1$ m/s, this equation simplifies to $\dfrac{\partial C}{\partial x} = \dfrac{\partial^2 C}{\partial y^2} + \dfrac{\partial^2 C}{\partial z^2}$.

(a) Separating the variables where $C(x, y, z) = C_1(x, y) \cdot C_2(x, z)$, show that two partial differential equations result: $\dfrac{\partial C_1}{\partial x} = \dfrac{\partial^2 C_1}{\partial y^2}$ and $\dfrac{\partial C_2}{\partial x} = \dfrac{\partial^2 C_2}{\partial z^2}$.

(b) Because of the similarity of these two equations, one need only consider a solution for one (since the other follows by symmetry). The boundary conditions for the given y region are $C_1(x, y) < M$ at $y = 0$, where M is a constant since the C_1 concentration must remain finite at $y = 0$, which implies that $\dfrac{\partial C_1}{\partial y}\Big|_{y=0} = 0$, $x > 0$. The second boundary condition $C_1(x, y) = 0$, $y = \infty$, $x > 0$, follows since far away from the source, the concentration must approach zero. Using a Fourier cosine transform method such that $c(x, w) = F_c\{C_1(x, y)\}$, show that the solution of this problem is $C_1(x, y) = \dfrac{A_1}{\sqrt{x}} \exp\left\{-\dfrac{y^2}{4x}\right\}$, where A_1 is an arbitrary constant. Note the following definite integral has been used in this derivation:
$$\int_{-\infty}^{\infty} e^{-aw^2} \cos(bw)\, dw = \frac{1}{2}\sqrt{\frac{\pi}{a}}\, e^{-\frac{b^2}{4a}}.$$

(c) The solution for the z direction follows on replacing the variable y with z in the solution of part (b) because of the symmetry of the problem $C_2(x, z) = \dfrac{A_2}{\sqrt{x}} \exp\left\{-\dfrac{z^2}{4x}\right\}$. The final solution for $C(x, y, z)$ therefore follows as $C(x, y, z) = C_1(x, y) \cdot C_2(x, z) = \dfrac{A}{x} \exp\left\{-\dfrac{1}{4x}\left[y^2 + z^2\right]\right\}$, where $A = A_1 \cdot A_2$. The arbitrary constant A follows from a conservation of mass using the source condition for the pollutant at $x = 0$. Hence, evaluate A by integrating the flux of material $[u \cdot C(x, y, z)]$ (where $u = 1$ m/s) over all space in the y and z directions and equating this result to the quantity S: $\displaystyle\lim_{x \to 0} \int_{-\infty}^{\infty} \int_{-\infty}^{\infty} [1 \cdot C(x, y, z)]\, dy\, dz = S$. Hint: Note the definite integral result $\displaystyle\int_{-\infty}^{\infty} e^{-a\xi^2}\, d\xi = \sqrt{\dfrac{\pi}{a}}$. This analysis therefore yields the steady-state Gaussian plume model for the dispersal of a pollutant with diffusion in the y and z directions and convective transport by the wind in the x direction.

5.6 Consider heat conduction in a semiinfinite slab, where the slab is initially ($t = 0$) at a constant temperature u_m and the end of the slab (at $x = 0$) is maintained at the constant temperature u_w. The partial differential equation for the temperature distribution $u(x, t)$ for thermal diffusivity κ is

$$\frac{\partial u}{\partial t} = \kappa \frac{\partial^2 u}{\partial x^2}, \quad x > 0,\ t > 0,$$

with $u(x, 0) = u_m$, $x > 0$, $u(0, t) = u_w$, $t > 0$, $|u(x, t)| < M$ (that is, u is finite as $x \to \infty$, $t > 0$). Using a Laplace transform method:

(a) Show that the transformed solution is $U(x, s) = (u_w - u_m)\dfrac{e^{-\sqrt{\frac{s}{\kappa}}x}}{s} + \dfrac{u_m}{s}$.

(b) Using the inverse Laplace transform $\mathcal{L}^{-1}\left\{\dfrac{e^{-a\sqrt{s}}}{s}\right\} = \text{erfc}\left(\dfrac{a}{2\sqrt{t}}\right)$, show that the solution is

$$u(x,t) = \mathcal{L}^{-1}\{U(x,s)\} = (u_w - u_m)\text{erfc}\left(\frac{x}{2\sqrt{\kappa t}}\right) + u_m.$$

5.7 A problem of interest to chemical and material engineers is the melting of a material. This problem can be solved as a moving-boundary heat conduction problem as a so-called "Stefan problem," where the melt front (that is, the liquid/solid interface) progresses with time. Consider a slab of material which is initially at the constant temperature u_m at $t = 0$. At time $t > 0$, the temperature at the wall $(x = 0)$ is raised to a value above the melting temperature of the material, u_w. At the liquid/solid interface, $x = s(t)$, the temperature is equal to u_m. These boundary conditions can be written mathematically as

$$u(0,t) = u_w, \quad t > 0,$$
$$u(x,t)|_{x=s(t)} = u_m, \quad t > 0.$$

As shown in Problem 5.6, the general solution of the heat conduction equation is of the form $A + B\,\text{erfc}\left(\dfrac{x}{2\sqrt{\kappa t}}\right)$, where A and B are constants determined from the boundary conditions.

(a) Show that by applying the boundary conditions, the solution of this moving boundary problem is $\dfrac{u(x,t) - u_w}{u_m - u_w} = \dfrac{\text{erf}\left(\dfrac{x}{2\sqrt{\kappa t}}\right)}{\text{erf}(\gamma)}$, where $s(t) = 2\gamma\sqrt{\kappa t}$. The relation for $s(t)$ indicates the position of the moving liquid front.

(b) Assuming that heat transfer occurs in the liquid phase by conduction only, the constant γ can be determined from an energy balance across the interface: $-K\dfrac{du}{dx}\bigg|_{x=s(t)} = \tau L\dfrac{ds}{dt}$. Using the definition of the thermal conductivity K, show that this energy balance becomes $-\kappa\sigma\dfrac{du}{dx}\bigg|_{x=s(t)} = L\dfrac{ds}{dt}$. Here τ is the density, κ is the thermal diffusivity, σ is the specific heat for the liquid phase, and L is the latent heat of melting/solidification. Show that by applying this condition with the solution for $u(x,t)$ one obtains the following transcendental equation for γ: $\gamma e^{\gamma^2}\text{erf}(\gamma) = \dfrac{\sigma(u_w - u_m)}{L\sqrt{\pi}}$. As shown in Problem 9.2, various approximate methods can be used to numerically solve this transcendental equation.

(c) To numerically solve for γ in part (b), one requires that the error function $\text{erf}(\gamma) = \dfrac{2}{\sqrt{\pi}}\displaystyle\int_0^\gamma e^{-u^2}\,du$ be evaluated. Using a Gaussian integration rule (with $n = 2$ terms) (Section 9.5.3), derive an analytic relation for $\text{erf}(\gamma)$.

5.8 Give a brief physical interpretation of the following partial differential equations. Classify each equation as elliptic, hyperbolic, or parabolic by comparing it to the general form of a linear partial differential equation with two independent variables.

(a) $\dfrac{\partial^2 u}{\partial x^2} + \dfrac{\partial^2 u}{\partial y^2} = 0.$

(b) $\dfrac{\partial u}{\partial t} = \kappa \dfrac{\partial^2 u}{\partial x^2}.$

(c) $\dfrac{\partial^2 u}{\partial t^2} = \alpha^2 \dfrac{\partial^2 u}{\partial x^2}.$

5.9 Consider the heat conduction in a thin insulated bar of length 3 m where the initial temperature at $t = 0$ is $f(x) = 15 - 10x\,°C$ and the ends of the bar are kept at $0°C$. The partial differential equation for the temperature distribution $u(x, t)$ at the distance x and time t in the bar is therefore given by $\dfrac{\partial u}{\partial t} = \dfrac{\partial^2 u}{\partial x^2}$. Here the thermal diffusivity is simply equal to $1\ m^2\ s^{-1}$. The boundary conditions for this problem are given mathematically by $u(0, t) = 0$, $t > 0$, $u(3, t) = 0$, $t > 0$, and the initial condition is $u(x, 0) = 15 - 10x$, for $0 < x < 3$, $t = 0$. Solve this problem using a separation of variables technique.

5.10 Solve Problem 5.9 using a Laplace transform method with the variable transformation $U(x, s) = \mathscr{L}\{u(x, t)\}$, where

(a) the transformed solution for $U(x, s)$ is given by

$$U(x, s) = 15 \left[\frac{\sinh\left(x\sqrt{s}\right)}{s \sinh\left(3\sqrt{s}\right)} + \frac{\sinh\left((x - 3)\sqrt{s}\right)}{s \sinh\left(3\sqrt{s}\right)} \right] + \frac{15 - 10x}{s}.$$

(b) By taking the inverse Laplace transform of $U(x, s)$ in part (a), what is the solution for $u(x, t)$?

5.11 Consider the heat conduction in a thin insulated bar of length 3 m where the initial temperature at $t = 0$ is $25°C$ and the ends of the bar are kept at $10°C$ at $x = 0$ and $40°C$ at $x = 3$. The partial differential equation for the temperature distribution $u(x, t)$ at the distance x and time t in the bar is given by $\dfrac{\partial u}{\partial t} = \dfrac{\partial^2 u}{\partial x^2}$. Here the thermal diffusivity is simply equal to $1\ m^2\ s^{-1}$. The boundary conditions for this problem are given by $u(0, t) = 10$, $t > 0$, $u(3, t) = 40$, $t > 0$, and the initial condition is $u(x, 0) = 25$, for $0 < x < 3$, $t = 0$. To solve this complicated boundary value problem, assume that $u(x, t) = v(x, t) + \psi(x)$, where $\psi(x)$ is to be suitably determined in part (a).

(a) Explain why $u(x, t)$ can be written as two separate functions. Show that the boundary value problem can be rewritten as two separate problems: $\psi''(x) = 0$, $\psi(0) = 10$, $\psi(3) = 40$, and $\dfrac{\partial v}{\partial t} = \dfrac{\partial^2 v}{\partial x^2}$ with $v(0, t) = 0$, $v(3, t) = 0$, $v(x, 0) = 15 - 10x$.

(b) What is the solution for $\psi(x)$? The solution for $\psi(x)$ is in fact needed to obtain the initial condition for $v(x, 0)$.

(c) The partial differential equation problem for the function $v(x, t)$ is in fact the same boundary value problem as given in Problems 5.9 and 5.10. Using this solution for $v(x, t)$, give the complete solution for $u(x, t)$.

(d) What is the physical significance of the solution for $\psi(x)$?

5.12 The slowing down of neutrons in the moderator of a nuclear reactor can be described by the slowing down density q (neutrons m^{-3} s^{-1}) in accordance with the so-called Fermi age equation, $\dfrac{\partial q}{\partial \tau} = \dfrac{\partial^2 q}{\partial x^2}$, where τ is the "age" of the neutron (in this theory, the unit of τ is m^2). This equation can be solved assuming a plane source of neutrons S (neutrons m^{-2} s^{-1}) in the yz plane at the origin of the coordinate system, where x is the distance from the plane source. The boundary conditions for this equation consider symmetry at $x = 0$ and require that: (i) q remains finite such that $\dfrac{\partial q}{\partial x}\bigg|_{x=0} = 0$, $\tau > 0$, and (ii) q is finite as $x \to \infty$, $\tau > 0$.

(a) Using a Fourier cosine transform on defining $Q(w, \tau) = F_c\{q(x, \tau)\}$, show that this problem yields the transformed solution: $Q(w, \tau) = A_1 e^{-w^2 \tau}$, where A_1 is an arbitrary constant.

(b) Taking the inverse Fourier cosine transform, show that the solution for $q(x, \tau)$ is $q(x, \tau) = \dfrac{A}{\sqrt{\tau}} e^{-x^2/(4\tau)}$, where A is a constant.

(c) Using a source condition that follows from the conservation of mass $S = \displaystyle\int_{-\infty}^{\infty} q(x, \tau)dx$, evaluate the constant A. This source condition replaces the initial condition for $\tau = 0$.

5.13 Consider the solution of d'Alembert's for the wave equation as mentioned in Section 1.1.1. Here a solution to the wave equation

$$\frac{\partial^2 u(x, t)}{\partial x^2} = \frac{1}{c^2} \frac{\partial^2 u(x, t)}{\partial t^2}$$

is given by

$$u(x, t) = F(x + ct) + G(x - ct),$$

where $F(x + ct)$ and $G(x - ct)$ are arbitrary functions. This solution is obtained by changing the variables x and t to $w = x + ct$ and $z = x - ct$, respectively, considering that $u(x, t) = \bar{u}(w, z)$ for the partial differential equation.

Difference numerical methods

Finite difference methods can be used to numerically solve first- and second-order ordinary differential equations (Section 6.1), as well as partial differential equations (Section 6.2), particularly when analytical methods are not feasible. This is a very common method used for the solution of engineering problems that commonly arise in real-world situations where the coefficients of the differential equation themselves may be a function of the dependent variable. For the solution method, there is a discretization of the differential equation using finite differences to approximate derivatives. This method essentially represents three (or less) spatial dimensions and perhaps a time dimension as arrays of mesh points, defined from a mesh spacing that is uniform or variable in size. This method can be alternatively compared to the finite element numerical method as described in Chapter 7. Numerical techniques developed here can also be used for the solution of nonlinear partial differential equations in Chapter 12 or utilized as an approach to linearize nonlinear equations.

6.1 Ordinary differential equations

6.1.1 First-order equations

Numerical methods can be used to solve the initial value problem

$$y' = f(x, y), \qquad y(x_0) = y_0.$$
(6.1)

One-step methods

The problem in Eq. (6.1) can be solved by truncating the Taylor series

$$y(x + h) = y(x) + hy'(x) + \frac{h^2}{2}y''(x) + ...,$$
(6.2)

where by Eq. (6.1), $y' = f$, $y'' = f' = \dfrac{\partial f}{\partial x} + \dfrac{\partial f}{\partial y}y'$, and so on. In a one-step method, one starts from $y_0 = y(x_0)$ and proceeds stepwise computing approximate values of the solution using a Taylor series approximation at given "mesh points":

$$x_1 = x_0 + h, x_2 = x_0 + 2h, x_3 = x_0 + 3h, ...,$$
(6.3)

Advanced Mathematics for Engineering Students. https://doi.org/10.1016/B978-0-12-823681-9.00014-9

where h is a fixed step size.

(i) *Euler method*

Truncating Eq. (6.2) after the term hy' yields the Euler method in which one computes in a step-by-step method:

$$y_1 = y_0 + hf(x_0, y_0),$$
$$y_2 = y_1 + hf(x_1, y_1),$$
$$\cdots$$

$$\boxed{y_{n+1} = y_n + hf(x_n, y_n)} \quad (n = 0, 1, ...). \tag{6.4}$$

This method is of first order since only constant terms are considered with a term containing the first power of h. The truncation of the series in Eq. (6.2) results in a local truncation error of order h^2. In addition, there are round-off errors in this calculation and other methods.

Example 6.1.1. Solve the initial value problem $y' = x + y + 1$, $y(0) = 0$ for $0 \leq x \leq 1$ with $h = 0.2$ using an Euler method.

Solution.
Here $f(x, y) = x + y + 1$ and Eq. (6.4) becomes

$$y_{n+1} = y_n + 0.2(x_n + y_n + 1).$$

The analytical solution is obtained with an integrating factor of e^{-x} as derived from the methodology in Chapter 2, yielding $y(x) = 2e^x - (x + 2)$. The numerical solution using this step-by-step equation is given in Table 6.1.

Table 6.1 Example calculation for the Euler method.

n	x_n	y_n	$0.2(x_n + y_n + 1)$	Exact value	Error
0	0.0	0.000	0.200	0.000	0.000
1	0.2	0.200	0.280	0.243	0.043
2	0.4	0.480	0.376	0.584	0.104
3	0.6	0.856	0.491	1.044	0.188
4	0.8	1.347	0.629	1.651	0.304
5	1.0	1.977		2.437	0.460

(ii) *Heun's method*

Taking one more term into account in Eq. (6.2) yields the Heun's method, which avoids the computation of calculating the derivatives of f with the use of an "auxiliary" equation, that is,

$$\boxed{y_{n+1}^* = y_n + hf(x_n, y_n),} \tag{6.5a}$$

$$y_{n+1} = y_n + \frac{1}{2}h[f(x_n, y_n) + f(x_{n+1}, y_{n+1}^*)]. \qquad (6.5b)$$

This method is a predictor-corrector method because in each step a value is predicted in Eq. (6.5a) and then corrected in Eq. (6.5b). The truncation error per step for this method is obtained as follows. Setting $\tilde{f}_n = f(x_n, y(x_n))$ and using Eq. (6.1) and Eq. (6.2),

$$y(x_n + h) - y(x_n) = h\tilde{f}_n + \frac{1}{2}h^2 \tilde{f}_n' + \frac{1}{6}h^3 \tilde{f}_n'' + ..., \qquad (6.6a)$$

and it follows that

$$y_{n+1} - y_n \simeq \frac{1}{2}h[\tilde{f}_n + \tilde{f}_{n+1}] = \frac{1}{2}h[\tilde{f}_n + (\tilde{f}_n + h\tilde{f}_n' + \frac{1}{2}h^2 \tilde{f}_n'' + ...)]. \qquad (6.6b)$$

Subtracting Eq. (6.6a) from Eq. (6.6b) gives the truncation error per step

$$\frac{h^3}{4}\tilde{f}_n'' - \frac{h^3}{6}\tilde{f}_n'' + ... = \frac{h^3}{12}\tilde{f}_n'' + ...$$

Thus, Heun's method is a second-order method because the local truncation error is of the order h^3.

Example 6.1.2. Solve the problem in Example 6.1.1 by Heun's method.

Solution.
Letting $k_1 = hf(x_n, y_n) = 0.2(x_n + y_n + 1)$ and $k_2 = hf(x_{n+1}, y_n + k_1) = 0.2(x_n + 0.2 + y_n + 0.2(x_n + y_n + 1) + 1)$, Eq. (6.5b) becomes

$$y_{n+1} = y_n + \frac{1}{2}(k_1 + k_2) = y_n + \frac{0.2}{2}(2.2x_n + 2.2y_n + 2.4)$$
$$= y_n + 0.22(x_n + y_n) + 0.24.$$

The results in Table 6.2 are more accurate than those in Table 6.1.

Table 6.2 Example calculation for the Heun method.

n	x_n	y_n	$0.22(x_n + y_n) + 0.24$	**Exact value**	**Error**
0	0.0	0.0000	0.2400	0.0000	0.0000
1	0.2	0.2400	0.3368	0.2428	0.0028
2	0.4	0.5768	0.4549	0.5836	0.0068
3	0.6	1.0317	0.5990	1.0442	0.0125
4	0.8	1.6307	0.7747	1.6511	0.0204
5	1.0	2.4054		2.4366	0.0311

(iii) *Runge–Kutta method*

Truncating after the term in h^4 gives the Runge–Kutta method. Again, in this method, the calculation of the derivatives in Eq. (6.2) is replaced by the evaluation of $f(x, y)$ at suitable points (x, y) with the following use of four auxiliary quantities:

$$k_1 = hf(x_n, y_n),$$

$$k_2 = hf(x_n + \frac{1}{2}h, y_n + \frac{1}{2}k_1),$$

$$k_3 = hf(x_n + \frac{1}{2}h, y_n + \frac{1}{2}k_2),$$

$$k_4 = hf(x_n + h, y_n + k_3),$$ (6.7a)

and then from these quantities we have the new value

$$y_{n+1} = y_n + \frac{1}{6}(k_1 + 2k_2 + 2k_3 + k_4).$$ (6.7b)

Example 6.1.3. Solve the problem in Example 6.1.1 by the Runge–Kutta method.

Solution.
For the present problem $f(x, y) = x + y + 1$ and $h = 0.2$. Therefore, Eq. (6.7a) gives

$$k_1 = 0.2(x_n + y_n + 1), k_2 = 0.2(x_n + 0.1 + y_n + 0.5k_1 + 1),$$

$$k_3 = 0.2(x_n + 0.1 + y_n + 0.5k_2 + 1), k_4 = 0.2(x_n + 0.2 + y_n + k_3 + 1).$$

The results using Excel analysis in Table 6.3 are more accurate than those in Table 6.1 and Table 6.2.

Since the error is of order h^5 for this method, in a switch from h to $2h$, the error is multiplied by $2^5 = 32$. However, one needs only half as many steps as before so that the error will only be multiplied by $32/2 = 16$. In fact, the error ϵ of an approximation $\tilde{y}$ obtained with step h equals about $1/15$ the difference $\delta = \tilde{y} - \tilde{\tilde{y}}$ of corresponding approximations obtained with steps h and $2h$, respectively:

$$\epsilon = \frac{1}{15}\left(\tilde{y} - \tilde{\tilde{y}}\right).$$ (6.8)

Multistep methods

The previous methods are known as one-step methods because each step uses only values obtained in the preceding step. In contrast, methods that use values from more than one preceding step are called multistep methods, which avoid computations such as Eq. (6.7a).

Table 6.3 Example calculation for the Runge–Kutta method.

n	x_n	y_n	k_1	k_2	k_3	k_4	$\frac{1}{6}(k_1 + 2k_2 + 2k_3 + k_4)$	Exact	Error $\times 10^6$
0	0.0	0.000000	0.200000	0.240000	0.244000	0.288800	0.242800	0.000000	0
1	0.2	0.242800	0.288560	0.337416	0.342302	0.397020	0.340836	0.242806	6
2	0.4	0.583636	0.396727	0.456400	0.462367	0.529201	0.460577	0.583649	14
3	0.6	1.044213	0.528843	0.601727	0.609015	0.690646	0.606829	1.044238	25
4	0.8	1.651042	0.690208	0.779229	0.788131	0.887835	0.785461	1.651082	40
5	1.0	2.436502						2.436564	61

By integrating cubic spline polynomials (see Section 9.3), one can obtain the important Adams–Moulton multistep method (Kreyszig, 1993). In this method, one first computes the predictor

$$y_{n+1}^* = y_n + \frac{h}{24}\left(55 f_n - 59 f_{n-1} + 37 f_{n-2} - 9 f_{n-3}\right), \tag{6.9a}$$

where $f_j = f(x_j, y_j)$, and then from it the corrector (the actual new value)

$$y_{n+1} = y_n + \frac{h}{24}\left(9 f_{n+1}^* + 19 f_n - 5 f_{n-1} + f_{n-2}\right), \tag{6.9b}$$

where $f_{n+1}^* = f(x_{n+1}, y_{n+1}^*)$. To get started, y_1, y_2, and y_3 must be computed by the Runge–Kutta method or some other method. This method is like the Runge–Kutta method that is of fourth order; however, it is faster because only two new values of f are needed per step in Eq. (6.9a) and Eq. (6.9b), as opposed to four values in Eq. (6.7a). Predictor-corrector methods can also provide an estimate of the error. If $|y_n - y_n^*|$ is large, then the error of y_n is probably large and h should be reduced. On the other hand, if y_n is small, then h may be increased (e.g., doubled in size).

Example 6.1.4. Solve the problem in Example 6.1.1 on the interval $0 \le x \le 2$ by the Adams–Moulton method with $h = 0.2$.

Solution.
Using Eq. (6.9a) and Eq. (6.9b) and the Runge–Kutta results in Table 6.3 for the starting values y_1, y_2, and y_3, the results are shown in Table 6.4 using Excel analysis. It is seen that the corrections improve the accuracy considerably.

6.1.2 Second-order equations

Numerical methods can be used to solve the initial value problem

$$y'' = f(x, y, y'), \qquad y(x_o) = y_o, \, y'(x_o) = y_o', \tag{6.10}$$

such as the Runge–Kutta–Nyström method. This is a fourth-order method that is an extension of the Runge–Kutta method in Section 6.1.1. Again, one computes four auxiliary quantities:

$$k_1 = \frac{1}{2} h f(x_n, y_n, y_n'),$$

$$k_2 = \frac{1}{2} h f(x_n + \frac{1}{2}h, y_n + K, y_n' + k_1) \text{ where } K = \frac{1}{2}h(y_n' + \frac{1}{2}k_1),$$

$$k_3 = \frac{1}{2} h f(x_n + \frac{1}{2}h, y_n + K, y_n' + k_2),$$

Table 6.4 Example calculation for the Adams–Moulton method.

n	x_n	Starting y_n	f_n	y_n^*	f_n^*	Corrected y_n	Exact	Error $\times 10^6$
0	0.0	0.000000	1.000000				0.000000	0
1	0.2	0.242800	1.442800				0.242806	6
2	0.4	0.583636	1.983636				0.583649	13
3	0.6	1.044213	2.644213				1.044238	25
4	0.8		3.451056	1.650720	3.450720	1.651056	1.651082	26
5	1.0		4.436537	2.436129	4.436129	2.436537	2.436564	26
6	1.2		5.640208	3.439708	5.639708	3.440208	3.440234	26
7	1.4		7.110377	4.709766	7.109766	4.710377	4.710400	23
8	1.6		8.906046	6.305300	8.905300	6.306046	6.306065	18
9	1.8		11.099284	8.298373	11.098373	8.299284	8.299295	10
10	2.0			10.777001	13.777001	10.778114	10.778112	−2

$$\boxed{k_4 = \frac{1}{2}hf(x_n + h, y_n + L, y_n' + 2k_3), \text{ where } L = h(y_n' + k_3),}$$ (6.11a)

and then from these quantities the new approximate value y_{n+1} of the solution y,

$$\boxed{y_{n+1} = y_n + h\left(y_n' + \frac{1}{3}(k_1 + k_2 + k_3)\right),}$$ (6.11b)

as well as an approximation of the derivative y' needed in the next step,

$$\boxed{y_{n+1}' = y_n' + \frac{1}{3}(k_1 + 2k_2 + 2k_3 + k_4).}$$ (6.11c)

Example 6.1.5. Solve the initial value problem $y'' = \frac{1}{2}(x + y + y' + 1)$, $y(0) = 0$, $y'(0) = 0$ for $0 \le x \le 1$ with $h = 0.2$, using the Runge–Kutta–Nyström method. Using the method of undetermined coefficients from Chapter 2, the particular solution of this differential equation is $y^p = -x$. The solution of the homogeneous equation is $y^h = Ae^x + Be^{-x/2}$. Hence, applying the boundary conditions, the exact solution is $y(x) = \frac{2}{3}\left[e^x - e^{-x/2}\right] - x$. The numerical solution can therefore be compared to this exact analytical solution.

Solution.
From Eq. (6.11a) we have

$$k_1 = 0.05(x_n + y_n + y_n' + 1),$$

$$k_2 = 0.05(x_n + 0.1 + y_n + K + y_n' + k_1 + 1), \quad K = 0.1(y_n' + \frac{1}{2}k_1),$$

$$k_3 = 0.05(x_n + 0.1 + y_n + K + y_n' + k_2 + 1),$$

$$k_4 = 0.05(x_n + 0.2 + y_n + L + y_n' + 2k_3 + 1), \quad L = 0.2(y_n' + k_3).$$

The computations based on Excel analysis are shown in Table 6.5, and compared to the exact solution. In this calculation, both truncation and round-off errors are present. The round-off error may accumulate so that it can affect the required solution (after many steps) resulting in a significant build-up error.

6.2 Partial differential equations

A partial differential equation is quasilinear if it is linear in the highest derivative. For example, a second-order quasilinear equation is

$$au_{xx} + 2bu_{xy} + cu_{yy} = F(x, y, u, u_x, u_y),$$ (6.12)

where u is an unknown function of the independent variables x and y. This equation can be further classified as

Table 6.5 Example calculation for the Runge–Kutta–Nyström method.

n	x_n	y_n	y'_n	k_1	k_2	k_3	k_4	I	II	Exact	Error
0	0.0	0.0000000	0.0000000	0.0500000	0.0576250	0.0580063	0.0663807	0.0110421	0.1158811	0.0000000	0
1	0.2	0.0110421	0.1158811	0.0663462	0.0754087	0.0758619	0.0858498	0.0376840	0.1515790	0.0110436	2
2	0.4	0.0487261	0.2674601	0.0858093	0.0966516	0.0971937	0.1091752	0.0721357	0.1942251	0.0487293	3
3	0.6	0.1208617	0.4616852	0.1091273	0.1221650	0.1228168	0.1372540	0.1159443	0.2454483	0.1208671	5
4	0.8	0.2368060	0.7071335	0.1371970	0.1529355	0.1537224	0.1711778	0.1710170	0.3072302	0.2368139	8
5	1.0	0.4078231								0.4078341	11

Note:
Column $I = 0.2\left[y'_n + (k_1 + k_2 + k_3)/3\right]$.
Column $II = (k_1 + k_2 + k_3 + k_4)/3$.
Error $= (Exact - y_n) \times 10^6$.

(i) *elliptic* if $ac - b^2 > 0$ (Laplace equation),
(ii) *parabolic* if $ac - b^2 = 0$ (heat equation),
(iii) *hyperbolic* if $ac - b^2 < 0$ (wave equation).

Here the coefficients a, b, and c may be functions of x and y. A numerical solution is obtained by replacing partial derivatives for the above types with difference quotients as shown in the following sections.

6.2.1 Poisson and Laplace difference equations

The Poisson equation is given by

$$\boxed{\nabla^2 u = u_{xx} + u_{yy} = f(x, y).}$$ (6.13)

If $f(x, y) = 0$, Eq. (6.13) reduces to the Laplace equation

$$\boxed{\nabla^2 u = u_{xx} + u_{yy} = 0.}$$ (6.14)

For a numerical solution, consider the following Taylor series formula:

$$u(x + h, y) = u(x, y) + h u_x(x, y) + \frac{1}{2} h^2 u_{xx}(x, y) + ...,$$ (6.15a)

$$u(x - h, y) = u(x, y) - h u_x(x, y) + \frac{1}{2} h^2 u_{xx}(x, y) + ...$$ (6.15b)

Adding Eq. (6.15a) and Eq. (6.15b) and neglecting higher-order terms,

$$u(x + h, y) + u(x - h, y) \simeq 2u(x, y) + h^2 u_{xx}(x, y),$$

or

$$u_{xx}(x, y) \simeq \frac{1}{h^2} [u(x + h, y) - 2u(x, y) + u(x - h, y)].$$ (6.16a)

Similarly,

$$u_{yy}(x, y) \simeq \frac{1}{k^2} [u(x, y + k) - 2u(x, y) + u(x, y - k)].$$ (6.16b)

Substituting Eq. (6.16a) and Eq. (6.16b) into Eq. (6.13), one obtains the Poisson difference equation for $k = h$,

$$\boxed{u(x + h, y) + u(x, y + h) + u(x - h, y) + u(x, y - h) - 4u(x, y) = h^2 f(x, y).}$$ (6.17)

Similarly, the Laplace difference equation is

$$u(x+h, y) + u(x, y+h) + u(x-h, y) + u(x, y-h) - 4u(x, y) = 0.$$ (6.18)

Here h is the mesh size. As a convenient notation for the mesh points (and the corresponding values of the solution), let

$$P_{ij} = (ih, jh), \quad u_{ij} = u(ih, jh).$$ (6.19)

Hence, for any mesh point P_{ij}, Eq. (6.18) can be written as

$$u_{i+1,j} + u_{i,j+1} + u_{i-1,j} + u_{i,j-1} - 4u_{i,j} = 0.$$ (6.20)

Example 6.2.1. The four sides of a square plate of side 12 cm made of a homogeneous material are kept at constant temperature 0°C and 100°C, as shown in Fig. 6.1(a). Using a wide grid of mesh 4 cm (see Fig. 6.1(b)), find the steady-state temperature at the mesh points.

Solution.
For steady-state conditions, the heat conduction equation reduces to the Laplace equation and Dirichlet problem (see Example 5.2.6).

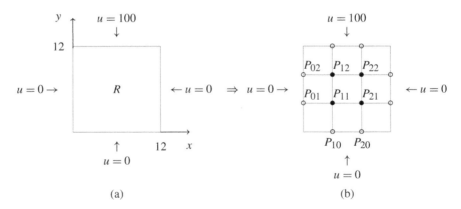

FIGURE 6.1

Heat conduction in (a) a square plate with (b) a meshed grid.

Using Eq. (6.20) for the mesh points P_{11}, P_{21}, P_{12}, and P_{22} in Fig. 6.1(b), one obtains the following system of equations. For example, for $P_{11}(i = 1, j = 1)$, one has $u_{21} + u_{12} + u_{01} + u_{10} - 4u_{11} = 0$. However, $u_{01} = 0$ and $u_{10} = 0$ (from the boundary conditions) so that

$$-4u_{11} + u_{21} + u_{12} = 0.$$

Thus, the system of equations for the four internal mesh points is

$$
\begin{array}{llllll}
-4u_{11} & +u_{21} & +u_{12} & & = 0 & (P_{11}) & (i=1, j=1), \\
u_{11} & -4u_{21} & & +u_{22} & = 0 & (P_{21}) & (i=2, j=1), \\
u_{11} & & -4u_{12} & +u_{22} & = -100 & (P_{12}) & (i=1, j=2), \\
& u_{21} & +u_{12} & -4u_{22} & = -100 & (P_{22}) & (i=2, j=2).
\end{array}
\tag{6.21}
$$

This small system of equations can be solved with matrix methods with Maple in Appendix A, yielding $u_{11} = u_{21} = 12.5°C$ and $u_{12} = u_{22} = 37.5°C$. More exact values to one decimal place can be obtained using Fourier series (see Example 5.2.6). Taking $f(x) = 100$ in Eq. (5.29) with $b = a = 12$, on integrating one obtains $A_n^* = \dfrac{400}{n\pi \sinh(n\pi)}$ with $n = 1, 3, 5\ldots$. Thus, the analytical solution in Eq. (5.28) is $u(x, y) = \dfrac{400}{\pi} \displaystyle\sum_{n=1,3,5\ldots} \dfrac{\sin \dfrac{n\pi x}{12} \sinh \dfrac{n\pi y}{12}}{n \sinh(n\pi)}$. Summing the series to 10 terms with Maple yields the analytical results $u_{11} = u_{21} = 11.9°C$ and $u_{12} = u_{22} = 38.1°C$. These latter values are identical to the numerical results obtained in Example 7.1.1 with COMSOL.   [answer]

The Gauss–Seidel (or Liebmann) method and the alternating direction implicit method can also be used to solve the system of equations in Eq. (6.21).

(i) *Gauss–Seidel method*

The system in Eq. (6.21) can also be written in the form

$$
\begin{array}{llll}
u_{11} = & & 0.25u_{21} & +0.25u_{12} & +0, \\
u_{21} = & 0.25u_{11} & & +0.25u_{22} & +0, \\
u_{12} = & 0.25u_{11} & & +0.25u_{22} & +25, \\
u_{22} = & & 0.25u_{21} & +0.25u_{12} & +25.
\end{array}
\tag{6.22}
$$

Using an iterative method, one starts from a poor approximation to the solution, for example, $u_{11}^{(0)} = 30$, $u_{21}^{(0)} = 30$, $u_{12}^{(0)} = 30$, and $u_{22}^{(0)} = 30$, and calculates from Eq. (6.22) a better approximation.

Use "old" values

↓

$$
\begin{array}{llll}
u_{11}^{(1)} = & 0.25u_{21}^{(0)} & +0.25u_{12}^{(0)} & +0 = 15.00, \\
u_{21}^{(1)} = & 0.25u_{11}^{(1)} & +0.25u_{22}^{(0)} & +0 = 11.25, \\
u_{12}^{(1)} = & 0.25u_{11}^{(1)} & +0.25u_{22}^{(0)} & +25 = 36.25, \\
u_{22}^{(1)} = & 0.25u_{21}^{(1)} & +0.25u_{12}^{(1)} & +25 = 36.88.
\end{array}
\tag{6.23}
$$

↑

Use "new" values

The next step yields

$$
\begin{aligned}
u_{11}^{(2)} &= & 0.25u_{21}^{(1)} &+0.25u_{12}^{(1)} & &+0 = 11.88, \\
u_{21}^{(2)} &= 0.25u_{11}^{(2)} & & &+0.25u_{22}^{(1)} &+0 = 12.19, \\
u_{12}^{(2)} &= 0.25u_{11}^{(2)} & & &+0.25u_{22}^{(1)} &+25 = 37.19, \\
u_{22}^{(2)} &= & 0.25u_{21}^{(2)} &+0.25u_{12}^{(2)} & &+25 = 37.49.
\end{aligned}
\tag{6.24}
$$

Further iteration yields the exact solution $u_{11} = u_{21} = \underline{12.5°C}$ and $u_{12} = u_{22} = \underline{37.5°C}$.

(ii) *Alternating direction implicit method*

The alternating direction implicit method uses the fact that the system of equations has a coefficient matrix that is <u>tridiagonal</u>, that is, the matrix has nonzero entries on the main diagonal and immediately adjacent to it. A tridiagonal matrix is obtained if there are only three points in a row where Eq. (6.20) can be written as

$$
u_{i-1,j} - 4u_{i,j} + u_{i+1,j} = -u_{i,j-1} - u_{i,j+1},
\tag{6.25a}
$$
$$
u_{i,j-1} - 4u_{i,j} + u_{i,j+1} = -u_{i-1,j} - u_{i+1,j}.
\tag{6.25b}
$$

In Eq. (6.25a), the left side belongs to the y-row and the right side to the x-column i. The Eq. (6.20) can alternatively be written as Eq. (6.25b). Thus, using an iterative method, on can substitute the mth iteration values on the right-hand side of Eq. (6.25a) and solve for the $(m + 1)$th values on the left-hand side for a <u>fixed row j</u>:

$$
\boxed{u_{i-1,j}^{(m+1)} - 4u_{i,j}^{(m+1)} + u_{i+1,j}^{(m+1)} = -u_{i,j-1}^{(m)} - u_{i,j+1}^{(m)}.}
\tag{6.26}
$$

This gives a system of N algebraic equations in N unknowns (where N is equal to the number of internal mesh points per row), which can be solved by Gauss elimination. One then proceeds to the next row, and so on. In the next step, the direction is alternated and a next approximation is computed, $u_{i,j}^{(m+2)}$, column by column, from the $u_{i,j}^{(m+1)}$ values and the given values for the boundary condition (BC). That is, from Eq. (6.25b),

$$
\boxed{u_{i,j-1}^{(m+2)} - 4u_{i,j}^{(m+2)} + u_{i,j+1}^{(m+2)} = -u_{i-1,j}^{(m+1)} - u_{i+1,j}^{(m+1)}}
\tag{6.27}
$$

for a <u>fixed column i</u>.

For the present example, one obtains first approximations $u_{11}^{(1)}$, $u_{21}^{(1)}$, $u_{12}^{(1)}$, and $u_{22}^{(1)}$ from starting values $u_{11}^{(0)} = 30, u_{21}^{(0)} = 30, u_{12}^{(0)} = 30$, and $u_{22}^{(0)} = 30$ and the appropriate boundary values in Fig. 6.1b (which can be written without an upper index). Thus, from Eq. (6.26) with $m = 0$, for a fixed $j = 1$ (first row), the system is

$$\underbrace{u_{01}}_{0(\text{BC})} \quad -4u_{11}^{(1)} \quad +u_{21}^{(1)} \qquad = -\underbrace{u_{10}}_{0(\text{BC})} \quad -\underbrace{u_{12}^{(0)}}_{30} \qquad (i = 1),$$

$$u_{11}^{(1)} \quad -4u_{21}^{(1)} \quad +\underbrace{u_{31}}_{0(\text{BC})} = -\underbrace{u_{20}}_{0(\text{BC})} \quad -\underbrace{u_{22}^{(0)}}_{30} \qquad (i = 2).$$

The solution is $u_{11}^{(1)} = u_{21}^{(1)} = 10$. For $j = 2$ (second row) one obtains from Eq. (6.26) the following system:

$$u_{02} \quad -4u_{12}^{(1)} \quad +u_{22}^{(1)} \qquad = -u_{11}^{(0)} \quad -u_{13} \qquad (i = 1),$$

$$u_{12}^{(1)} \quad -4u_{22}^{(1)} \quad +u_{32} \quad = -u_{21}^{(0)} \quad -u_{23} \qquad (i = 2).$$

The solution is $u_{12}^{(1)} = u_{22}^{(1)} = 43.33$.

Second approximations $u_{11}^{(2)}$, $u_{21}^{(2)}$, $u_{12}^{(2)}$, and $u_{22}^{(2)}$ are now obtained from Eq. (6.27) with $m = 0$ by using the first approximations just computed and the boundary values. For $i = 1$ (first column) one obtains from Eq. (6.27) the following system:

$$u_{10} \quad -4u_{11}^{(2)} \quad +u_{12}^{(2)} \qquad = -u_{01} \quad -u_{21}^{(1)} \qquad (j = 1),$$

$$u_{11}^{(2)} \quad -4u_{12}^{(2)} \quad +u_{13} \quad = -u_{02} \quad -u_{22}^{(1)} \qquad (j = 2).$$

The solution is $u_{11}^{(2)} = 12.22$, $u_{12}^{(2)} = 38.89$.

For $i = 2$ (second column) one obtains from Eq. (6.27) the following system:

$$u_{20} \quad -4u_{21}^{(2)} \quad +u_{22}^{(2)} \qquad = -u_{11} \quad -u_{31} \qquad (j = 1),$$

$$u_{21}^{(2)} \quad -4u_{22}^{(2)} \quad +u_{23} \quad = -u_{12}^{(1)} \quad -u_{32} \qquad (j = 2).$$

The solution is $u_{21}^{(2)} = 12.22$, $u_{22}^{(2)} = 38.89$.

Thus, in the second approximation, $u_{11}^{(2)} = u_{21}^{(2)} = 12.22°\text{C}$ and $u_{12}^{(2)} = u_{22}^{(2)} = 38.89°\text{C}$. These values are about the same as those of two Gauss–Seidel steps in (i).

Neumann and mixed problems

In Neumann and mixed problems, there are boundary points at which the normal derivative $u_n = \dfrac{\partial u}{\partial n}$ is given but not u.

Example 6.2.2. Solve the mixed boundary value problem (see Fig. 6.2) for the Poisson equation:

$$\nabla^2 u = u_{xx} + u_{yy} = 16xy.$$

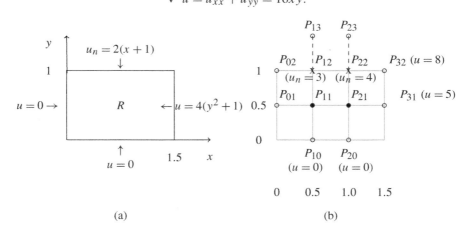

(a)   (b)

FIGURE 6.2

Heat conduction in (a) a rectangular plate with mixed boundaries with (b) a meshed grid.

Solution.

For the grid in Fig. 6.2(b), $h = 0.5$. From the boundary conditions $u = 4(y^2 + 1)$ and $u_n = 2(x + 1)$ so that

$$u_{31} = 4[(0.5)^2 + 1] = 5, u_{32} = 4[(1)^2 + 1] = 8,$$

$$\frac{\partial u_{12}}{\partial n} = \frac{\partial u_{12}}{\partial y} = 2[0.5 + 1] = 3, \frac{\partial u_{22}}{\partial y} = 2[1 + 1] = 4. \tag{6.28}$$

The interior points P_{11} and P_{21} are obtained from Eq. (6.17) (with $h^2 = 0.25$ and $f(x, y) = 12xy$) and the given boundary values. Thus,

$$-4u_{11} + u_{21} + u_{12} \quad = \underbrace{-0}_{-u_{01}} \quad \underbrace{-0}_{-u_{10}} \quad +\underbrace{0.25}_{h^2} \cdot \underbrace{(16)(0.5)(0.5)}_{(16)\,(x)\,(y)} = 1,$$

$$u_{11} - 4u_{21} + u_{22} \quad = \quad -5 + 2 = -3.$$

$$\tag{6.29}$$

For the points P_{12} and P_{22}, one extends the region R (to $y = 1.5$), and Eq. (6.17) yields

$$u_{11} \qquad -4u_{12} + u_{22} + u_{13} \qquad\qquad = 2,$$

$$u_{21} + u_{12} - 4u_{22} \qquad + u_{23} = -8 + 4 = -4. \tag{6.30}$$

From Eq. (6.28),

$$3 = \frac{\partial u_{12}}{\partial y} \approx \frac{u_{13} - u_{11}}{2h} = u_{13} - u_{11} \Rightarrow u_{13} = u_{11} + 3,$$

$$4 = \frac{\partial u_{22}}{\partial y} \approx \frac{u_{23} - u_{21}}{2h} = u_{23} - u_{21} \Rightarrow u_{23} = u_{21} + 4.$$

Substituting these results into Eq. (6.30) gives

$$\begin{array}{llll} 2u_{11} & -4u_{12} & +u_{22} & = 2 - 3 = -1, \\ 2u_{21} & +u_{12} & -4u_{22} & = -4 - 4 = -8. \end{array} \tag{6.31}$$

Thus, Eq. (6.29) and Eq. (6.31) in matrix form is

$$\begin{bmatrix} -4 & 1 & 1 & 0 \\ 1 & -4 & 0 & 1 \\ 2 & 0 & -4 & 1 \\ 0 & 2 & 1 & -4 \end{bmatrix} \begin{bmatrix} u_{11} \\ u_{21} \\ u_{12} \\ u_{22} \end{bmatrix} = \begin{bmatrix} 1 \\ -3 \\ -1 \\ -8 \end{bmatrix}. \tag{6.32}$$

The solution of Eq. (6.32) using Maple to solve this matrix system of equations in Appendix A is $u_{11} = 0.484$, $u_{21} = 1.658$, $u_{12} = 1.280$, $u_{22} = 3.149$. The more exact values as evaluated with a complete numerical analysis with COMSOL in Example 7.1.2 are $u_{11} = 0.507$, $u_{21} = 1.717$, $u_{12} = 1.466$, $u_{22} = 3.269$.

6.2.2 Parabolic equations

By an appropriate transformation, the heat equation and corresponding boundary conditions can be written as

$$\boxed{\begin{array}{ll} u_t = u_{xx}, & 0 \le x \le 1, t \ge 0, \\ u(x, 0) = f(x) & \text{(initial condition)}, \\ u(0, t) = u(1, t) = 0 & \text{(boundary conditions)}. \end{array}} \tag{6.33}$$

Using Eq. (6.16a), a simple finite difference approximation of Eq. (6.33) is

$$\frac{1}{k}\left(u_{i,j+1} - u_{i,j}\right) = \frac{1}{h^2}\left(u_{i+1,j} - 2u_{i,j} + u_{i-1,j}\right) \tag{6.34}$$

for a mesh size h in the x direction and k in the t direction. Solving for $u_{i,j+1}$ (which corresponds to time row $j + 1$ in Fig. 6.3),

$$\boxed{u_{i,j+1} = (1 - 2r)u_{i,j} + r(u_{i+1,j} + u_{i-1,j}),} \qquad r = \frac{k}{h^2}. \tag{6.35}$$

It can be shown for <u>convergence</u> that

$$r = \frac{k}{h^2} \le \frac{1}{2}, \tag{6.36}$$

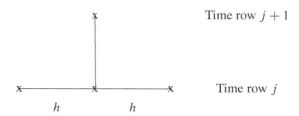

FIGURE 6.3

The four points in Eq. (6.34) and Eq. (6.35).

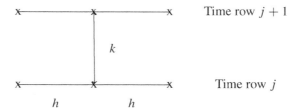

FIGURE 6.4

The six points in the Crank–Nicolson formulas in Eq. (6.38) and Eq. (6.39).

The restriction in Eq. (6.36) can be avoided with the <u>Crank–Nicolson method</u>, which replaces the difference quotient on the right-hand side of Eq. (6.34) by $\frac{1}{2}$ times the sum of two such difference quotients at two time rows (see Fig. 6.4), that is,

$$
\begin{aligned}
\frac{1}{k}\left(u_{i,j+1}-u_{i,j}\right) = &\frac{1}{2h^2}\left(u_{i+1,j}-2u_{i,j}+u_{i-1,j}\right) \\
&+\frac{1}{2h^2}\left(u_{i+1,j+1}-2u_{i,j+1}+u_{i-1,j+1}\right),
\end{aligned}
\tag{6.37}
$$

or equivalently (with $r = k/h^2$)

$$(2+2r)u_{i,j+1}-r(u_{i+1,j+1}+u_{i-1,j+1}) = (2-2r)u_{i,j}+r(u_{i+1,j}+u_{i-1,j}). \tag{6.38}$$

If $r = 1$, Eq. (6.38) becomes simply

$$4u_{i,j+1}-u_{i+1,j+1}-u_{i-1,j+1}=u_{i+1,j}+u_{i-1,j}. \tag{6.39}$$

Example 6.2.3 (Heat conduction in an insulated bar). Consider a laterally insulated metal bar of unit length and $\kappa = 1$. The ends of the bar are kept at a temperature of $u = 0°C$. At $t = 0$, the temperature distribution is given by the triangular distribution in Fig. 2.8 with $L = 1$ such that

$$
f(x) = \begin{cases} 2x & \text{if } 0 < x < 0.5, \\ 2(1-x) & \text{if } 0.5 \le x < 1. \end{cases}
$$

For $h = 0.2$, find the temperature in the bar for $0 \leq t \leq 0.08$.

Solution.

(i) *Crank–Nicolson method*

Taking $r = 1$, we have $k = h^2 = 0.04$ (see grid in Fig. 6.5). The initial values are $u_{10} = 0.4$ and $u_{20} = 0.8$. Also by symmetry, $u_{30} = u_{20}$ and $u_{40} = u_{10}$. In addition, $u_{31} = u_{21}$ and $u_{41} = u_{11}$. Using Eq. (6.39), for $j = 0$,

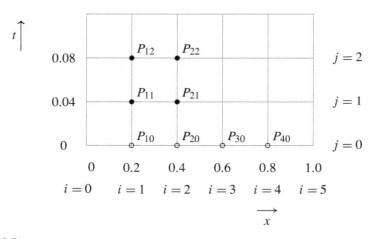

FIGURE 6.5

Mesh for the numerical solution of an insulated metal bar.

$(i = 1)$:
$$4u_{11} \quad -u_{21} \quad - \underbrace{u_{01}}_{0(\text{BC})} \ = \underbrace{u_{20}}_{0.8} \ + \underbrace{u_{00}}_{0(\text{BC})} \quad = 0.8,$$

$(i = 2)$:
$$-u_{11} \ +4u_{21} \quad - \underbrace{u_{31}}_{u_{21}(\text{sym.})} \ = u_{10} \ + \underbrace{u_{30}}_{u_{20}(\text{sym.})} \ = 0.4 + 0.8 = 1.2.$$

Therefore,

$$\begin{aligned}
4u_{11} \ -u_{21} \quad\quad &= 0.8, \\
-u_{11} \ +4u_{21} \ -u_{21} &= 1.2.
\end{aligned} \tag{6.40}$$

The solution of Eq. (6.40) is

$$\boxed{u_{11} = 0.327 \text{ and } u_{21} = 0.509.}$$

Similarly, for $j = 1$,

$$(i=1) \quad 4u_{12} - u_{22} \quad \underbrace{- u_{02}}_{0(BC)} \quad = u_{01} + u_{21} \quad = 0 + 0.509 = 0.509,$$

$$(i=2) \quad -u_{12} + 4u_{22} \quad \underbrace{- u_{32}}_{u_{22}(\text{symmetry})} = u_{11} + \underbrace{u_{31}}_{u_{21}} = 0.327 + 0.509 = 0.836.$$

The system then becomes

$$\begin{aligned} 4u_{12} \;-u_{22} \;\;\;\;\;\;&= 0.509091, \\ -u_{12} + 4u_{22} - u_{22} &= 0.836364, \end{aligned} \qquad (6.41)$$

with the following solution for Eq. (6.41):

$$\boxed{u_{12} = 0.215 \ \text{ and } \ u_{22} = 0.350.}$$

The temperature distribution in the bar is given in Table 6.6. The dashed vertical line in Table 6.6 indicates the line of symmetry.

Table 6.6 Temperature in a thin bar using a Crank–Nicolson method.

t	$x = 0$	$x = 0.2$	$x = 0.4$	$x = 0.6$	$x = 0.8$	$x = 1$	j
0.00	0	0.4	0.8	0.8	0.4	0	0
0.04	0	0.327	0.509	0.509	0.327	0	1
0.08	0	0.215	0.350	0.350	0.215	0	2
		$i = 1$	$i = 2$				

(ii) *Direct method*

Letting $r = 0.25$ $(< \frac{1}{2})$ yields $k = rh^2 = 0.25(.2)^2 = 0.01$. As such, this value requires four times as many time steps as the Crank–Nicolson method. The formula in Eq. (6.35) with $r = 0.25$ gives

$$u_{i,j+1} = 0.25(u_{i-1,j} + 2u_{ij} + u_{i+1,j}). \qquad (6.42)$$

For $j = 0$, $u_{00} = 0$ (boundary condition), $u_{10} = 0.4$, $u_{20} = u_{30} = 0.8$ (by symmetry) so that

$$\begin{aligned} (i=1) \quad u_{11} &= 0.25(u_{00} + 2u_{10} + u_{20}) = 0.4, \\ (i=2) \quad u_{21} &= 0.25(u_{10} + 2u_{20} + u_{30}) = 0.25(u_{10} + 3u_{20}) = 0.7. \end{aligned}$$

For $j = 1$ and noting that $u_{01} = u_{02} = \ldots = 0$ from the boundary conditions,

$$\begin{aligned} (i=1) \quad u_{12} &= 0.25(u_{01} + 2u_{11} + u_{21}) = 0.375, \\ (i=2) \quad u_{22} &= 0.25(u_{11} + 3u_{21}) = 0.625. \end{aligned}$$

and so on. As shown in Table 6.7, the accuracy is comparable to that of the Crank–Nicolson method. The exact solution is obtained from the general solution in Example 5.2.4 where $L = \kappa = 1$. Thus, substituting

$$f(x) = \begin{cases} 2x & \text{if } 0 < x < 0.5, \\ 2(1-x) & \text{if } 0.5 \le x < 1. \end{cases}$$

into Eq. (5.24) and integrating with Maple yields $B_n = \dfrac{8}{(n\pi)^2} \sin\left(\dfrac{n\pi}{2}\right)$. Hence, the analytical solution is given by Eq. (5.23)

$$u(x,t) = \frac{8}{\pi^2} \sum_{n=1}^{\infty} = \frac{\sin(n\pi/2)}{n^2} \sin(n\pi x) e^{-(n\pi)^2 t}.$$

Using Maple to sum the series to 10 terms yields the exact solution also shown to three decimal places. In addition, the COMSOL result using a finite element technique with an error tolerance of 1×10^{-6} in Example 7.1.3 is also shown in Table 6.7, which is in perfect agreement with the analytical solution.

Table 6.7 Temperature in a thin bar using a direct method.

t	$x = 0.2$				$x = 0.4$			
	CN	Direct	Exact	COMSOL	CN	Direct	Exact	COMSOL
0	0.4	0.4	0.4	0.4	0.8	0.8	0.8	0.8
0.04	0.327	0.313	0.319	0.319	0.509	0.508	0.521	0.521
0.08	0.215	0.210	0.216	0.216	0.350	0.340	0.350	0.350

CN – Crank–Nicolson method.

6.2.3 Hyperbolic equations

By an appropriate transformation, the wave equation and corresponding boundary conditions can be written as

$$\begin{aligned} u_{tt} &= u_{xx}, & 0 \le x \le 1, t \ge 0, \\ u(x,0) &= f(x) & \text{(initial displacement)}, \\ u_t(x,0) &= g(x) & \text{(initial velocity)}, \\ u(0,t) &= u(1,t) = 0 & \text{(boundary conditions)}. \end{aligned} \qquad (6.43)$$

Using Eq. (6.16a) and Eq. (6.16b), a finite difference approximation of Eq. (6.43) is

$$\frac{1}{k^2}\left(u_{i,j+1} - 2u_{ij} + u_{ij-1}\right) = \frac{1}{h^2}\left(u_{i+1,j} - 2u_{i,j} + u_{i-1,j}\right), \qquad (6.44)$$

where h is the mesh size in the x direction and k in the mesh size in the t direction. Similarly, solving for $u_{i,j+1}$ (which corresponds to time row $j+1$ in Fig. 6.6), where $r^* = \dfrac{k^2}{h^2} = 1$,

$$u_{i,j+1} = u_{i-1,j} + u_{i+1,j} + u_{i,j-1}. \qquad (6.45)$$

It can be shown for that for $0 < r^* \le 1$, the present method is stable.

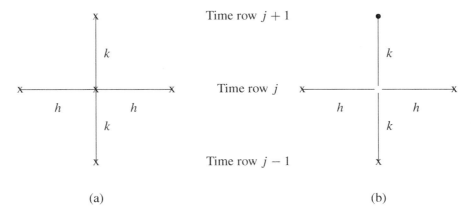

Time row $j+1$

Time row j

Time row $j-1$

(a)   (b)

FIGURE 6.6

Mesh points used in Eq. (6.44) and Eq. (6.45).

For the condition $u_t(x,0) = g(x)$ in Eq. (6.43), one can derive the difference formula

$$\frac{1}{2k}\left(u_{i,1} - u_{i,-1}\right) = g(i) \quad \text{or} \quad u_{i,-1} = u_{i,1} - 2kg_i, \qquad (6.46)$$

where $g_i = g(ih)$. For $t = 0$ (that is, $j = 0$), Eq. (6.45) and Eq. (6.46) yield

$$u_{i,1} = \frac{1}{2}\left(u_{i-1,0} + u_{i+1,0}\right) + kg_i \qquad (j = 0) \qquad (6.47)$$

(which expresses $u_{i,1}$ in terms of the initial data).

Example 6.2.4 (Vibrating string). Consider a string of unit length and $c = 1$. The grid is the same as in Fig. 6.5, where the string is fixed at both ends, and has the same initial distribution: $f(x) = \begin{cases} 2x & \text{if } 0 < x < 0.5, \\ 2(1-x) & \text{if } 0.5 \le x < 1, \end{cases}$ with an initial velocity $g(x) = 0$. For $h = k = 0.2$, solve the wave equation for the deflection of a string $u(x,t)$. This problem is analogous to Example 6.2.3.

Solution.
The grid is the same as in Fig. 6.5, except for the t values which are now at $0.2, 0.4, \ldots$ (instead of $0.04, 0.08\ldots$). The initial values are the same as Example 6.2.3. From Eq. (6.47) with $g(x) = 0$,

$$u_{i,1} = \frac{1}{2}\left(u_{i-1,0} + u_{i+1,0}\right)$$

so that

$$(i = 1) \quad u_{11} = \frac{1}{2}\left(u_{00} + u_{20}\right) = \frac{1}{2}[0 + 0.8] = \underline{0.4},$$

$$(i = 2) \quad u_{21} = \frac{1}{2}\left(u_{10} + u_{30}\right) = \frac{1}{2}[0.4 + 0.8] = \underline{0.6}.$$

As in the previous example, by symmetry, $u_{31} = u_{21}$ and $u_{41} = u_{11}$.

From Eq. (6.47) with $j = 1$ (where $u_{01} = u_{02} = \ldots = 0$ from the boundary conditions):

$$(i = 1) \quad u_{12} = u_{01} + u_{21} - u_{10} = 0 + 0.6 - 0.4 = \underline{0.2},$$
$$(i = 2) \quad u_{22} = u_{11} + u_{31} - u_{20} = 0.4 + 0.6 - 0.8 = \underline{0.2}.$$

In the computation for $(i = 2)$, the symmetry $u_{31} = u_{21}$ has been used. Also by symmetry, $u_{32} = u_{22}$ and $u_{42} = u_{12}$. Proceeding in this manner, the values in Table 6.8 are obtained. The line of symmetry is shown by the dashed line. An analytical solution is derived in Example 5.2.3, where for this problem $L = c^2 = 1$ and $g(x) = 0$ (so that $A_n = 0$ in Eq. (5.19)). Moreover, the same integral arises for B_n in Eq. (5.18), which is evaluated as $B_n = \dfrac{8}{(n\pi)^2} \sin\left(\dfrac{n\pi}{2}\right)$. Hence, from Eq. (5.17),

$$u(x,t) = \frac{8}{\pi^2} \sum_{n=1}^{\infty} = \frac{\sin(n\pi/2)}{n^2} \sin(n\pi x)\cos(n\pi t). \text{ Maple can be used to sum this}$$

series, where 100 terms are needed since the analytic series converges more slowly than the one in Example 6.2.3. The numerical values in Table 6.8 are in fact identical with the analytic solution as well as the COMSOL solution in Example 7.1.4.

Table 6.8 Distribution of a vibrating string.

t	$x = 0$	$x = 0.2$	$x = 0.4$	$x = 0.6$	$x = 0.8$	$x = 1$
0.00	0	0.4	0.8	0.8	0.4	0
0.2	0	0.4	0.6	0.6	0.4	0
0.4	0	0.2	0.2	0.2	0.2	0
0.6	0	−0.2	−0.2	−0.2	−0.2	0
0.8	0	−0.4	−0.6	−0.6	−0.4	0
1.0	0	−0.4	−0.8	−0.8	−0.4	0

Problems

6.1   Explain the advantages and disadvantages of the Adam–Moulton, Runge–Kutta, and Euler methods for the numerical solution of first-order ordinary differential equations.

6.2   Consider the first-order ordinary differential equation $y' = \dfrac{dy}{dx} = f(x, y)$. Derive the predictor and corrector equations for the Heun method by using a truncated Taylor series for y_{n+1}.

6.3   Given the initial value problem $y' = y - y^2$, $y(0) = 0.5$:

(a)   Solve the problem exactly using standard analytical methods for first-order ordinary differential equations.

(b)   Apply the Runge–Kutta method (of fourth order) and solve the problem for $0 \le x \le 1$, with a step size of $h = 0.2$. Compute the error, using the exact values in part (a).

6.4   Using the Maple software package:

(a)   Consider the square plate in Example 6.2.1 for a mesh spacing of $h = 4$ cm (that is, $h = a/n$, where $a = 12$ cm and $n = 3$). However, the bottom and side boundaries each have a temperature of $100°C$ and the top boundary is at $0°C$. Therefore show that $u_{11} = u_{21} = 87.5°C$ and $u_{12} = u_{22} = 62.5°C$.

(b)   Repeat the calculation for a mesh spacing of $h = 3$ cm (that is, $n = 4$) and find the steady-state temperature at the internal mesh points. Hint: The number of mesh points can be reduced if symmetry is considered.

(c)   The analytical solution is given by

$$u(x, y) = \frac{400}{\pi} \sum_{n=0,1...}^{\infty} \frac{1}{(2n + 1)\sinh(2n + 1)\pi}$$

$$\times \left\{ \frac{\sin(2n + 1)\pi x}{a} \sinh \frac{(2n + 1)\pi(a - y)}{a} \right.$$

$$\left. + \frac{\sin(2n + 1)\pi y}{a} \left[\sinh \frac{(2n + 1)\pi(a - x)}{a} + \sinh \frac{(2n + 1)\pi x}{a} \right] \right\}.$$

Show that this analytical solution yields the values $u_{11} = u_{21} = 88.1°C$ and $u_{12} = u_{22} = 61.9°C$ for the given mesh in part (a). Compare the numerical results of part (b) with the values obtained from the analytical solution.

6.5   Write the steady-state heat conduction equation (that is, the Laplace equation) for a Cartesian coordinate system for two spatial dimensions. During the winter (in which the air has an ambient temperature of $-10°C$), a thin car window with dimensions 1 m high by 1.5 m wide is completely covered with ice. A poor quality electric heater is only able to apply heat to the bottom edge of the window at $23°C$. The other edges of the window remain at the ambient temperature of the air. Determine if one can see through the middle third of the bottom half of the window if one waits for a long enough time by performing a numerical solution of the steady-state heat conduction equation with stated boundary conditions (that is, calculate the temperature at the mesh points u_{11} and u_{21} with a given mesh size of $h = 0.5$ m).

6.6 Consider the temperature distribution $u(x, t)$ in an insulated bar of length L as governed by the heat conduction equation $\frac{\partial u}{\partial t} = \kappa \frac{\partial^2 u}{\partial x^2}$, subject to the boundary conditions $\frac{\partial u}{\partial x}\Big|_{x=0} = 0$ $t > 0$, $u(L, t) = 0$, $t > 0$, and initial condition $u(x, 0) = u_o$, $0 < x < L$, $t = 0$.

(a) Consider the transformation of the above parabolic heat conduction equation. Using the nondimensional variables $v = \frac{u}{u_o}$, $\eta = \frac{x}{L}$, and $\tau = \frac{\kappa t}{L^2}$, give the transformed partial differential equation for $v(\eta, \tau)$ with the corresponding boundary and initial conditions.

(b) Consider the direct method for the numerical solution of the parabolic partial differential equation. Show that with $\kappa = 1$ and $L = 1$, with a reflexive boundary condition at $i = 0$, the following formula results: $u_{0, j+1} = (1 - 2r)u_{0, j} + 2ru_{1, j}$, where $r = k/h^2$.

6.7 Explain what methodology could be used to determine if a numerical solution for a given problem is correct and accurate.

6.8 Consider the heat conduction in a thin thermally insulated bar of length L which is initially at a temperature of $0°C$. At $t > 0$, heat is applied to the ends of the bar, which are kept at a constant temperature of $55°C$. The temperature distribution is therefore described by the partial differential equation: $\frac{\partial u}{\partial \tilde{t}} = \kappa \frac{\partial^2 u}{\partial \tilde{x}^2}$, with boundary conditions $u(0, \tilde{t}) = u(L, \tilde{t}) = 55$, $\tilde{t} > 0$, and initial conditions $u(\tilde{x}, 0) = 0$, for $0 < \tilde{x} < L$, $\tilde{t} = 0$.

(a) What transformation can be applied to the independent variables in order to reduce the problem to $\frac{\partial u}{\partial t} = \frac{\partial^2 u}{\partial x^2}$ or $u_t = u_{xx}$ with the corresponding boundary and initial conditions $u(0, t) = u(1, t) = 55$, $t > 0$, and $u(x, 0) = 0$, for $0 < x < 1$, $t = 0$?

(b) Using a Crank–Nicolson method, solve the two resulting finite difference equations for the first time step for the unknowns u_{11} and u_{21} with $r = k/h^2 = 1$ and $h = 0.2$. What is the advantage of the Crank–Nicolson method over the conventional direct method?

(c) By transforming the dependent variable and using the more general solution in Example 5.2.4, give the analytical solution for the problem in part (a).

Finite element methods

7

Finite elements provide an alternative numerical technique in contrast to finite difference techniques for the solution of ordinary and partial differential equations.

COMSOL MULTIPHYSICS (COMSOL Inc., 2018) is a commercial software package that can be used to numerically solve real-world ordinary and partial differential equations arising in various engineering applications. The Finite Element Method (FEM) or equivalently Finite Element Analysis (FEA) is widely utilized in well known engineering simulation packages such as ANSYS, ABAQUS, LS-DYNA, NASTRAN and a variety of free and open source tools. This tool is particularly useful if the problems are nonlinear where the coefficients of the differential equation may be a function of the independent variable (such as for the thermal conductivity depending on temperature in a heat conduction problem).

The application of this tool is detailed in Section 7.1. All numerical problems presented in Chapter 6 are solved again. In this discussion, the use of the tool is demonstrated with an annotation of the graphical user interface for solution of a given problem in a step-by-step manner. This description therefore provides the procedure to solve different types of problems for both ordinary and partial differential equations. This is an extremely important software tool that an engineer would use for a real-world problem rather than requiring the writing of their own computer program for the numerical solution. Thus, COMSOL can be used to solve very complex and coupled ordinary and partial differential equations as "multiphysics" type analysis. This terminology means that one can couple all types of physical phenomena together, including, for example, heat and mass transfer, chemical reactions, material structural effects of stress and strain, and thermophysical properties of materials that may depend on independent solution variables as a single problem. A more complex multiphysics problem in given as Problem 7.7.

The general theory of finite elements, which underscores the development of COMSOL, is also detailed in Section 7.2. This discussion involves the covering of a domain with finite elements of various dimensions as well as the development of linear and higher-order shape functions. This technique approximates the solution to the differential equation with appropriate weightings and numerical techniques to minimize the error (for example, the Galerkin method). A sample finite element problem is worked through by hand in Problems 7.8 to 7.18.

Advanced Mathematics for Engineering Students. https://doi.org/10.1016/B978-0-12-823681-9.00015-0

7.1 COMSOL application methodology

The COMSOL methodology specifically involves the following general steps to solve a given problem, with a selection of

- model wizard (or an import of a previous simulation);
- space dimension (dimensional representation and symmetry of the problem);
- type of physics to be solved (for example, heat conduction or wave equation);
- type of solver: stationary (steady-state) or time-dependent problem;
- model builder options and solver, including: (i) geometry/physical object(s) being simulated, (ii) coupled multiphysics, (iii) material properties, and (iv) boundary and initial conditions;
- finite element mesh size for the object;
- problem solution and output/process of results.

7.1.1 Solved problems with COMSOL

The step-by-step methodology for solution with COMSOL 5.4 of various problem examples are described below. Problems taken from Chapter 6 are identified in each of the following examples.

Partial differential equations

Example 7.1.1 (see Example 6.2.1). As shown in Fig. 6.1(a), four sides of a square plate (12 cm × 12 cm) are kept at constant temperature 0°C and 100°C. Find the steady-state temperature for this Laplace heat conduction problem with Dirichlet boundary conditions with COMSOL.

Solution. Select the following choices.

Step 1: start model application.
New: click "Model Wizard" (see Fig. 7.1).

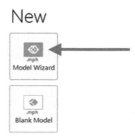

FIGURE 7.1

Start of the COMSOL 5.4 software tool.

Step 2: space dimension.
Space dimension: click "2D" (see Fig. 7.2) since the plate can be considered two-dimensional if the temperature doesn't change significantly through the thickness of the plate.

Select Space Dimension

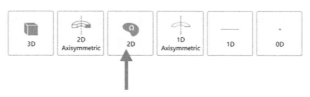

FIGURE 7.2

Selection of a space dimension.

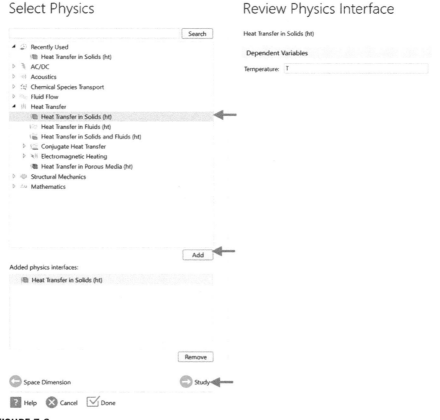

FIGURE 7.3

Selection of the physics for a heat transfer analysis.

Step 3: physics.
Physics: "Heat Transfer in Solids (ht)" → click "Add" → click "Study" (see Fig. 7.3).
The dependent variable is the temperature T and the independent variables are the
positions x and y.

Step 4: study.
Study: click "Stationary" → click "☑ Done" (see Fig. 7.4).

Select Study

FIGURE 7.4

Selection of the time dependence for the heat transfer model. A time-independent (stationary) model is chosen.

Step 5: model builder.
The following Model Builder page is generated; choose the "Length unit" as cm (see Fig. 7.5). During the creation of an object (especially a complex one), COMSOL automatically repairs any generated geometry by removing small edges and faces through the "Default repair tolerance" selection. The default value is relative to the overall size of the geometry but can be edited to a smaller tolerance value as required

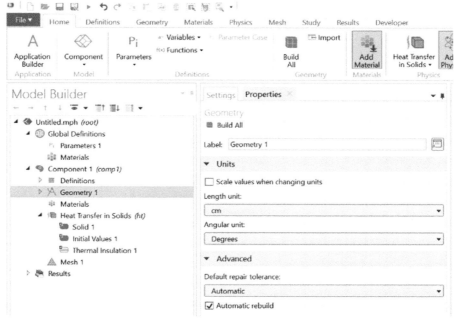

FIGURE 7.5

Model builder page and selection of the unit length of the square plate for the heat transfer analysis.

if one needs extra fine detail. For the simple plate in this example, an "Automatic" choice is selected.

Using this page perform the following actions. Right click "Geometry 1" (choose "Square □" in the drop down menu → in the resulting figure enter "12" cm for the "Side length" and select "Build All Objects") (see Fig. 7.6).

Right click "Materials" (choose "Blank Material" → thermal conductivity value for k of "1" (see Fig. 7.7). As shown for the Laplace equation in Eq. (5.25) for heat conduction in a plate, this problem is independent of the value of k but a value needs to be entered or an error will occur while the computation is performed in Step 7). This issue can be avoided by selecting "Mathematics" ⟶ "Classical PDEs" ⟶ "Laplace equation" for the Physics option in Step 3 (see, for example, Example 7.1.2), where a material entry is no longer required.

Right click: "Heat Transfer in Solids (ht)" and select "Temperature" (do this four times to add the four temperature subheadings under "Heat Transfer in Solids (ht)":

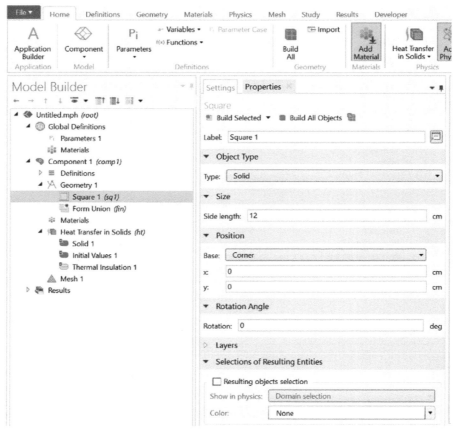

FIGURE 7.6

Selection of the size of the square plate for the heat transfer analysis.

"Temperature 1," "Temperature 2," "Temperature 3," "Temperature 4") (see Fig. 7.8). Click on "Temperature 1," click on one side of the square in the picture, and then enter one of the four boundary condition values in the box under "Temperature:" as a "User defined" value. Repeat this step for the other three temperature headings to assign the four Dirichlet boundary conditions on each side of the square. Note that calculations are performed in SI units with temperatures in K, but the boundary temperatures can be input instead in degrees Celsius as 0 [degC] and 100 [degC], accordingly.

Step 6: create a finite element mesh for the object.
Click on "Mesh 1," select the "Element size:" of the mesh (such as "Normal" or, if desired, the mesh can be finer for better accuracy) and click on "Build All." A

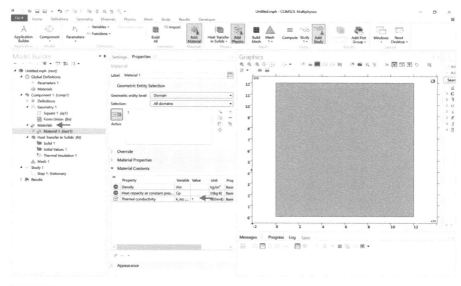

FIGURE 7.7

Selection of the material properties of the square plate for the heat transfer analysis.

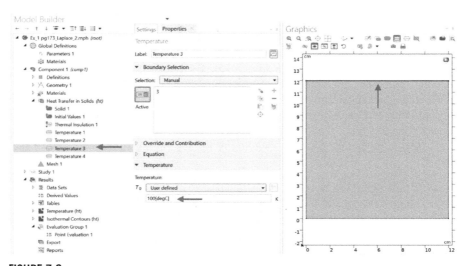

FIGURE 7.8

Selection of boundary conditions for the heat transfer analysis for the square plate.

"Physics-controlled mesh" is automatically generated of triangular shape as shown in the picture on the right-hand side of the page (see Fig. 7.9).

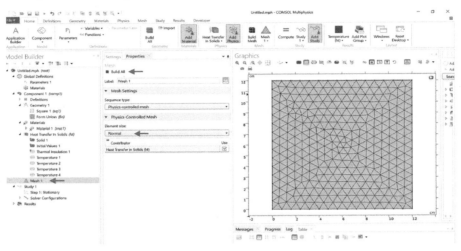

FIGURE 7.9

Meshing of the square plate.

Step 7: solve the problem and output results.

Finally, the problem can be solved by clicking on the "= Compute" at the top banner or by right clicking "Study 1" and choosing "Compute" at the top of the pop-up box.

The results are shown as a color scale plot on the right-hand side by clicking the "Temperature (ht)" entry (see Fig. 7.10). By clicking on a specific point in the color plot one can obtain the temperature T in K under the plot in a table for the given position of "X:" and "Y:" that is clicked. The results can be given in degrees Celsius by selecting "Surface" under the "Temperature (ht)" entry and changing the "Unit:" to "degC" in the drop down box and clicking the "Plot" button under the "Surface" heading at the top on the right-hand side of the page.

However, this method is more difficult when one wishes to obtain a value of the temperature at an exact position of interest on the plot. Thus, to obtain a temperature at a precise coordinate location:

(i) Right click "Data Sets" and choose "Cut Point 2D ⊡." A new item is created as "⊡ Cut Point 2D 1." Enter the coordinates of the desired point, for example: "X:" 4 cm and "Y:" 4 cm for evaluation of the temperature at this location (see Fig. 7.11).

(ii) Right click "Results" and click "Evaluation Group," resulting in a new entry in the menu along the left-hand side: "Evaluation Group 1." Right click on "Evaluation Group 1" and choose "Point Evaluation," which results in another new entry

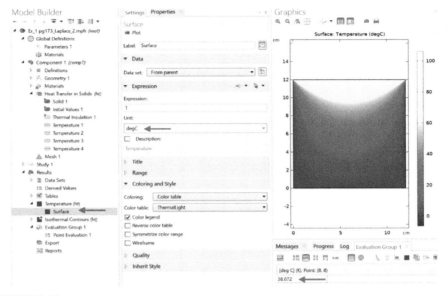

FIGURE 7.10

Solution output of the temperature distribution in the square plate.

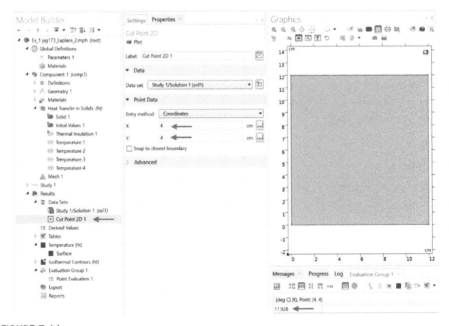

FIGURE 7.11

Selection of the coordinate location on the square plate for evaluation of the temperature.

in the menu along the left-hand side of "Point Evaluation 1." Click on "Evaluation Group 1" and select "Cut Point 2D 1" in the selection box for "Data set:." Click on the "Point Evaluation 1" item and in the box for "Data set:" add under the "Expression" heading: "T(K)-273.15," delete the "K" symbol under the "Unit" heading, and add "[degC]" under the "Description" heading (see Fig. 7.12). This relabeling provides values in degrees Celsius (vice K). By selecting "Evaluate" on the "Point Evaluation 1" page, the result at position $(4, 4)$ is shown at the bottom of the table under the figure as 11.928 [degC]. Likewise, the temperature at position $(4, 8)$ is 38.072 [degC]. This latter value is obtained again by clicking on the "Cut Point 2D 1" item in step (i) and choosing coordinates $(4, 8)$ for "X:" and "Y:." These two results are in agreement with the analytical Fourier series solution.

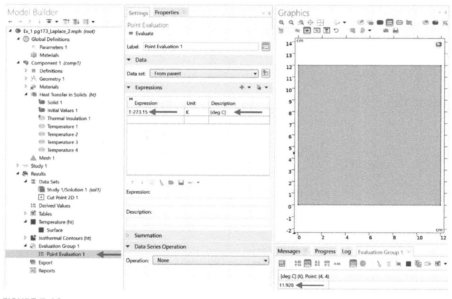

FIGURE 7.12

Evaluation of the temperature at a particular coordinate location on the square plate (in °C).

Example 7.1.2 (see Example 6.2.2). Similarly, the Poisson equation

$$\nabla^2 u = 16xy \tag{7.1}$$

can be solved with COMSOL for the boundary value problem involving both Dirichlet and Neumann conditions, as shown in Fig. 6.2(a).

Solution. The same methodology as used in Example 7.1.1 is applied, except that the physics, geometry, and boundary conditions are different.

In **Step 3**, "Poisson's equation" is chosen under "Classical PDEs" in place of the physics model "Heat Transfer in Solids (ht)" (see Fig. 7.13).

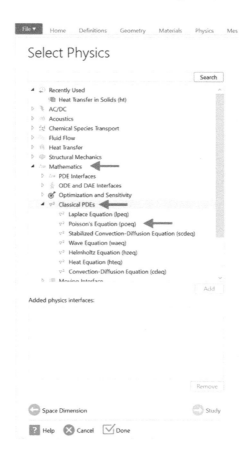

FIGURE 7.13

Selection of the physics for solution of a Poisson equation.

In **Step 5** a "Rectangle" is chosen for "Geometry 1" (with a "Width:" 1.5 and "Height:" 1.0) (see Fig. 7.14).

Right click "Poisson's equation (poeq)" and select "Dirichlet Boundary Condition" (three times) and "Flux Source" for the four mixed boundary conditions (see Fig. 7.15). Click on "Dirichlet Boundary Condition 1" and "Dirichlet Boundary Condition 2" and enter a "0" in the box for each of these "r" entries under the "Dirichlet Boundary Condition" heading after clicking on each of the two sides of the figure. However, for the more complicated "Dirichlet Boundary Condition 3," click on the

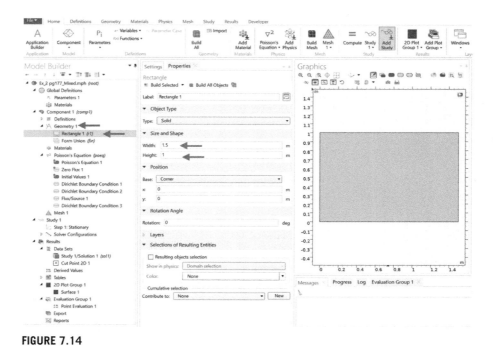

FIGURE 7.14

Selection of the size of the rectangle for the geometry input for solution of the Poisson equation.

far side of the square in the picture and then enter "4*(y^2+1)" for "r" (see bottom pane of Fig. 7.15). Finally, for the last boundary with the Neumann condition, click on "Flux/Source 1" and enter "2*(x+1)" in the box for "g" after clicking on the top boundary in the figure (see top pane of Fig. 7.15).

Click on the "Poisson Equation 1" and enter "-16*x*y" for the parameter "f" to complete the form of Eq. (7.1). Use an "Element size:" of "Extra fine" for "Mesh 1" to obtain higher accuracy.

The COMSOL solution gives $u(0.5, 0.5) = 0.507$, $u(1, 0.5) = 1.717$, $u(0.5, 1) = 1.466$, and $u(1, 1) = 3.269$ (see Fig. 7.16). The value of the temperature at $x = 0.5$ and $y = 0.5$ is shown in the table under the graphic in the top pane of Fig. 7.16, and the value at $x = 1.0$ and $y = 1.0$ is further shown in the bottom table under the graphic in the bottom pane of Fig. 7.16. For this problem with more complicated boundary conditions, the numerical solution of Example 6.2.2 with a course grid spacing is in reasonable agreement with the accurate solution provided by COMSOL.

Example 7.1.3 (see Example 6.2.3). Similarly, the parabolic equation

$$\frac{\partial u}{\partial t} = \frac{\partial^2 u}{\partial x^2} \tag{7.2}$$

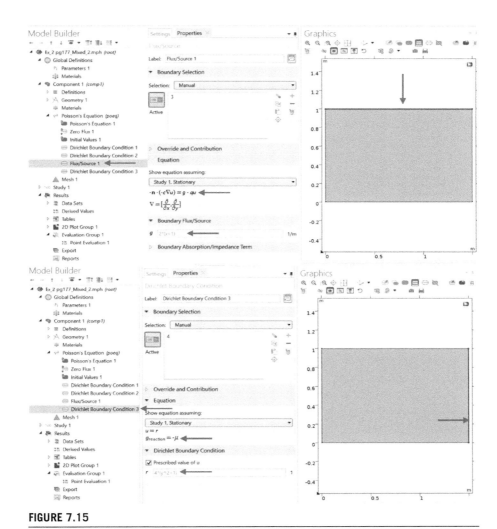

FIGURE 7.15

Entry of boundary conditions for the rectangle for solution of the Poisson equation.

can be solved for the time-dependent heat conduction problem for an insulated metal bar of unit length, where the ends of the bar are kept at a temperature of $0°C$. An initial temperature distribution is applied:

$$f(x) = \begin{cases} 2x & \text{if } 0 < x < 0.5, \\ 2(1-x) & \text{if } 0.5 \leq x < 1. \end{cases}$$

Solution. The same methodology can be used as in the previous examples. The changes to the physics, geometry, and boundary conditions are detailed below. A

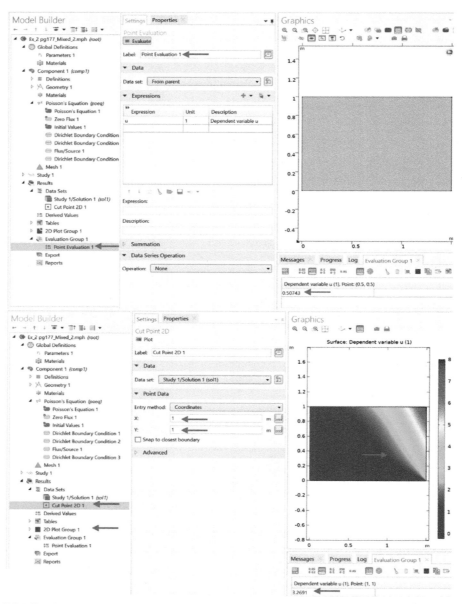

FIGURE 7.16

Solution of the Poisson equation.

Select Physics

FIGURE 7.17

Selection of the heat equation in COMSOL for the time-dependent heat conduction in a thin metal bar.

one-dimensional model is chosen for the thin metal bar. In the model setup, "Heat equation(hteq)" is chosen for the physics under "Mathematics $\longrightarrow$ "Classical PDEs" (see Fig. 7.17).

Under "Geometry 1," "Interval" is selected, resulting in a menu item of "Interval 1," where the "Coordinates" 0 and 1 can be input representing the bar of unit length. By right clicking "∇^2 Heat Equation (hteq)" in the left-side menu, the "Dirichlet Boundary Condition 1" and "Dirichlet Boundary Condition 2" items can be created in order to apply the boundary conditions of zero at each end of the bar. Eq. (7.2) can be input by clicking on "Heat Equation 1" and choosing $c = 1$, $f = 0$, and $d_a = 1$ (see Fig. 7.18). The initial value of $(x < 0.5) * (2 * x) + (x \geq 0.5) * (1 - x)$ can be input into the box for the "Initial value for u" on clicking the "Initial Values 1" menu item (see Fig. 7.19). Here the inequalities $(x < 0.5)$ and $(x \geq 0.5)$ are interpreted by COMSOL as conditional statements which evaluate as 1 if the inequality is true, and 0 if the inequality is false. The solid is then meshed and built.

The needed accuracy for the problem requires a suitable mesh size and time step. The object can be meshed with a finer mesh size by clicking on the "Mesh 1" menu item and selecting the "Element size." The tolerance for the time step can be entered

FIGURE 7.18

Input of the heat equation physics for conduction of heat in the thin metal bar.

FIGURE 7.19

Input of the initial condition for the problem.

in the "Step 1: Time Dependent" menu item (see Fig. 7.20), and choosing an input value in the box for the "User Controlled" "Relative tolerance" (that is, 1E−6 in this example). The output time range (that is, the time at a reported step and the total time of the problem simulation) is also input by the user on this page.

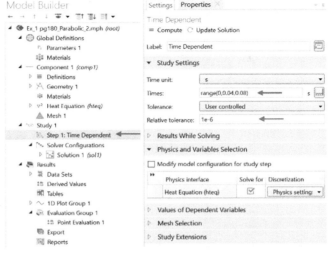

FIGURE 7.20

Selection of time-dependent range and relative tolerance.

A solution can then be computed with a selection of the "= Compute" button after right clicking the "Study 1" menu item.

For comparison to the values in Table 6.7, an output step size of 0.04 s is chosen with a total simulation time of 0.08 s. To output the results at a specific x location, right click "Results" and choose "Evaluation Group" in the drop-down box, which creates the submenu item "Evaluation Group 1." Right click "Data Sets" and select "Cut Point 1 D." A new submenu item "Cut Point 1 D 1" is created, where one can enter the value of the x coordinate of interest. Click the new "Evaluation Group 1" item and select "Data Set:" "Cut Point 1 D 1" and then right click "Evaluation Group 1" and choose "Point Evaluation." Click on the new created "Point Evaluation 1" and the "Evaluate" button. Click on the "1 D Plot Group 1" menu item, thereby display-ing the solution at different time steps and as a function of position along the bar (see Fig. 7.21). The initial distribution at $t = 0$ is shown in Fig. 7.21. At the bottom of the figure, values are also shown at the specified position as chosen (that is, at 0.2 m). The two values shown in the bottom table match the exact values in Table 6.7 for $x = 0.2$ for times of 0.04 and 0.08 s.

Example 7.1.4 (see Example 6.2.4). For this example, a hyperbolic equation is ap-plied for a vibrating string of unit length, where the ends of the string are fixed. We have

$$\frac{\partial^2 u}{\partial t^2} = \frac{\partial^2 u}{\partial x^2},$$

(7.3)

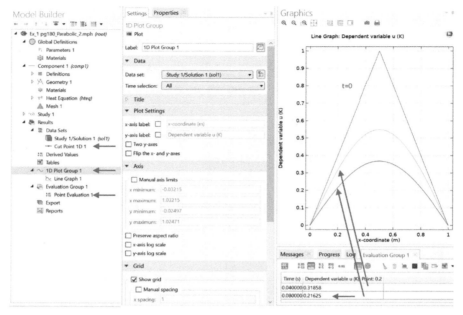

FIGURE 7.21

Output solution of results and values at $x = 0.2$ m.

$$u(x, 0) = \begin{cases} 2x & \text{if } 0 < x < 0.5, \\ 2(1-x) & \text{if } 0.5 \leq x < 1, \end{cases}$$

$$\frac{\partial u}{\partial t}(x, 0) = 0,$$

$$u(0, t) = u(1, t) = 0.$$

Here, there is an initial distribution of $f(x) = \begin{cases} 2x & \text{if } 0 < x < 0.5, \\ 2(1-x) & \text{if } 0.5 \leq x < 1 \end{cases}$ and a zero initial velocity.

Solution. Once again, the same methodology can be used as in the previous examples. A one-dimensional model is again chosen for the string with the same boundary conditions as in Example 7.1.3. Instead, for this model, "Wave Equation (waeq)" is chosen for the physics under "Mathematics $\longrightarrow$ "Classical PDEs." The same geometry as in the last example is used as well as the Dirichlet boundary conditions.

Eq. (7.3) can be input by clicking on "Wave Equation 1" and choosing $c = 1$, $f = 0$, and $d_a = 1$ (see Fig. 7.22). The initial value of $(x < 0.5) * 2 * x + (x \geq 0.5) * 2 * (1 - x)$ is input for the "Initial value for u" on clicking the "Initial Values 1" menu item. The string is then meshed and built.

FIGURE 7.22

Input of the physics for the vibrating string.

A "Relative tolerance" of 1E−6 can be chosen in the "Step 1: Time Dependent" menu item. An output time range with a step size of 0.2 s and total time of 1 s is chosen in accordance with Table 6.8. A solution can now be computed with the "= Compute" button after right clicking the "Study 1" menu item.

Output results are shown for the string as a function of distance along the string and as a function of time (see Fig. 7.23). Also shown are values of the string displacement at the location $x = 0.4$ m for times of 0.2 s and 0.4 s in the bottom table below the output figure. The solutions match the exact solutions in Table 6.8.

Ordinary differential equations

COMSOL is a powerful tool typically used to solve complicated partial differential equations, particularly when several equations are coupled when they represent an underlying multiphysics nature (such as both heat and mass transfer). For example, see the multiphysics question in Problem 7.7. However, COMSOL can also be used to solve ordinary differential equations of first and second order, (or higher order decomposed into multiple coupled lower-order ordinary differential equations), as also detailed in Chapter 6. Again, this section draws on the examples presented in Chapter 6 for these types of problems.

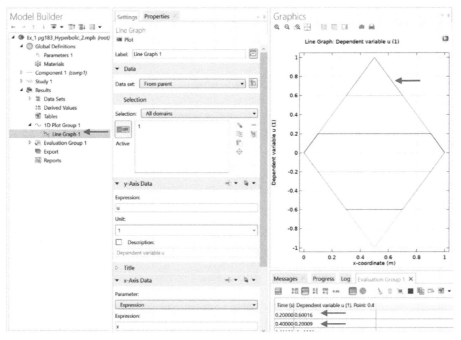

FIGURE 7.23

Output solution for the vibrating string with results also shown at $x = 0.4$ m for several times.

Example 7.1.5 (see Example 6.1.1 (Euler's method)). The same problem is also solved with different solution methods of increasing accuracy, including Example 6.1.2 (Heun's method), Example 6.1.3 (Runge–Kutta's method), and Example 6.1.4 (Adams–Moulton's method). This example demonstrates the solution method with COMSOL for a first-order ordinary differential equation. We have

$$\frac{dy'}{dx} = x + y + 1 \quad y(0) = 0 \quad \text{for } 0 \le x \le 2. \tag{7.4}$$

The COMSOL solution can be compared to the most accurate values based on Adams–Moulton's method in Table 6.4.

Solution. The same methodology can be used as in the previous examples. A one-dimensional geometry is chosen with the "Interval" option selected by right clicking the "Geometry 1" menu entry, thereby creating a submenu item "Interval 1 (i1)." In this model, a "Coefficient Form PDE(c)" is chosen for the physics under "Mathematics $\longrightarrow$ "PDE Interfaces." A "Dirichlet Boundary Condition" is selected on right

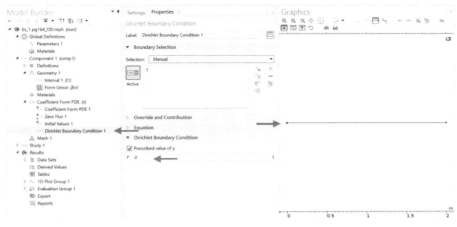

FIGURE 7.24

Input of the Dirichlet boundary condition for the first-order ordinary differential equation in Eq. (7.4).

clicking the "Coefficient Form PDE(c)." Similarly, as in previous examples, a new submenu "Dirichlet Boundary Condition 1" is created, from which one assigns "r:" equal to 0 in the box for the "Prescribed value of r" to apply the boundary condition $y(0) = 0$ at the point $x = 0$ on the interval (see Fig. 7.24).

Eq. (7.4) can be input by clicking on "Coefficient Form PDE 1" and choosing $c = 1$, $a = -1$, $f = x + 1$, $e_a = 0$, $d_a = 0$, $\alpha = 0$, $\beta = 1$, and $\gamma = 0$ (see Fig. 7.25). The object is then meshed and built. A solution can now be computed with the "= Compute" button after right clicking the "Study 1" menu item.

Output results for $y(x)$ are shown in Fig. 7.26. Also shown is the specific value of y at $x = 2.0$ in the bottom table below the output figure using the same method as before for selecting (cutting) a specific one-dimensional point. The solution matches the exact solution in Table 6.4.

Example 7.1.6 (see Examples 6.1.5 (Runge–Kutta–Nyström's method)). This example demonstrates the solution method with COMSOL for the following second-order ordinary differential equation:

$$\frac{d^2y}{dx^2} = \frac{1}{2}\left(x + y + \frac{dy}{dx} + 1\right) \quad y(0) = 0, \quad \frac{dy}{dx}(0) = 0 \quad \text{for } 0 \le x \le 1. \quad (7.5)$$

The COMSOL solution can be compared to the exact values in Table 6.5.

Solution. The same methodology can be used as in Example 7.1.5, except that the problem can be recast as a one-dimensional "Time Dependent" study (in contrast to

FIGURE 7.25

Input of the physics for the first-order ordinary differential equation.

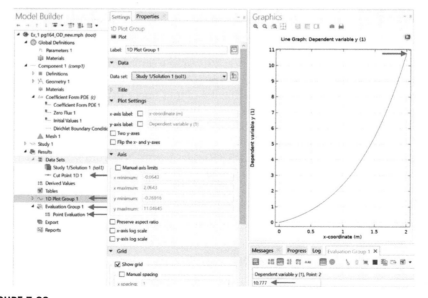

FIGURE 7.26

Output solution for $y(x)$ for Eq. (7.4) with results also shown at $x = 2.0$.

FIGURE 7.27

Input of the physics for the second-order ordinary differential equation in Eq. (7.5).

a "Stationary" study as in the previous example). In this problem, one can simply replace x by t in the ordinary differential equation to better capture the two conditions in Eq. (7.5) as initial conditions: $y(t = 0) = 0$ and $\dfrac{dy}{dt}(t = 0) = 0$. Similarly, a "Coefficient Form PDE(c)" is chosen for the physics under "Mathematics ⟶ "PDE Interfaces." However, no Dirichlet boundary conditions need to be considered in this problem. Although a one-dimensional geometry is chosen again with "Interval 1 (i1)," this assignment is somewhat arbitrary as there is no relevant spatial dimension in this problem so that default values can be assumed for the "Coordinates." Instead, the initial conditions are captured in the "Initial Values 1" submenu that is created on right clicking "Coefficient Form PDE(c)" and selecting the option "Initial Values."

Eq. (7.5) can now be input by clicking on "Coefficient Form PDE 1" and choosing $c = 0$, $a = 0$, $f = 0.5(t + y + 1)$, $e_a = 1$, $d_a = -0.5$, $\alpha = 0$, $\beta = 0$, and $\gamma = 0$ (see Fig. 7.27). A step size of 0.2 and a range from 0 to 1 can be input in the Step 1: "Time Dependent" submenu with a "User controlled" input for a "Relative tolerance" of 1E−6. The object is then meshed and built. A solution can now be computed with the "= Compute" button after right clicking the "Study 1" menu item.

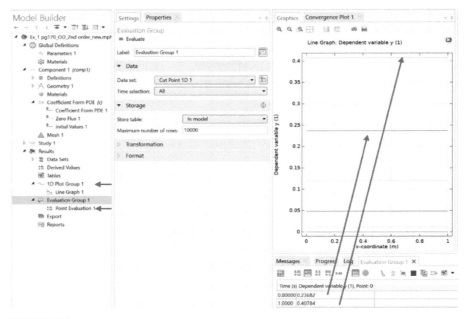

FIGURE 7.28

Output solution for y with results also shown at several time steps.

Output results for y as a function of x yield a different flat line at each time step since the problem has been recast as a time-dependent one (see Fig. 7.28). The value of y is also shown at several time steps in the bottom table. Using the same method as before for selecting (cutting) a specific one-dimensional point, a table is generated of y versus t. Here one can choose any arbitrary value of x within the problem range (since the line is flat) to capture the solution.

The solution in Fig. 7.28 matches the exact values shown in Table 6.5 (where the time step value is recognized as a value of x_n).

7.2 General theory of finite elements

The finite element method, or equivalently finite element analysis, is a set of numerical techniques combined to approximate solutions to differential equations on arbitrary domains. It is similar to the finite difference method encountered in Section 6.2, with the primary advantage being easier handling of complex geometries and variable spatial resolution. At a high level, the finite element method obtains a piecewise approximation to the solution of a differential equation with a controllable level of accuracy.

First, the discretization is achieved by dividing the domain into a finite number of smaller, geometrically simple subdomains called elements, from which the technique derives its name. The individual elements together are referred to as the finite element "mesh" and the process of subdividing the domain is called "meshing." The solution is approximated as the sum of polynomial basis functions with unknown coefficients in each element. The unknown coefficients, called degrees of freedom, are determined such that they minimize the residual (a measure of error) when the approximate solution is utilized in the differential equation. This process converts the differential equation into a multidimensional minimization problem.

This minimum of the error estimate is found by differentiating the error with respect to each unknown leading to a set of nonlinear equations which can be solved numerically by techniques such as a damped Newton's method described in Section 9.1.2, resulting in a series of linear equations. The optimal combination of the unknown coefficients which minimizes the error over the whole domain is known as the solution vector. Decreasing the size of the elements and/or increasing the number of basis functions in each element increases the number of degrees of freedom, such that the approximate solution becomes closer to the true solution. This approach also increases the computational costs since more unknowns must be determined.

In the following sections this process is described in greater detail. Many finite element packages are available commercially (such as COMSOL MULTIPHYSICS, demonstrated in Section 7.1) or as open source software. These packages provide tools to perform these steps with varying levels of user input. Typically the user defines the model in the form of: (1) the relevant physics or mathematical equations, (2) the geometry and finite element mesh, (3) boundary conditions, and (4) solution techniques and tolerances for solving the nonlinear and linear systems of equations. The user is also required to provide judgment to ensure that the approximations at each step are sufficiently accurate for the intended application.

7.2.1 Finite elements and shape functions

This section introduces the concept of finite elements and their corresponding shape functions which can be utilized to approximate the solution to partial differential equations. The process for creating elements of arbitrary order is demonstrated in one dimension, which is then extended to higher dimensions. In Section 7.2.2 these elements are used to discretize differential equations and obtain solutions.

One dimension
Linear elements

Suppose we wish to approximate a solution on a one-dimensional domain as a continuous piecewise linear function. For linear elements the approximate solution $\tilde{u}$ in each element is $\tilde{u} = a_0 + a_1 x$, where a_0 and a_1 are the unknown coefficients defined independently in each element. We define nodes at both ends of each element, numbering the left node in each element as 1 and the right node as 2. We can rewrite $\tilde{u}$ in terms of the value of $\tilde{u}$ at these nodes (that is, $\tilde{u}_1 = \tilde{u}(x_1)$ and $\tilde{u}_2 = \tilde{u}(x_2)$) as

$$\begin{bmatrix} \tilde{u}_1 \\ \tilde{u}_2 \end{bmatrix} = \begin{bmatrix} 1 & x_1 \\ 1 & x_2 \end{bmatrix} \begin{bmatrix} a_0 \\ a_1 \end{bmatrix}, \tag{7.6}$$

which can be solved for the unknown coefficients

$$\begin{bmatrix} a_0 \\ a_1 \end{bmatrix} = \begin{bmatrix} 1 & x_1 \\ 1 & x_2 \end{bmatrix}^{-1} \begin{bmatrix} \tilde{u}_1 \\ \tilde{u}_2 \end{bmatrix} = \frac{1}{x_2 - x_1} \begin{bmatrix} x_2 & -x_1 \\ -1 & 1 \end{bmatrix} \begin{bmatrix} \tilde{u}_1 \\ \tilde{u}_2 \end{bmatrix}. \tag{7.7}$$

Using these values of a_0 and a_1 the approximate solution can be written as

$$\tilde{u} = \frac{x_2\tilde{u}_1 - x_1\tilde{u}_2}{x_2 - x_1} + \frac{-\tilde{u}_1 + \tilde{u}_2}{x_2 - x_1} x. \tag{7.8}$$

Collecting $\tilde{u}_1$ and $\tilde{u}_2$ yields

$$\tilde{u} = \frac{x_2 - x}{x_2 - x_1}\tilde{u}_1 + \frac{-x_1 + x}{x_2 - x_1}\tilde{u}_2 = N_1(x)\tilde{u}_1 + N_2(x)\tilde{u}_2. \tag{7.9}$$

The functions $N_1(x)$ and $N_2(x)$ are known as shape functions, interpolation functions, or basis functions. The approximate solution is written as a sum of the product of the basis functions times the nodal values (that is, the solution is a linear combination of basis functions). The method of determining the nodal values for a particular problem is discussed in the next section. The $N_1(x)$ shape function decreases linearly from $N_1(x_1) = 1$ to $N_1(x_2) = 0$ across the element, while the $N_2(x)$ shape function increases from $N_2(x_1) = 0$ to $N_2(x_2) = 1$. Outside the element in which they are defined, their value is zero.

The shape functions are defined independently on each element, but neighboring elements can share the node between them (the right node of the left element is the left node of the right element), resulting in a continuous solution across elements. The linear basis functions corresponding to a hypothetical finite element mesh with elements with ranges $[0, 1.5]$, $[1.5, 3]$, and $[3, 4]$ are shown in Fig. 7.29(a). In Fig. 7.29(b) these elements and additional elements with ranges $[4, 4.5]$, $[4.5, 4.75]$, $[4.75, 4.875]$, and $[4.875, 5]$ are used to approximate $y = \sin(x)$ by interpolation.

Note that the error in the approximation is nonuniform. The error is lowest when the function is approximately linear on the scale of the element. If the function deviates significantly from linear on the scale of the element, the approximation cannot reproduce the nonlinearity. More smaller elements produce less error compared to fewer larger elements. The approximation is continuous, but the derivatives at the nodes are not continuous. This is referred to as c^0 continuity, the zero here referring to the number of continuous derivatives. Although not covered in this work, higher-order continuity can be achieved with more advanced finite elements such as Hermite, but these elements come at additional computational cost and are usually reserved for special applications.

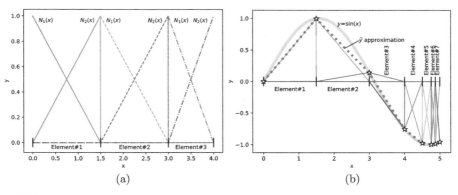

FIGURE 7.29

(a) One-dimensional linear shape functions for element number 1 [0, 1.5] (solid), element number 2 [1.5, 3] (dashed), and element number 3 [3, 4] (dash-dot). (b) $y = \sin(x)$ (thick solid line) and a linear finite element approximation (thick dotted line) along with the contribution from individual shape functions for a nonuniform finite element mesh (thin solid lines).

Quadratic elements

Suppose now one wants to approximate a solution on a one-dimensional domain as a continuous piecewise quadratic function. The above process is repeated with $\tilde{u} = a_0 + a_1 x + a_2 x^2$ with the additional unknown coefficient a_2. Since there are now three unknown coefficients we position three nodes in each element, one at each end and one located inside the element. Numbering these nodes from left to right, the unknown coefficients in terms of the values at the nodes are

$$\begin{bmatrix} a_0 \\ a_1 \\ a_2 \end{bmatrix} = \begin{bmatrix} 1 & x_1 & x_1^2 \\ 1 & x_2 & x_2^2 \\ 1 & x_3 & x_3^2 \end{bmatrix}^{-1} \begin{bmatrix} \tilde{u}_1 \\ \tilde{u}_2 \\ \tilde{u}_3 \end{bmatrix}. \tag{7.10}$$

As before, one: (1) calculates the inverse matrix to obtain the unknown coefficients, (2) substitutes the unknown coefficients into the polynomial form, and (3) collects nodal values ($\tilde{u}_1$, $\tilde{u}_2$, and $\tilde{u}_3$) to obtain

$$\tilde{u} = N_1(x)\tilde{u}_1 + N_2(x)\tilde{u}_2 + N_2(x)\tilde{u}_2 \tag{7.11}$$

in terms of the quadratic shape functions

$$N_1(x) = \frac{x_2 x_3 (x_3 - x_2) + \left(x_2^2 - x_3^2\right) x + (x_3 - x_2) x^2}{-x_1^2 x_2 + x_1 x_2^2 + x_1^2 x_3 - x_2^2 x_3 - x_1 x_3^2 + x_2 x_3^2}, \tag{7.12a}$$

$$N_2(x) = \frac{x_1 x_3 (x_1 - x_3) + \left(x_3^2 - x_1^2\right) x + (x_1 - x_3) x^2}{-x_1^2 x_2 + x_1 x_2^2 + x_1^2 x_3 - x_2^2 x_3 - x_1 x_3^2 + x_2 x_3^2}, \tag{7.12b}$$

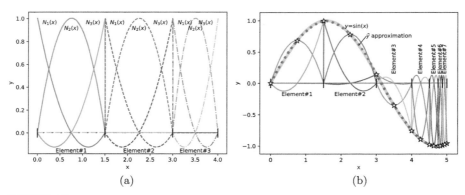

FIGURE 7.30

(a) One-dimensional quadratic shape functions for element number 1 [0, 1.5] (solid), element number 2 [1.5, 3] (dashed), and element number 3 [3, 4] (dash-dot). (b) $y = \sin(x)$ (thick solid line) and a linear finite element approximation (thick dotted line) along with the contribution from individual shape functions for a nonuniform finite element mesh (thin solid lines).

$$N_3(x) = \frac{x_1 x_2 (x_2 - x_1) + \left(x_1^2 - x_2^2\right)x + (x_2 - x_1)x^2}{-x_1^2 x_2 + x_1 x_2^2 + x_1^2 x_3 - x_2^2 x_3 - x_1 x_3^2 + x_2 x_3^2}. \tag{7.12c}$$

This is a general expression where x_2 can be located anywhere $x_1 < x_2 < x_3$, which can be simplified if x_2 is located at the center of the element such that $x_2 = (x_1 + x_3)/2$ so that the shape functions become

$$N_1(x) = \frac{(x - x_3)(2x - x_1 - x_3)}{(x_1 - x_3)^2}, \tag{7.13a}$$

$$N_2(x) = -4\frac{(x - x_1)(x - x_3)}{(x_1 - x_3)^2}, \tag{7.13b}$$

$$N_3(x) = \frac{(x - x_1)(2x - x_1 - x_3)}{(x_1 - x_3)^2}. \tag{7.13c}$$

Again, adjacent elements share the connecting node such that $\tilde{u}_3$ of the left element is $\tilde{u}_1$ of the right element, which produces a continuous approximation across elements. These quadratic basis functions are shown in Fig. 7.30(a) for the same nonuniform mesh, and applied to approximate $y = \sin(x)$ in Fig. 7.30(b). Note the approximation more closely resembles $y = \sin(x)$ compared to the example with linear elements and now includes some of the curvature.

Higher-order elements

A feature of basis functions of any order constructed in this manner is that they always have a value of one at their corresponding node and zero at all other nodes. These correspond to Lagrange interpolation polynomials of order n (see Section 9.2.1) from which we can write the ith shape function, $l_i^n(x) = N_i(x)$, as

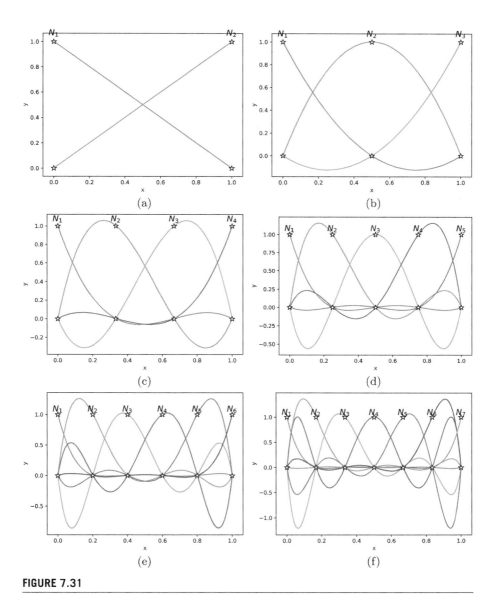

FIGURE 7.31

Comparison of shape functions of orders 1–6.

$$N_i(x) = l_i^n(x) = \prod_{j=1, j\neq i}^{n} \frac{x - x_j}{x_i - x_j}. \tag{7.14}$$

These are plotted in Fig. 7.31 for element orders 1–6 for evenly spaced nodes. An nth-order Lagrange element in one dimension has $n + 1$ nodes corresponding to

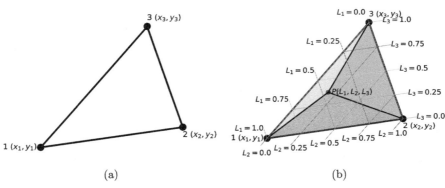

FIGURE 7.32

(a) A linear triangle element with three nodes. (b) Diagram of area coordinates for triangular element in part (a).

$n + 1$ shape functions. The higher-order elements have more degrees of freedom per element and are thus able to more closely approximate an arbitrary function and achieve lower error per element at increased computational cost. The computational cost per degree of freedom is also higher since there are more overlapping shape functions leading to more complicated systems of linear equations. The optimal balance of element order and mesh density is problem-specific. Finite element implementations which focus on controlling error by using high-order elements are known as "p-methods," while those that instead focus on reducing the size of relatively low-order elements are known as "h-methods." The combination of the two approaches is called "hp-methods."

Two dimensions
Linear triangular elements

For the two-dimensional case, we can begin by decomposing our domain into triangular elements with a node at each vertex, as shown in Fig. 7.32(a). With three nodes, we have three degrees of freedom to approximate $u(x, y)$. The lowest-order polynomials with three degrees of freedom are

$$\tilde{u}(x, y) = a_0 + a_1 x + a_2 x^2, \tag{7.15a}$$

$$\tilde{u}(x, y) = a_0 + a_1 x + a_2 y, \tag{7.15b}$$

$$\tilde{u}(x, y) = a_0 + a_1 y + a_2 y^2. \tag{7.15c}$$

However, it is usually desirable to maintain equal powers of both x and y so that the quality of the approximation is independent of the rotation of the element. Therefore, one can proceed with Eq. (7.15b).

Following the same procedure as for the one-dimensional case, one can write $\tilde{u}$ in terms of the nodal values to obtain the coefficients a_0 to a_2,

$$\begin{bmatrix} a_0 \\ a_1 \\ a_2 \end{bmatrix} = \begin{bmatrix} 1 & x_1 & y_1 \\ 1 & x_2 & y_2 \\ 1 & x_3 & y_3 \end{bmatrix}^{-1} \begin{bmatrix} \tilde{u}_1 \\ \tilde{u}_2 \\ \tilde{u}_3 \end{bmatrix}$$

$$= \frac{1}{2A} \begin{bmatrix} x_2 y_3 - x_3 y_2 & x_3 y_1 - x_1 y_3 & x_1 y_2 - x_2 y_1 \\ y_2 - y_3 & -y_1 + y_3 & y_1 - y_2 \\ -x_2 + x_3 & x_1 - x_3 & -x_1 + x_2 \end{bmatrix} \begin{bmatrix} \tilde{u}_1 \\ \tilde{u}_2 \\ \tilde{u}_3 \end{bmatrix}, \qquad (7.16)$$

where A is the area of the element (and the determinant of the matrix) given by $A = x_1 (y_2 - y_3) + x_2 (-y_1 + y_3) + x_3 (y_1 - y_2)$. By collecting the terms the shape functions are

$$N_1 (x, y) = \frac{x_2 y_3 - x_3 y_2 + (y_2 - y_3) x + (-x_2 + x_3) y}{2A}, \qquad (7.17a)$$

$$N_2 (x, y) = \frac{-x_1 y_3 + x_3 y_1 + (-y_1 + y_3) x + (x_1 - x_3) y}{2A}, \qquad (7.17b)$$

$$N_3 (x, y) = \frac{x_1 y_2 - x_2 y_1 + (y_1 - y_2) x + (-x_1 + x_2) y}{2A}. \qquad (7.17c)$$

Area coordinates for triangles

The linear shape functions derived in Eq. (7.17) are also known as "area coordinates" or "triangle coordinates." They are a useful coordinate system for working with triangles that do not have orthogonal edges and will be used to derive the higher-order triangular elements. In two dimensions, Cartesian coordinates have two independent components (e.g., $P(x, y)$), while area coordinates also have two independent components plus a third dependent component (e.g., $P(L_1, L_2, L_3)$), as depicted in Fig. 7.32(b). A point within an arbitrary triangle can be uniquely defined by

$$L_1 = N_1 (x, y), \qquad (7.18a)$$

$$L_2 = N_2 (x, y), \qquad (7.18b)$$

$$L_3 = N_3 (x, y) \qquad (7.18c)$$

or by recognizing that the constraint $N_1 + N_2 + N_3 = 1 = L_1 + L_2 + L_3$ can produce the third dependent coordinate from any two independent coordinates. The name "area coordinates" arises because their value relates to the area of the shaded subtriangles in Fig. 7.32(b) to the area of the whole triangle

$$L_1 = \frac{\text{area} (P23)}{\text{area} (123)}, \qquad (7.19a)$$

$$L_2 = \frac{\text{area} (P13)}{\text{area} (123)}, \qquad (7.19b)$$

$$L_3 = \frac{\text{area} (P12)}{\text{area} (123)}. \qquad (7.19c)$$

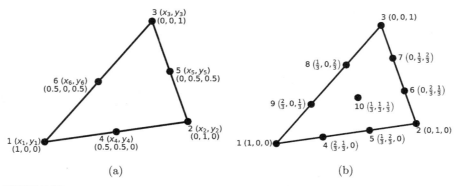

FIGURE 7.33

(a) A quadratic triangle element with six nodes with global Cartesian coordinates and local area coordinates. (b) A cubic triangle element with 10 nodes with local area coordinates.

To return to Cartesian coordinates from area coordinates the relationship is

$$x = L_1 x_1 + L_2 x_2 + L_3 x_3, \tag{7.20a}$$

$$y = L_1 y_1 + L_2 y_2 + L_3 y_3. \tag{7.20b}$$

Quadratic and cubic triangular elements

The second-order triangle, shown in Fig. 7.33(a), has six nodes (three vertex nodes and three edge nodes) such that $\tilde{u}(x, y) = a_0 + a_1 x + a_2 y + a_3 x^2 + a_4 x y + a_5 y^2$. The same process outlined throughout this chapter can be utilized to determine the coefficients a_0 to a_5 and therefore the shape functions. However, since there are now six nodes, this method requires inverting a 6×6 matrix for each element to obtain the shape functions. The growing complexity of the shape functions, as well as the eventual need to efficiently integrate functions over each element, motivates an alternate approach to derive high-order and high-dimensional elements. Instead, one can write the shape functions in terms of the area coordinates L_1, L_2, L_3 defined by Eq. (7.17) and Eq. (7.18). One can write the shape functions as products, which is one at the corresponding node and zero at all others:

$$N_1 = (2L_1 - 1) L_1, \tag{7.21a}$$

$$N_2 = (2L_2 - 1) L_2, \tag{7.21b}$$

$$N_3 = (2L_3 - 1) L_3, \tag{7.21c}$$

$$N_4 = 4L_1 L_2, \tag{7.21d}$$

$$N_5 = 4L_2 L_3, \tag{7.21e}$$

$$N_6 = 4L_1 L_3. \tag{7.21f}$$

Similarly, the shape functions of the 10-node cubic triangle shown in Fig. 7.33(b) may be written as

$$N_1 = \frac{9}{2}\left(L_1 - \frac{1}{3}\right)\left(L_1 - \frac{2}{3}\right)L_1, \tag{7.22a}$$

$$N_2 = \frac{9}{2}\left(L_2 - \frac{1}{3}\right)\left(L_2 - \frac{2}{3}\right)L_2, \tag{7.22b}$$

$$N_3 = \frac{9}{2}\left(L_3 - \frac{1}{3}\right)\left(L_3 - \frac{2}{3}\right)L_3, \tag{7.22c}$$

$$N_4 = \frac{27}{2}L_1 L_2\left(L_1 - \frac{1}{3}\right), \tag{7.22d}$$

$$N_5 = \frac{27}{2}L_1 L_2\left(L_1 - \frac{2}{3}\right), \tag{7.22e}$$

$$N_6 = \frac{27}{2}L_2 L_3\left(L_2 - \frac{1}{3}\right), \tag{7.22f}$$

$$N_7 = \frac{27}{2}L_2 L_3\left(L_2 - \frac{2}{3}\right), \tag{7.22g}$$

$$N_8 = \frac{27}{2}L_1 L_3\left(L_1 - \frac{2}{3}\right), \tag{7.22h}$$

$$N_9 = \frac{27}{2}L_1 L_3\left(L_1 - \frac{1}{3}\right). \tag{7.22i}$$

This process can be repeated to efficiently produce triangular Lagrange elements of arbitrary order.

Linear quadrilateral elements

The other common element shape for discretizing in two dimensions are quadrilaterals. The simplest quadrilateral with one node at each vertex, as depicted in Fig. 7.34(a), is given by $\tilde{u}(x, y) = a_0 + a_1 x + a_2 y + a_3 xy$. Again, the general process outlined in the previous sections could be applied in which the coefficients a_0 to a_3 are determined in terms of the coordinates (x_0 to x_3 and y_0 to y_3) by solving a linear equation in order to obtain the shape functions. However, like the higher-order triangles, an alternate approach is to use local element coordinates (ξ, η) to more easily account for potential nonparallel sides. First, one considers the simple quadrilateral in Fig. 7.34(b) with nodes at $\xi = \pm 1$ and $\eta = \pm 1$. For this simple element geometry, one can formulate the shape functions in terms of Lagrange interpolation polynomials (Eq. (7.14)) along the edges, which are easily defined in (ξ, η) (see Fig. 7.35(a) and Fig. 7.35(b)),

$$N_1(\xi, \eta) = \frac{1}{4}(1 - \xi)(1 - \eta), \tag{7.23a}$$

$$N_2(\xi, \eta) = \frac{1}{4}(1 + \xi)(1 - \eta), \tag{7.23b}$$

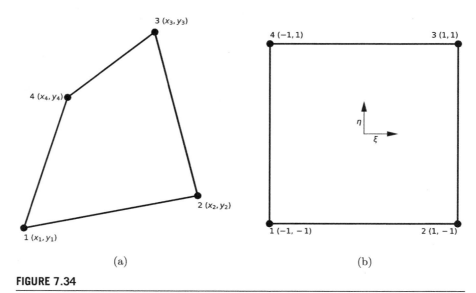

FIGURE 7.34

(a) A quadrilateral element with four nodes in global (x, y) Cartesian coordinates. (b) A quadrilateral element with four nodes in local (ξ, η) coordinates.

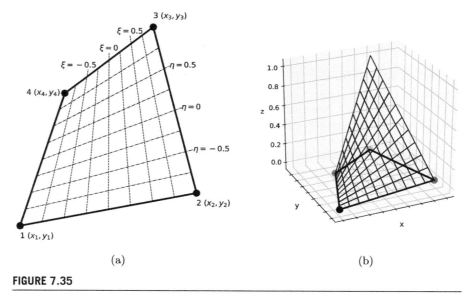

FIGURE 7.35

(a) A quadrilateral element with lines of constant local (ξ, η) coordinates. (b) Three-dimensional visualization of the N_3 shape function over the sample element.

$$N_3 (\xi, \eta) = \frac{1}{4} (1 + \xi) (1 + \eta), \tag{7.23c}$$

$$N_4 (\xi, \eta) = \frac{1}{4} (1 - \xi) (1 + \eta). \tag{7.23d}$$

The global coordinates (x, y) corresponding to local coordinates (ξ, η) are a linear combination of the nodal global coordinates and local shape functions

$$x (\xi, \eta) = N_1 (\xi, \eta) x_1 + N_2 (\xi, \eta) x_2 + N_3 (\xi, \eta) x_3 + N_4 (\xi, \eta) x_4, \tag{7.24a}$$

$$y (\xi, \eta) = N_1 (\xi, \eta) y_1 + N_2 (\xi, \eta) y_2 + N_3 (\xi, \eta) y_3 + N_4 (\xi, \eta) y_4. \tag{7.24b}$$

This relationship allows arbitrary quadrilateral elements without parallel sides to be integrated or plotted parametrically using the easy to define domain in local (ξ, η) coordinates. Spatial derivatives such as $\frac{\partial \tilde{u}}{\partial x}$ requires the derivatives of the shape functions in terms of the global (x, y) coordinates which can be obtained by applying the chain rule,

$$\frac{\partial N (\xi, \eta)}{\partial \xi} = \frac{\partial N (\xi, \eta)}{\partial x} \frac{\partial x}{\partial \xi} + \frac{\partial N (\xi, \eta)}{\partial y} \frac{\partial y}{\partial \xi}, \tag{7.25a}$$

$$\frac{\partial N (\xi, \eta)}{\partial \eta} = \frac{\partial N (\xi, \eta)}{\partial x} \frac{\partial x}{\partial \eta} + \frac{\partial N (\xi, \eta)}{\partial y} \frac{\partial y}{\partial \eta}, \tag{7.25b}$$

which may also be written in terms of the Jacobian matrix

$$\begin{bmatrix} \frac{\partial N(\xi,\eta)}{\partial \xi} \\ \frac{\partial N(\xi,\eta)}{\partial \eta} \end{bmatrix} = \begin{bmatrix} \frac{\partial x}{\partial \xi} & \frac{\partial y}{\partial \xi} \\ \frac{\partial x}{\partial \eta} & \frac{\partial y}{\partial \eta} \end{bmatrix} \begin{bmatrix} \frac{\partial N(\xi,\eta)}{\partial x} \\ \frac{\partial N(\xi,\eta)}{\partial y} \end{bmatrix}. \tag{7.26}$$

Higher-order quadrilateral elements

Using a local coordinate system in which the nodes are conveniently arranged, higher-order elements can be easily created from products of Lagrange interpolation polynomials in terms of ξ, η. A nine-node element supports a polynomial of the form $\tilde{u} (x, y) = a_0 + a_1 x + a_2 y + a_3 xy + a_4 x^2 + a_5 y^2 + a_6 x^2 y + a_7 xy^2 + a_8 x^2 y^2$. The nodes form an evenly spaced grid local at $\xi = -1, 0, 1$ and $\eta = -1, 0, 1$. Examples of the shape functions corresponding to a corner, edge, and central node are shown in Fig. 7.36.

Similarly, a 16-node element supports a polynomial of the form $\tilde{u} (x, y) = a_0 + a_1 x + a_2 y + a_3 xy + a_4 x^2 + a_5 y^2 + a_6 x^2 y + a_7 xy^2 + a_8 x^2 y^2 + a_9 x^3 + a_{10} y^3 + a_{11} x^3 y + a_{12} xy^3 + a_{13} x^3 y^2 + a_{14} x^2 y^3 + a_{15} x^3 y^3$. The nodes form an evenly spaced grid local at $\xi = -1, \frac{-1}{3}, \frac{1}{3}, 1$ and $\eta = -1, \frac{-1}{3}, \frac{1}{3}, 1$. Examples of the shape functions corresponding to a corner, edge, and central node are shown in Fig. 7.37.

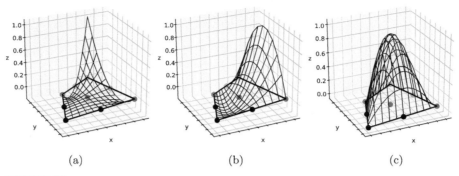

FIGURE 7.36

Shape functions for a nine-node quadrilateral corresponding to a node located: (a) at the corner ($\xi = 1, \eta = 1$), (b) along the edge ($\xi = 1, \eta = 0$), and (c) in the middle ($\xi = 0, \eta = 0$).

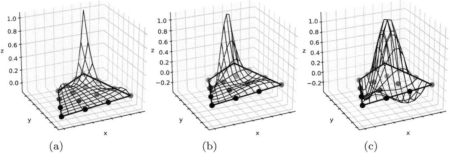

FIGURE 7.37

Shape functions for a 16-node quadrilateral corresponding to the nodes located: (a) at the corner ($\xi = 1, \eta = 1$), (b) along the edge ($\xi = 1, \eta = \frac{1}{3}$), and (c) in the middle ($\xi = \frac{1}{3}, \eta = \frac{1}{3}$).

7.2.2 Solution of partial differential equations
Weighted error

In this section, one utilizes the finite element basis developed in Section 7.2.1 to find the approximate solution to a partial differential equation which minimizes the error. Once the mesh and shape functions are defined, the approximate solution to a partial differential equation can be written as the sum of the product of the shape functions times the magnitude at each of the nodes:

$$\tilde{u}(\vec{x}) = \sum_{i=1}^{n} N(\vec{x})u_i = [\vec{N}] \cdot [\vec{u}]. \tag{7.27}$$

To demonstrate this process, consider the example of Poisson's differential:

$$\nabla^2 u - f(\vec{x}) = 0. \tag{7.28}$$

Since $\tilde{u}$ is approximate rather than exact, it may not perfectly satisfy the differential and one expects to have a spatially varying residual error $R(\vec{x})$ leading to

$$\nabla^2 \tilde{u} - f(\vec{x}) = R(\vec{x}). \tag{7.29}$$

This error can be converted to a global measure of error by introducing a currently undefined weighting function $W(\vec{x})$ and integrating over the domain where the error is defined as

$$Error = \int_V W(\vec{x}) R(\vec{x}) dV = \int_V W(\vec{x}) \left(\nabla^2 \tilde{u} - f(\vec{x}) \right) dV \tag{7.30}$$

$$= \int_V (W(\vec{x}) \nabla^2 \vec{N}(\vec{x}) \vec{u} - W(\vec{x}) f(\vec{x})) dV. \tag{7.31}$$

For a smooth, well-behaved weighting function, it is advantageous to manipulate this error considering the divergence of a scalar vector product $a\vec{b}$,

$$\nabla \cdot \left(a\vec{b} \right) = \nabla a \cdot \vec{b} + a \nabla \cdot \vec{b}. \tag{7.32}$$

If $a = W(\vec{x})$ and $\vec{b} = \nabla N(\vec{x})\vec{u}$, the $W(\vec{x})\nabla^2 \vec{N}(\vec{x})\vec{u}$ term in the error may be written as

$$Error = \int_V (\nabla \cdot W(\vec{x}) \nabla \vec{N}(\vec{x}) \vec{u} - \nabla W(\vec{x}) \cdot \nabla \vec{N}(\vec{x}) \vec{u} - W(\vec{x}) f(\vec{x})) dV. \tag{7.33}$$

The divergence theorem in Eq. (4.16) may be applied to the first term on the right-hand side to convert it to a surface integral,

$$Error = \oint_S \vec{n} \cdot W(\vec{x}) \nabla \vec{N}(\vec{x}) \vec{u} \, dS - \int_V (\nabla W(\vec{x}) \cdot \nabla \vec{N}(\vec{x}) \vec{u} + W(\vec{x}) f(\vec{x})) dV. \tag{7.34}$$

The surface integral term is associated with the boundary conditions, and is often but not always zero (i.e., no flux boundary conditions). This manipulation has effectively transferred one of the derivatives from the shape functions to the weighting functions, which can be advantageous for obtaining numerical solutions.

Weak form

If the error is zero and the weighting function is well behaved, this formulation is known as the "weak form", because it is equivalent to the "strong form" shown in Eq. (7.31), but has weaker (relaxed) smoothness requirements on the solution (and therefore the shape functions). In the weak form, the weighting functions are known as test functions. The general process for obtaining the weak form starting with the strong form follows the same approach: (1) multiply the strong form by a test function, (2) integrate the equation over the whole domain, and (3) manipulate the equation to transfer some of the derivatives to the test function.

Galerkin method

The choice of the weighting and shape functions produces different numerical methods: least squares, collocation (i.e., the sum of fixed points), and the Galerkin method, among other techniques. For finite element methods the most common is the Galerkin method, which sets the weighting function equal to the shape functions $W(\vec{x}) = N_i(\vec{x})$ and searches for an approximate solution with error equal to zero. For each node this approach leads to equations of the form

$$\int_V (\nabla N_i(\vec{x}) \cdot \nabla \vec{N}(\vec{x})\vec{u} + N_i(\vec{x})f(\vec{x}))dV = 0, \tag{7.35}$$

which collectively produce a system of equations for the unknown components of $\vec{u}$. The components of $\vec{u}$ are the unknown nodal values $\tilde{u}_i$, which are not spatially varying and can therefore be factored from the integrals, leading to

$$
\begin{bmatrix}
u_1 \int \nabla N_1 \nabla N_1 dV + u_2 \int \nabla N_1 \nabla N_2 dV + \cdots + u_n \int \nabla N_1 \nabla N_n dV \\
u_1 \int \nabla N_2 \nabla N_1 dV + u_2 \int \nabla N_2 \nabla N_2 dV + \cdots + u_n \int \nabla N_2 \nabla N_n dV \\
\vdots \\
u_1 \int \nabla N_n \nabla N_1 dV + u_2 \int \nabla N_n \nabla N_2 dV + \cdots + u_n \int \nabla N_n \nabla N_n dV
\end{bmatrix}
$$
$$
= -
\begin{bmatrix}
\int N_1 f dV \\
\int N_2 f dV \\
\vdots \\
\int N_n f dV
\end{bmatrix}. \tag{7.36}
$$

Finally, factoring $\vec{u}$ yields a matrix equation

$$
\begin{bmatrix}
\int \nabla N_1 \nabla N_1 dV & \int \nabla N_1 \nabla N_2 dV & \ldots & \int \nabla N_1 \nabla N_n dV \\
\int \nabla N_2 \nabla N_1 dV & \int \nabla N_2 \nabla N_2 dV & \ldots & \int \nabla N_2 \nabla N_n dV \\
& & \vdots & \\
\int \nabla N_n \nabla N_1 dV & \int \nabla N_n \nabla N_2 dV & \ldots & \int \nabla N_n \nabla N_n dV
\end{bmatrix}
\begin{bmatrix}
u_1 \\
u_2 \\
\vdots \\
u_n
\end{bmatrix}
$$
$$
= -
\begin{bmatrix}
\int N_1 f dV \\
\int N_2 f dV \\
\vdots \\
\int N_n f dV
\end{bmatrix}, \tag{7.37}
$$

or in matrix form

$$[K]\vec{u} = \vec{F}. \tag{7.38}$$

Here the matrix K is termed the "stiffness" matrix. The elements of the stiffness matrix have the form

$$K_{i,j} = \int \nabla N_i \nabla N_j dV. \tag{7.39}$$

Typically most of the entries in this matrix evaluate to zero because the shape functions are only nonzero in the elements which include the node (i.e., nonoverlapping shape functions yield zero for this integral). This produces a "sparse" matrix where most entries are zero and only the nonzero values are stored, which reduces the computational costs compared to "dense" matrices.

The calculation of the nonzero elements of the matrix K and the vector $\vec{F}$ requires evaluation of the integrals on each finite element. In most finite element implementations, this calculation is usually performed numerically with a "quadrature" method that approximates an integral as a weighted sum of the integrand evaluated at specific points:

$$\int g(\vec{x})dV = \sum_{i=1}^{n} w_i g\vec{x}_i. \tag{7.40}$$

Here w_i are the quadrature weights applied to the integrand $g\vec{x}$ evaluated at $\vec{x}_i$ and n is the number of points evaluated. The most widely used quadrature method is "Gaussian quadrature," which is exact for polynomials $2n - 1$ or less (see, for example, Section 9.5.3). Other quadrature techniques such as "Gauss–Kronrod" are sometimes utilized when estimates of the integration error are required. In practice it is often sufficient to integrate to a higher degree but skip the error test.

The element formulations are normally precalculated for standard equations and element types. The stiffness matrix is assembled by adding the contributions from each element in the mesh. This "stiffness" matrix depends on the partial differential equation, the boundary conditions, and the mesh.

Boundary conditions need to be applied that will constrain the system to obtain a nonsingular solution. If the system is nonlinear, an iterative approach is utilized in which the stiffness of each element is calculated using the previous solution. The stiffness is updated on each iteration along with the solution vector and the calculation proceeds until convergence is achieved. For example, a nonlinear system may involve the calculation of temperature, where the thermal conductivity itself is dependent on the temperature.

Once the linear system of equations from Eq. (7.38) is assembled, standard solution techniques can be applied as discussed in Chapter 4. There are two main approaches used for finite elements:

- Direct matrix factorization. Techniques such as lower-upper factorization are more stable/robust and simpler to use and configure. However, it is slower for large problems and requires more memory.
- Iterative methods. These methods solve the system by updating an initial guess; the most common are Krylov methods such as GMRES and BiCSTAB. They often

require less memory and are faster than the direct factorization method for large (usually three-dimensional) problems. However, it is less stable for poorly/ill-conditioned problems.

Additional considerations for finite element method analysis

It is important to recognize that the finite element mesh does not have a physical meaning since the true solution does not depend on the mesh itself, and an infinite number of meshes can provide equivalent results. Changing the mesh a small amount should not substantially change the answer. If it does, the result is an artifact of the solution process and is not representative of the true solution. It is good practice to perform a mesh convergence study to ensure that refining the mesh does not strongly alter the solution. The mesh is simply a part of the finite element method, which must be small enough in order to "accurately" represent the solution.

On the other hand, the boundary conditions are extremely important and are physically relevant to the problem. They can be simple (e.g., a fixed displacement or constant temperature) or extremely complicated (e.g., with the contact of adjoining surfaces, friction between surfaces, or no slip flow).

One must also ensure that the element order is sufficient to capture the desired properties of the solution. For example, consider a displacement field which satisfies a solid mechanics model. Components of the strain (ϵ) and stress (σ) are related to the derivative of the displacement u such that

$$\epsilon_x = \frac{\partial u}{\partial x}, \tag{7.41}$$

$$\sigma_x = E\epsilon_x = E\frac{\partial u}{\partial x}, \tag{7.42}$$

where E is the Young's modulus. For a linear triangular element, $\tilde{u}(x, y) = a_0 + a_1 x + a_2 y = N_1(x, y)u_1 + N_2(x, y)u_2 + N_3(x, y)u_3$ from Eq. (7.15b) and Eq. (7.17), and therefore the strain is

$$\frac{\partial \tilde{u}(x, y)}{\partial x} = a_1 = u_1\frac{\partial N_1(x, y)}{\partial x} + u_2\frac{\partial N_2(x, y)}{\partial x} + u_3\frac{\partial N_3(x, y)}{\partial x} \tag{7.43}$$

$$= \frac{(y_2 - y_3)u_1 + (-y_1 + y_3)u_2 + (y_1 - y_2)u_3}{2A}, \tag{7.44}$$

which is no longer a function of the spatial coordinates. Therefore, the strain and stress field within linear triangular shape functions cannot vary smoothly between elements. By contrast, the derivatives of higher-order triangular elements, or linear quadrilateral elements, are not constant and therefore better able to resolve the strain fields. In general there is no clear "best" element for all applications, i.e., this choice depends on the application. One needs to consider what features are expected/needed to be captured. Can rectangular elements be used? How significant are the gradients? Is there a region of interest? How many spatial derivatives are required?

Similarly, to achieve convergence one requires an acceptable error tolerance that depends on the quantity being solved for, the required accuracy needed for the solution, and the accuracy of the underlying physics and known material properties. Other considerations include (i) the available computing resources (i.e., the random-access memory [RAM] or central processing unit [CPU]); (ii) the degree of nonlinearity, which has a strong effect on the solution convergence (i.e., large deformation and geometric stiffness); (iii) the type of problem being solved (e.g., stationary, time-dependent, eigenvalue, frequency response); and (iv) whether the problem involves single physics or multiphysics (with one- or two-directional coupling). A more detailed introduction to finite elements is provided in Hutton (2004) and Zienkiewicz et al. (2013).

Problems

7.1 Solve the time-independent Laplace heat conduction equation in Example 6.2.1 with the COMSOL software package.

7.2 Solve the Poisson equation in Example 6.2.2 with the COMSOL software package.

7.3 Solve the time-dependent heat conduction equation in Example 6.2.3 with the COMSOL software package.

7.4 Solve the wave equation in Example 6.2.4 with the COMSOL software package.

7.5 Solve the first-order ordinary differential equation in Example 6.1.4 with the COMSOL software package.

7.6 Solve the second-order ordinary differential equation in Example 6.1.5 with the COMSOL software package.

7.7 A time-dependent fission product diffusion model in the nuclear fuel matrix can be coupled with a mass balance in the fuel-to-sheath gap and coolant to predict the coolant activity behavior from a defective fuel element for variable power operation in a nuclear reactor.[1] The diffusion equation for the fission product concentration C in an idealized fuel grain sphere of radius a, accounting for radioactive decay λ with a time-dependent diffusion coefficient D and fission product generation rate B ($=$ fission rate F times the fission product yield Y), is

$$\frac{\partial C(r,t)}{\partial t} = \frac{D(t)}{r^2} \frac{\partial}{\partial r}\left(r^2 \frac{\partial C(r,t)}{\partial r}\right) - \lambda C(r,t) + \frac{F(t)Y}{V}. \tag{7.45}$$

---

[1] B.J. Lewis, P.K. Chan, A. El-Jaby, F.C. Iglesias, and A. Fitchett, "Fission product release modeling for application of fuel-failure monitoring and detection – An overview," J. Nucl. Mater. 489 (2017) 64–83.

Defining the dimensionless variable $x = r/a$ and multiplying through by the volume V yields

$$\frac{\partial u(x,t)}{\partial t} = \frac{D'(t)}{x^2} \frac{\partial}{\partial x} \left(x^2 \frac{\partial u(x,t)}{\partial x} \right) - \lambda u(x,t) + F(t)Y, \qquad (7.46)$$

where $u = CV$ and $D' = D/a^2$. The initial and boundary conditions are given as

$$u(x, 0) = 0, \quad 0 < x < 1, \quad t = 0, \qquad (7.47a)$$

$$\frac{\partial u}{\partial x} = 0, \quad x = 0, \quad t > 0, \qquad (7.47b)$$

$$u(1, t) = 0 = 0, \quad x = 1, \quad t > 0. \qquad (7.47c)$$

A zero concentration is assumed at $t = 0$. For the boundary conditions, symmetry is assumed at the center of the sphere (i.e., a zero flux) and a zero concentration at the fuel grain surface with fission product release from the grain surface. The diffusional release-to-birth rate ratio derived from the Fick's law of diffusion is

$$\left(\frac{R}{B} \right) = \frac{4\pi a^2}{FY(4\pi a^3/3)} \left(-D \frac{\partial CV}{\partial r} \bigg|_{r=a} \right) = -\frac{3D'}{FY} \frac{\partial u}{\partial x} \bigg|_{x=1}. \qquad (7.48)$$

Equivalently, the release rate R (atoms s^{-1}) from the fuel matrix is

$$R = -3D' \frac{\partial u}{\partial x} \bigg|_{x=1}. \qquad (7.49)$$

Thus, the time-dependent diffusion equation in Eq. (7.46) can be solved by numerical methods subject to the conditions in Eq. (7.47a) to Eq. (7.47c). The derivative of this solution (at $x = 1$) is subsequently used in Eq. (7.49), which is the source release rate from the fuel matrix into the fuel-to-sheath gap. As such, the mass balance for the number of atoms N_g in the gap is

$$\frac{dN_g(t)}{dt} = R(t) - (\lambda + v(t))N_g(t), \qquad (7.50a)$$

with initial condition

$$N_g(t) = 0, \quad t = 0. \qquad (7.50b)$$

A first-order loss-rate process is assumed from the gap characterized by an escape rate coefficient v (s^{-1}) which can also be considered as a function of time with element deterioration. The mass balance in the coolant is similarly given by

$$\frac{dN_c(t)}{dt} = v(t)N_g(t) - (\lambda + \beta(t))N_c(t) \qquad (7.51a)$$

with initial condition

$$N_c(t) = 0, \quad t = 0. \qquad (7.51b)$$

Here a time-dependent coolant purification rate constant $\beta(t)$ for the coolant is considered. For this problem, the empirical diffusion coefficient D' (s^{-1}) as a function of the element linear power P (kW m^{-1}) is derived from historical coolant activity data in the commercial stations as

$$D' = 8.6 \times 10^{-12} \left[\frac{10^{(0.114 \cdot P)}}{P^2} \right]. \tag{7.52}$$

In addition, assuming 180 MeV fission^{-1} and a fuel stack length of 0.48 m, the fission rate F (fissions s^{-1}) $= 1.489 \times 10^{13}$ P (kW m^{-1}).

(a) The solution of the coupled Eq. (7.46) to Eq. (7.52) provides a prediction of both the number of atoms in the gap and coolant as a function of time for a variable fuel element linear rating and coolant purification history. Solve the time-dependent governing transport equations using the commercial COMSOL software package (version 5.4) employing a finite element technique following the methodology in Section 7.1. For this calculation, evaluate the I-131 activity concentration $A_c(t)$ (μCi kg^{-1}) $= \lambda N_c(t)/(3.7 \times 10^{-4}$ Bq μCi$^{-1})/244{,}000$ kg in the reactor coolant over an irradiation period of 30 days at a constant linear power rating of $P = 40$ kW m^{-1}. For I-131, the following data apply: $Y = 0.0288$ atoms fission^{-1} and $\lambda = 9.98 \times 10^{-7}$ s^{-1}. One can take $\nu = 3 \times 10^{-6}$ s^{-1} as a typical escape rate coefficient for iodine with a reactor coolant purification constant of $\beta = 5 \times 10^{-5}$ s^{-1}.

7.8 Find the weak form of $\nabla \cdot \nabla u + 2u + exp(x) = 0$.

7.9 Starting with the weak form from Problem 7.8, use the Galerkin method to find the contribution to the element load vector from each shape function for a one-dimensional linear Lagrange finite element on the range x_1 to x_2.

7.10 Evaluate the contributions to the load vector from Problem 7.9 for a mesh consisting of one-dimensional linear Lagrange finite elements with four elements with nodes positioned at $x = 0$, 1, 1.5, 1.75, and 2.

7.11 Assemble the global load vector from Problem 7.10.

7.12 Starting from the weak form from Problem 7.8, use the Galerkin method to find the contribution to the element stiffness matrix from each shape function for a one-dimensional linear Lagrange finite element on the range x_1 to x_2.

7.13 Evaluate the contributions to the stiffness matrix from Problem 7.12 for the mesh described in Problem 7.10.

7.14 Assemble the global stiffness matrix from Problem 7.13.
7.15 Solve the linear system $Ku = F$ for the unknown nodal values of u using K from Problem 7.14 and F from Problem 7.11.
7.16 Graph the approximate solution from Problem 7.15 for $u(x)$ over the range $x = 0$ to 2. Also graph $\frac{du}{dx}$ over the range $x = 0$ to 2.

7.17 Repeat the steps in Problems 7.8 to 7.16 for the same mesh using second-order Lagrange elements with the internal node placed at the center of each element. Compare with the approximate solution obtained using linear finite elements.

7.18 Solve the differential equation in Problem 7.8 with the COMSOL software package with a fine mesh and compare to the results obtained in Problems 7.16 and 7.17.

Treatment of experimental results

Every physical measurement is subject to a degree of uncertainty. *"An engineer or scientist can ill afford to waste time in the indiscriminate pursuit of the ultimate in accuracy when such is not needed."*

8.1 Definitions

A measurement is normally taken several times. The "best" value of a measurement set may be determined from either the *mean (average, $\bar{x}$)* or the *median*. The mean is obtained by dividing the sum of a set of replicate measurements by the number of individual results in that set. The median of a set is the result about which all others are equally distributed (for an odd number of measurements, it is the middle number of the set when all values are arranged in order of increasing magnitude; for an even number of measurements, it is the average of the central pair).

Example 8.1.1. Calculate the mean and median for the measurement set 10.06, 10.20, 10.08, 10.10.

Solution:
The mean $= \bar{x} = (10.06 + 10.20 + 10.08 + 10.10)/4 = 10.11$, and the median $= (10.08 + 10.10)/2 = 10.09$ (since the set has an even number of measurements).

The *precision* is used to describe the reproducibility of results and resolution of a measurement when measured again with the same instrument. The *accuracy* of a measurement denotes its nearness to a true or accepted value. A measurement that is highly accurate has an associated small error. In a number of circumstances, the true value for an experiment is not necessarily known so that only an estimated accuracy can be realized. Although very precise equipment can consistently measure small differences in a physical measurement, it may still be inaccurate as a high-precision measurement is not necessarily a sufficient condition for high accuracy.

The deviation from the mean in absolute terms, $|x_i - \bar{x}|$, is a common method for describing precision. The *spread* or *range* ($w = x_{max} - x_{min}$) in a set of data is the numerical difference between the highest and lowest results. Sometimes, it is worthwhile to indicate the precision relative to the mean in terms of a percentage

Advanced Mathematics for Engineering Students. https://doi.org/10.1016/B978-0-12-823681-9.00016-2

$\left(\dfrac{|x_i - \bar{x}|}{\bar{x}}\right) \times 100$. The most important measures of precision are the *standard deviation* and the *variance* (see below). The accuracy of a measurement is often described in terms of an *absolute error*, $E = x_i - x_t$, defined as the difference between the observed value x_i and the true value x_t. The *relative error* can also be expressed as a percentage, $\left(\dfrac{x_i - x_t}{x_t}\right) \times 100\%$, where, in this case, the sign of the error is retained to indicate whether the result is high or low.

Experimental uncertainties that can be revealed by repeating measurements are called *random errors*. Errors that cannot be revealed in this way are called *systematic errors*. Thus, uncertainties that arise in a given measurement can be classified into two categories: (i) *determinate errors* (*systematic errors*) and (ii) *indeterminate errors* (random errors). Determinate errors arise from: (i) an instrument bias due to a calibration error or uncompensated drift for example (most of these are unidirectional); (ii) method errors (e.g., analysis based on nonideal chemical or physical behavior); and (iii) personal errors (for example, judgment bias in reading a scale, gross arithmetical mistakes). Instrumental errors are usually found and corrected by calibration. Indeterminate errors arise from an inability to discriminate between readings differing by less than a small amount. This type of error cannot be controlled by an experimenter and eliminated, producing a random scatter of results for replicate measurements. As an example, Table 8.1 shows all possible ways that four uncertainties

Table 8.1 Possible ways four equal-sized uncertainties can combine.

Combination of uncertainties	Indeterminate error	Frequency
$+U_1 + U_2 + U_3 + U_4$	$+4U$	1
$-U_1 + U_2 + U_3 + U_4$		
$+U_1 - U_2 + U_3 + U_4$	$+2U$	4
$+U_1 + U_2 - U_3 + U_4$		
$+U_1 + U_2 + U_3 - U_4$		
$-U_1 - U_2 + U_3 + U_4$		
$+U_1 + U_2 - U_3 - U_4$		
$+U_1 - U_2 + U_3 - U_4$	0	6
$-U_1 + U_2 - U_3 + U_4$		
$-U_1 + U_2 + U_3 - U_4$		
$+U_1 - U_2 - U_3 + U_4$		
$+U_1 - U_2 - U_3 - U_4$		
$-U_1 + U_2 - U_3 - U_4$	$-2U$	4
$-U_1 - U_2 + U_3 - U_4$		
$-U_1 - U_2 - U_3 + U_4$		
$-U_1 - U_2 - U_3 - U_4$	$-4U$	1

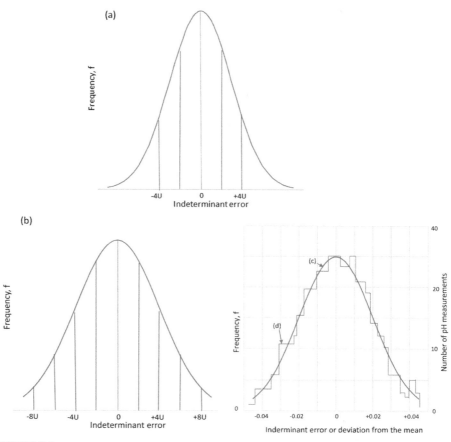

FIGURE 8.1

Theoretical distribution of indeterminate error arising from (a) four, (b) eight, and (c) a large number of uncertainties. Curve (c) shows the theoretical Gaussian distribution and (d) is an experimental distribution curve (adapted from Skoog and West, 1982, p. 50).

(of equal magnitude) can combine to provide an indeterminate error. The theoretical frequency distribution of errors is shown in Fig. 8.1(a). Similarly, Fig. 8.1(b) shows the theoretical distribution for 8 equal-sized uncertainties. Extending this to a very large number of uncertainties of smaller and smaller size, the continuous "bell-shaped" distribution curve of Fig. 8.1(c) results, called a *Gaussian* or *normal error curve*. Indeterminate errors in chemical analysis (e.g., deviations of the mean of a large number of repetitive pH measurements on a single sample) most commonly have a discontinuous distribution that approaches a Gaussian curve (see Fig. 8.1(d)). The Gaussian distribution represents the theoretical distribution of experimental results to be expected as the number of analyses involved approaches infinity. The

Gaussian distribution of most analytical data permits the use of statistical techniques to estimate the limits of indeterminate error from the precision of such data.

8.1.1 Standard normal distribution

The Gaussian distribution or normal distribution function is a continuous probability distribution for a real-valued random variable. This distribution function is symmetric about the mean, which shows that data near the mean are more frequent in occurrence than data away from it. Since it is a probability, the area under the curve is unity. It can be described mathematically as (see Fig. 8.1(a))

$$f(x) = \frac{e^{\frac{-(x-\mu)^2}{2\sigma^2}}}{\sigma\sqrt{2\pi}}, \tag{8.1}$$

where mathematically μ is the center of the distribution and σ is the width of the distribution. Specifically, the random variable x represents values of individual measurements, which are centered about the arithmetic mean μ for an infinite number of such measurements. The parameter f is the frequency of occurrence for each value of the deviation from the mean $(x - \mu)$, and σ, which is called the *standard deviation*, characterizes the width of the curve. A property of f includes its normalization, $\int_{-\infty}^{\infty} f(x)\,dx = 1$. Also, by definition from calculus theory, the expected values $\bar{x} = \dfrac{\int_{-\infty}^{\infty} xf(x)\,dx}{\int_{-\infty}^{\infty} f(x)\,dx} = \mu$ and $s^2 = \dfrac{\int_{-\infty}^{\infty} (x - \bar{x})^2 f(x)\,dx}{\int_{-\infty}^{\infty} f(x)\,dx} = \sigma^2$, where μ and σ are the theoretical mean and standard deviation, respectively, obtained after many measurements. A new random variable,

$$z = \frac{x - \mu}{\sigma}, \tag{8.2}$$

can be introduced to transform the entire family of normal curves into just one curve called "*the standard normal distribution*" for all values of σ (see Fig. 8.2). Thus, the unit for z in Fig. 8.2 is $\pm\sigma$.

The normal error curve permits an estimate of the probable magnitude of the indeterminate error in a given measurement provided that the standard deviation is known. For instance, the probability that a measurement will fall within t standard deviations is P (with $t\sigma$) $= \int_{\mu-t\sigma}^{\mu+t\sigma} f(x)\,dx = \dfrac{1}{\sqrt{2\pi}} \int_{-t}^{t} e^{-z^2/2}\,dz = normal\ error$ *integral*. For the second integral, the transformation in Eq. (8.2) has been used. In particular, P (with $t\sigma$) $= 68.3\%$, that is, the probability that a measurement will fall within one standard deviation of the true answer is 68.3%. Thus, regardless of the standard deviation, the normal curve in Eq. (8.1) and Fig. 8.2 has the property that 68.3% of the area beneath the curve lies within one standard deviation ($\pm1\sigma$) of the

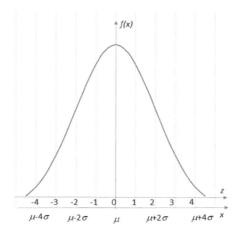

FIGURE 8.2

Normal error curve and standard normal error curve obtained with the transformation in Eq. (8.2).

mean, μ, (in other words, 68.3% of the values lie within these boundaries). Likewise, 95.5% of all values lie within $\pm 2\sigma$, and 99.7% of the values lie within $\pm 3\sigma$. Thus, if σ is available, the chances are 68.3 out of 100 that the indeterminate error associated with any given measurement is smaller than $\pm 1\sigma$, and 95.5 out of 100 that the error is less than $\pm 2\sigma$, and so forth.

For a very large set of data consisting of N measurements, the *standard deviation*, σ, is given in terms of the squares of the individual deviations from the mean:

$$\sigma = \sqrt{\frac{\sum\limits_{i=1}^{N}(x_i - \mu)^2}{N}}. \tag{8.3}$$

The *variance* is equal to σ^2. Eqs. (8.1) and (8.3) are not valid for a small number of replicate measurements (2 to 20) because the mean of an infinitely large number of measurements, μ (that is, the true value of a *population*) is never known and, instead, one is forced to approximate μ by the average of a small number of measurements in the *sample*, $\bar{x}$. With a small data set, not only is $\bar{x}$ likely to differ from μ, but equally important the estimate of the standard deviation. This negative bias for the small data set can be eliminated by substituting the *number of degrees of freedom* ($N - 1$) for N in Eq. (8.3), so that the approximation, s, of the true value of the standard deviation is given by

$$s = \sqrt{\frac{\sum_{i=1}^{N}(x_i - \bar{x})^2}{N - 1}}. \qquad (8.4)$$

Since μ is unknown, the mean must be calculated from the set of replicate data which removes one degree of freedom in Eq. (8.4). For 20 or more data, the difference between Eq. (8.3) and Eq. (8.4) is generally negligible.

Example 8.1.2. Calculate the standard deviation for the data set in Example 8.1.1.

Solution: With $\bar{x} = 10.11$ from the calculation in Example 8.1.1, the standard deviation from Eq. (8.4) (that is, since $N = 4 < 20$) is evaluated as

$$s = \sqrt{\frac{\sum_{i=1}^{4}(x_i - \bar{x})^2}{4 - 1}}$$

$$= \sqrt{\frac{(10.06 - 10.11)^2 + (10.08 - 10.11)^2 + (10.10 - 10.11)^2 + (10.20 - 10.11)^2}{3}}$$

$$= 0.06 = \pm 0.06.$$

Note that the data are not rounded until the end of the calculation (if the data were rounded too early, many of the differences from $\bar{x}$ would be zero).

For analyses that are time consuming, data from a series of samples can often be pooled to provide an estimate of s which is superior to the value of any given subset. In this calculation, the number of degrees of freedom for the pooled s is equal to the total number of measurements minus the number of subsets (see Example 8.1.3).

Example 8.1.3. The mercury concentration in samples of seven fish was determined by the method of absorption of radiation by elemental mercury (see Table 8.2). Calculate a standard deviation for this method, based upon the pooled precision data.

Solution:
The values in columns 4 and 5 of Table 8.2 were calculated as follows:

x_i	$|(x_i - \bar{x}_1)|$	$(x_i - \bar{x}_1)^2$
1.80	0.127	0.0161
1.58	0.093	0.0086
1.64	0.033	0.0011

$$3 | \underline{5.02} \qquad\qquad SS = 0.0258$$

$$\bar{x}_1 = 1.673 = 1.67$$

Table 8.2 Pooled sample data analysis.

Sample No.	No. of meas.	Results Hg content (ppm)	Mean ppm	SS of deviation from mean
1	3	1.80, 1.58, 1.64	1.67	0.0258
2	4	0.96, 0.98, 1.02, 1.10	1.02	0.0116
3	2	3.13, 3.35	3.24	0.0242
4	6	2.06, 1.93, 2.12, 2.16, 1.89, 1.95	2.02	0.0611
5	4	0.57, 0.58, 0.64, 0.49	0.57	0.0114
6	5	2.35, 2.44, 2.70, 2.48, 2.44	2.48	0.0685
7	4	1.11, 1.15, 1.22, 1.04	1.13	0.0170
Total = 28			Total sum of squares (SS) = 0.2196	

The other data in column 5 of Table 8.2 were obtained similarly. Then $s = \sqrt{\dfrac{0.2196}{28 - 7}} = 0.10$ ppm Hg. Because the number of degrees is greater than 20, this estimate of s is a good approximation of σ.

8.2 **Uses of statistics**

Some applications of statistics which are used by experimentalists concerning the effects of indeterminate error include:

(i) the confidence interval around the mean within which the true mean can be expected to be found (with a certain probability),

(ii) the number of times a measurement should be repeated in order for the experimental mean to be included (with a certain probability) within a predetermined interval around the true mean,

(iii) guidance concerning whether or not an outlying value in a set of replicated results should be retained or rejected,

(iv) estimation of the probability that two samples analyzed by the same method are significantly different in composition,

(v) estimation of the probability that a difference in precision exists between two methods of analysis.

8.2.1 **Confidence interval**

The area under the normal error curve encompassed by $z = \pm 1.96\sigma$ corresponds to 95% of the total area. In this case, the confidence level is 95%, so that 95 times out of 100 the calculated value of $(x - \mu)$ for a large number of measurements will be equal to or less than $\pm 1.96\sigma$ (see Table 8.3). Thus, using Eq. (8.2) and solving for μ:

$$\text{confidence limit for } \mu = x \pm z\sigma. \tag{8.5}$$

Table 8.3 Confidence limits for various values of z.

Confidence level	z	Confidence level	z
50	0.67	96	2.00
68	1.00	99	2.58
80	1.29	99.7	3.00
90	1.64	99.9	3.29
95	1.96		

Example 8.2.1. Calculate the 50% and 95% confidence limits for the first entry (1.80 ppm Hg) in Example 8.1.3.

Solution:

In Example 8.1.3, $s = 0.10$ ppm Hg and there were sufficient data to assume $s \to \sigma$. From Table 8.3, $z = \pm 0.67$ and ± 1.96 for the confidence levels of 50 and 95%, respectively. Thus, from Eq. (8.5):

- the 50% confidence limit for $\mu = 1.80 \pm 0.67 \times 0.10 = 1.80 \pm 0.07$ ppm Hg;
- the 95% confidence limit for $\mu = 1.80 \pm 1.96 \times 0.10 = 1.80 \pm 0.20$ ppm Hg.

For example, the second result reveals that the chances are 95 in 100 that μ, the true mean (in the absence of determinate error) will be in the interval between 1.60 and 2.00 ppm Hg.

Eq. (8.5) applies to the result of a single measurement. As previously discussed, σ characterizes the uncertainty of the separate measurements $x_1, ...x_N$. However, the answer $x_{best} = \bar{x}$, where

$$\bar{x} = \frac{x_1 + x_2 ... + x_N}{N}, \tag{8.6}$$

represents a judicious combination of all measurements, which is more reliable than any one of the measurements considered separately. In fact, if one were to make many determinations of the average of N measurements, these would be normally distributed about the true value μ. The width of the distribution for $\bar{x}$ is called the *standard error* or *standard deviation of the mean*, $\sigma_{\bar{x}}$. This quantity follows from the general formula in (see Section 8.4) for the propagation of random errors:

$$\sigma_{\bar{x}} = \sqrt{\left(\frac{\partial \bar{x}}{\partial x_1} \sigma_{x_1} \right)^2 + ... + \left(\frac{\partial \bar{x}}{\partial x_N} \sigma_{x_N} \right)^2}. \tag{8.7}$$

Since all $x_1, ..., x_N$ are measurements of the same quantity, their widths are all equal, that is, $\sigma_{x_1} = ... = \sigma_{x_N} = \sigma$. Also from Eq. (8.6), the partial derivatives are the same:

$$\frac{\partial \bar{x}}{\partial x_1} = \ldots = \frac{\partial \bar{x}}{\partial x_N} = \frac{1}{N}.$$ Therefore, using these results, Eq. (8.7) yields

$$\sigma_{\bar{x}} = \sqrt{\left(\frac{1}{N}\sigma\right)^2 + \ldots + \left(\frac{1}{N}\sigma\right)^2} = \sqrt{N\left(\frac{\sigma}{N}\right)^2} = \frac{\sigma}{\sqrt{N}}. \quad (8.8)$$

Thus, for the average of N replicate measurements, the more general form of Eq. (8.5) is

$$\text{confidence limit for } \mu = \bar{x} \pm \frac{z\sigma}{\sqrt{N}}. \quad (8.9)$$

Example 8.2.2. Calculate the 50% and 95% confidence limits for the mean value (1.67 ppm Hg) for sample 1 in Example 8.1.3.

Solution:
Again, $s \to \sigma = 0.10$ ppm Hg. For the three measurements using Eq. (8.9):

- the 50% confidence limit for $\mu = 1.67 \pm \dfrac{0.67 \times 0.10}{\sqrt{3}} = 1.67 \pm 0.04$ ppm Hg;

- the 95% confidence limit for $\mu = 1.67 \pm \dfrac{1.96 \times 0.10}{\sqrt{3}} = 1.67 \pm 0.11$ ppm Hg.

Example 8.2.3. Based on the previous calibration of a 10-mL pipette in Table 8.4, calculate the number of replicate measurements needed to decrease the 95% confidence interval for the calibration of a 10-mL pipette to 0.005 mL, assuming a similar procedure has been followed.

Table 8.4 Replicate measurements from the calibration of a 10-mL pipette.

Trial	Volume of water delivered (mL)	Trial	Volume of water delivered (mL)	Trial	Volume of water delivered (mL)
1	9.990	9	9.988	17	9.977
2	9.993[a]	10	9.976	18	9.982
3	9.973	11	9.981	19	9.974
4	9.980	12	9.974	20	9.985
5	9.982	13	9.970[b]	21	9.987
6	9.988	14	9.989	22	9.982
7	9.985	15	9.981	23	9.979
8	9.970[b]	16	9.985	24	9.988

Mean volume = 9.9816 = 9.982 mL; median volume = 9.982 mL.
Average deviation from mean = 0.0052 mL = 0.005 mL.
Spread = 9.993 – 9.970 = 0.23 mL; standard deviation = 0.0065 mL = 0.006 mL.
[a] *Maximum value.*
[b] *Minimum value.*

Solution:
The standard deviation for the measurement is 0.0065 mL (Table 8.5). Because s is based on 24 values, one may assume $s \rightarrow \sigma = 0.0065$. Using Eq. (8.9), the confidence interval is given by

$$\text{confidence interval} = \pm \frac{z\sigma}{\sqrt{N}}. \tag{8.10}$$

Therefore, $0.005 \text{ mL} = 1.96 \times \dfrac{0.0065}{\sqrt{N}}$, which yields $N = 6.5$. Thus, by employing the mean of seven measurements, there would be a somewhat better than 95% chance of knowing the true mean volume delivered by the pipette to ± 0.005 mL.

Table 8.5 Standard deviation calculations.

Entry	Trial number (Table 8.4)	Mean (mL)	Standard deviation (mL) (σ) (Eq. (8.3))	Standard deviation (mL) (s) (Eq. (8.4))
1	1–24	9.982	0.0063	0.0065
2	1–12	9.982	0.0070	0.0074
	13–24	9.982	0.0056	0.0058
	mean	9.982	0.0063	0.0066
	spread	0.000	0.0014	0.0016
3	1–3	9.985[a]	0.0088[a]	0.0108[a]
	4–6	9.983	0.0034	0.0042
	7–9	9.981	0.0079	0.0096
	10–12	9.977[b]	0.0029[b]	0.0036[b]
	13–15	9.980	0.0078	0.0095
	16–18	9.981	0.0033	0.0041
	19–21	9.982	0.0057	0.0070
	22–24	9.983	0.0037	0.0046
	mean	9.982	0.0054	0.0067
	spread	0.008	0.0059	0.0072

[a] Maximum value.
[b] Minimum value.

Limitations in the amount of time or the number of available samples may preclude an accurate estimate of σ. Thus, it may be necessary that a single set of replicate measurements must provide not only a mean value but also a precision estimate of s. In this case, the t parameter is used in place of z, where (compare with Eq. (8.2))

$$t = \frac{x - \mu}{s}. \tag{8.11}$$

In general, the t distribution (or *Student's t distribution*) is flatter than the normal probability distribution, causing more area to be contained in the tails of the distribution (see Fig. 8.3).

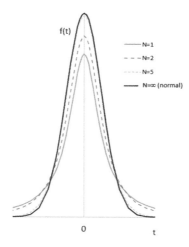

f(t)

——— N=1

– – – N=2

· · · · · N=5

——— N=∞ (normal)

0   t

FIGURE 8.3

Comparison of the Student t distribution with the normal distribution.

The parameter t is dependent not only on the desired confidence level but also on the number of degrees of freedom. As the number of degrees of freedom becomes infinite (more specifically, ≥ 30), the values of t in Table 8.6 approach the values of z in Table 8.3. The t distribution yields a larger confidence interval in order to account for the greater variability in s. The *"central limit theorem"* guarantees the sample means distribution will always be normally distributed if $N \geq 30$ regardless of the *population* distribution; however, care must be taken in applying the t distribution to small sample sizes, where it must be ensured that the measured data are normally distributed. Thus, the confidence limit analogous to Eq. (8.9) is given by

$$\text{confidence limit for } \mu = \bar{x} \pm \frac{ts}{\sqrt{N}}. \qquad (8.12)$$

Example 8.2.4. A chemist obtained the following data for the alcohol content in a sample of blood: ethanol concentration $= 0.084\%$, 0.089%, and 0.079%. Calculate the 95% confidence limit for the mean assuming (a) no additional data about the precision of the method and (b) that on the basis of previous experience $s \rightarrow \sigma = 0.006\%$ ethanol.

Solution:

Table 8.6 Values of t for various levels of probability.

Degrees of freedom	Factor for confidence interval (%)				
	80	90	95	99	99.9
1	3.08	6.31	12.7	63.7	637
2	1.89	2.92	4.30	9.92	31.6
3	1.64	2.35	3.18	5.84	12.9
4	1.53	2.13	2.78	4.60	8.60
5	1.48	2.02	2.57	4.03	6.86
6	1.44	1.94	2.45	3.71	5.96
7	1.42	1.90	2.36	3.50	5.40
8	1.40	1.86	2.31	3.36	5.04
9	1.38	1.83	2.26	3.25	4.78
10	1.37	1.81	2.23	3.17	4.59
11	1.36	1.80	2.20	3.11	4.44
12	1.36	1.78	2.18	3.06	4.32
13	1.35	1.77	2.16	3.01	4.22
14	1.34	1.76	2.14	2.98	4.14
∞	1.29	1.64	1.96	2.58	3.29

(a) We have $\bar{x} = \dfrac{(0.084 + 0.089 + 0.079)}{3} = 0.084$ and

$$s = \sqrt{\frac{(0.000)^2 + (0.005)^2 + (0.005)^2}{3 - 1}} = 0.005.$$

Table 8.6 indicates that $t = \pm 4.30$ for two degrees of freedom at a 95% confidence. Hence, the 95% confidence limit is $0.084 \pm 4.3 \times 0.005/\sqrt{3} = 0.084 \pm 0.012$. The confidence limit at 95% can also be easily calculated with the Excel software package (Microsoft Office 365, build version 2018). Here one enters the three data points into a single column A (that is, A1:A3). By selecting "Formulas" in the top menu of Excel → "More Functions" → "Statistical" → "STDEV.S" and highlighting the three numbers in the **Number 1** text box (A1:A3), a standard deviation value of 0.005 is similarly calculated. Subsequently, choosing the function "CONFIDENCE.T" and entering in the three pop-up text boxes (i) a value of **Alpha** $= 1 - 0.95 = 0.05$ (for a 95% confidence limit), (ii) the previous **Standard_dev** $= 0.005$ and a **Size** $= 3$, a confidence interval for a population mean using a Student's t distribution of 0.01242 is evaluated, identical to the previous hand calculation. Finally, using the function "AVERAGE," a value of 0.084 is calculated for the three numbers. This analysis thereby yields the same previous result for the 95% confidence limit of $0.084 \pm 0.012\%$.

(b) Because a good value of σ is available, the 95% confidence limit $= 0.084 \pm$
$\dfrac{z\sigma}{\sqrt{N}} = 0.084 \pm \dfrac{1.96 \times 0.0060}{\sqrt{3}} = 0.084 \pm 0.007\%$. Note that a more accurate knowledge of σ decreased the confidence interval by almost half.

8.2.2 Rejection of data

When an outlying result occurs in a data set, a decision must be made whether to reject it. In this case, the Q test can be used (although a cautious approach to the rejection of data should be adopted). Here the difference between the questionable result and its *next* nearest neighbor is divided by the spread of the entire data set; the resulting ratio is then compared to a critical value for a particular degree of confidence (see Table 8.7) (Dixon, 1951).

Table 8.7 Critical values for rejection values for Q.

Number of observations	Q_{crit} (reject if $Q_{exp} > Q_{crit}$)		
	90% Confidence	96% Confidence	99% Confidence
3	0.94	0.98	0.99
4	0.76	0.85	0.93
5	0.64	0.73	0.82
6	0.56	0.64	0.74
7	0.51	0.59	0.68
8	0.47	0.54	0.63
9	0.44	0.51	0.60
10	0.41	0.48	0.57

Example 8.2.5. The analysis of a calcite sample yielded CaO contents of 55.95%, 56.00%, 56.04%, 56.08%, and 56.23%. Should the last value that appears anomalous be retained?

Solution:
The difference between 56.23 and 56.08 is 0.15. The spread is $(56.23 - 55.95) = 0.28$. Thus, $Q_{exp} = 0.15/0.28 = 0.54$. For five measurements, Q_{crit} is 0.64 (at 90% confidence). Because $0.54 < 0.64$, the last value should be retained.

8.2.3 Hypothesis testing for experimental design

Hypothesis testing can be used in scientific and engineering analysis processes as a basis for experimental testing. In this approach, a null hypothesis is employed, which assumes that the numerical quantities being compared are the same. The probabilities of the observed differences appearing as a result of indeterminate error are then computed from statistical theory. Usually, if the observed difference is as large as or larger than a difference, which would occur at a specific probability level (for example, 5 times out of 100 for a 5% probability level), the null hypothesis is rejected and the difference is judged to be significant.

Comparison of an experimental mean with a true value

A judgment may have to be made whether the difference between an experimental mean and a true value μ is the consequence of indeterminate (random) error in the measurement or of the presence of determinate (systematic) error in the method. The statistical treatment for this type of problem involves comparing the difference $(\bar{x} - \mu)$. If this value is larger than the critical value expected for indeterminate error, it may be assumed that the difference is real (that is, rejection of the null hypothesis) and that a determinate error exists. The critical value for rejection of the null hypothesis can be obtained from Eq. (8.12) (or equivalently from Eq. (8.9) if a good estimate of σ is available) such that

$$\bar{x} - \mu = \pm\frac{ts}{\sqrt{N}}, \tag{8.13}$$

where N is the number of replicate measurements employed in the test.

Example 8.2.6. A new procedure for the rapid analysis of sulfur in kerosenes was tested by the analysis of a sample which was known from its method of preparation to contain 0.123% S. The results were as follows: sulfur content = 0.112%, 0.118%, 0.115%, and 0.119%. Do the data indicate the presence of a negative determinate error in the new method?

Solution:

$$\bar{x} = \frac{(0.112 + 0.118 + 0.115 + 0.119)}{4} = 0.116 \quad \text{and}$$

$$s = \sqrt{\frac{(0.004)^2 + (0.002)^2 + (0.001)^2 + (0.003)^2}{4 - 1}} = 0.0033.$$

Also $\bar{x} - \mu = 0.116 - 0.123 = -0.007$. Table 8.6 indicates that $t = 3.18$ for three degrees of freedom and 95% confidence. Hence, $\dfrac{ts}{\sqrt{N}} = \dfrac{3.18 \times 0.0033}{\sqrt{4}} = \pm0.0052$, as compared to $\bar{x} - \mu = -0.007$. This analysis indicates that five times out of 100, an experimental mean can be expected to deviate by ±0.0052 or more. Thus, if one concludes that -0.007 is a significant difference and that a determinate error is present, one will, on average, be right 95 times out of 100 judgments (or equivalently, wrong five times out of 100). On the other hand, using a 99% confidence level (with $t = 5.84$), $\dfrac{ts}{\sqrt{N}} = \pm0.0096$. Thus, if one insists on being wrong no more than one time out of 100, one would have to say that no difference has been demonstrated between the original and new method (note this statement is different from saying that no determinate error exists).

Comparison of two experimental means

A chemical engineer frequently employs analytical data to establish whether two materials are identical. If N_1 replicate analyses are made on material 1, yielding an experimental mean $\bar{x}_1$, and N_2 analyses on material 2, yielding a mean $\bar{x}_2$, Eq. (8.13) gives

$$\mu_1 = \bar{x}_1 \pm \frac{ts}{\sqrt{N_1}} \text{ and } \mu_2 = \bar{x}_2 \pm \frac{ts}{\sqrt{N_2}} \Rightarrow \bar{x}_1 - \bar{x}_2 = \pm ts\sqrt{\frac{N_1 + N_2}{N_1 N_2}}. \quad (8.14)$$

Equivalently, in accordance with Eq. (8.9), t and s can be replaced by z and σ if a good estimate of σ is available. The last relation follows if one considers the null hypothesis that μ_1 and μ_2 are identical. The number of degrees of freedom for finding t in Table 8.6 is defined as $N_1 + N_2 - 2$. If the experimental difference $\bar{x}_1 - \bar{x}_2$ is smaller than the computed value at a given confidence level, the null hypothesis is accepted and no significant difference between the means has been demonstrated.

Example 8.2.7. The composition of a flake of paint found on the clothes of a hit-and-run victim was compared with paint from a car suspected of causing the accident. Do the following data for the spectroscopic analysis of titanium in the paint suggest a difference in composition between the two materials? From previous experience, the standard deviation for the analysis is known to be 0.35% Ti (that is, $s \rightarrow \sigma = 0.35\%$ Ti). The following data were obtained. In paint from the clothes, titanium content $= 4.0\%$ and 4.6%; and in paint from the car, titanium content $= 4.5\%$, 5.3%, 5.5%, 5.0%, and 4.9%.

Solution:
We have $\bar{x}_1 = \dfrac{(4.0 + 4.6)}{2} = 4.3$ and $\bar{x}_2 = \dfrac{(4.5 + 5.3 + 5.5 + 5.0 + 4.9)}{5} = 5.0$. Hence, $\bar{x}_1 - \bar{x}_2 = 4.3 - 5.0 = -0.7\%$ Ti. Using z and σ in Eq. (8.14), with z values from Table 8.3 for 95% and 99% confidence levels, yields

$$\pm z\sigma\sqrt{\frac{N_1 + N_2}{N_1 N_2}} = \begin{cases} \pm 1.96 \times 0.35\sqrt{\dfrac{2+5}{2\times 5}} = \pm 0.57 \text{ for 95\% confidence,} \\ \\ \pm 2.58 \times 0.35\sqrt{\dfrac{2+5}{2\times 5}} = \pm 0.76 \text{ for 99\% confidence.} \end{cases}$$

Thus, 5 out of 100 data should differ by 0.57% or greater and only one out of 100 should differ by as much as 0.76% Ti. Thus it seems reasonably probable (between 95% and 99% certain) that the observed difference of -0.7% does not arise from indeterminate error but in fact is caused by a real difference between the two paint samples. Hence, the suspected vehicle was probably not involved in the accident.

Comparison of precision measurements

An F test may be used to determine whether (i) Method 1 is more precise than Method 2 or (ii) there is a difference in the precision of two methods. The null hypothesis for the F test assumes that the precisions are identical. In this test, the ratio of the variances of the two measurements is compared with a critical value of F given in Table 8.8; if the experimental value exceeds the critical value, the null hypothesis is rejected. For procedure (i), the variance of the supposedly more precise procedure is always placed in the denominator; for procedure (ii), the larger variance always appears in the numerator.

Table 8.8 Critical F values at the 5% level.

Degrees of freedom (denominator)	Degrees of freedom (numerator)							
	2	3	4	5	6	12	20	∞
2	19.00	19.16	19.25	19.30	19.33	19.41	19.45	19.50
3	9.55	9.28	9.12	9.01	8.94	8.74	8.66	8.53
4	6.94	6.59	6.39	6.26	6.16	5.91	5.80	5.63
5	5.79	5.41	5.19	5.05	4.95	4.68	4.56	4.36
6	5.14	4.76	4.53	4.39	4.28	4.00	3.87	3.67
12	3.89	3.49	3.26	3.11	3.00	2.69	2.54	2.30
20	3.49	3.10	2.87	2.71	2.60	2.28	2.12	1.84
∞	3.00	2.60	2.37	2.21	2.10	1.75	1.57	1.00

Example 8.2.8. A standard method for the determination of carbon monoxide levels in gaseous mixtures is known from many hundreds of measurements to have a standard deviation of 0.21 ppm CO. A modification of the method has yielded an s of 0.15 ppm CO for a pooled set of data with 12 degrees of freedom. A second modification, also based on 12 degrees of freedom, has a standard deviation of 0.12 ppm CO. (a) Is either of the modifications significantly more precise than the original? (b) Is the precision of the second method significantly better than the first?

Solution:

(a) Because an improvement is claimed, the variances of the modifications are placed in the denominator. Method 1 gives $F_1 = \dfrac{s_{std}^2}{s_1^2} = \dfrac{(0.21)^2}{(0.15)^2} = 1.96$, and

Method 2 gives $F_2 = \dfrac{s_{std}^2}{s_2^2} = \dfrac{(0.21)^2}{(0.12)^2} = 3.06$. For the standard procedure, $s \rightarrow \sigma$, and the number of degrees of freedom for the numerator can be taken as infinite, yielding a critical F value of 2.30 from Table 8.8. Thus, the F value for the first method is less than this critical value, and the null hypothesis is accepted at the 95% probability level (that is, Method 1 has the same precision as the original). However, the second method does appear to have a significantly greater precision.

(b) On the other hand, the ratio $F_1 = \dfrac{s_1^2}{s_2^2} = \dfrac{(0.15)^2}{(0.12)^2} = 1.56$ is lower than the critical value of 2.69 (for 12 degrees of freedom in the numerator and denominator of Table 8.8), indicating no significant difference between the two methods.

Analysis of variance (ANOVA)

In the previous section, two populations were compared for an estimation of the precision between two methods with an F statistical test. For populations greater than two, an *analysis of variance (ANOVA)* can be used to examine differences among group means. This test provides a way to assess if each mean likely came from a larger overall population. The ANOVA test evaluates a *variance ratio*. As depicted in Fig. 8.4, this ratio considers the distances from the sampled distribution means to the "overall" mean (that is, the variance *between*) relative to the internal spread around each mean (the variance *within*). The ratio of these two quantities is then compared to a critical F statistic (see Table 8.8). This statistic allows a test of the null hypothesis assuming the means (μ) are from a larger population where, for example, $H_0 : \mu_1 = \mu_2 = \mu_3$. If the ratio of the variances for the *between* to *within* values is significantly greater than F_{crit}, then one rejects the null hypothesis H_0. As such, this would indicate that at least one mean is an outlier and each distribution is narrow and distinct from one another. On the other hand, if this ratio is much less than F_{crit}, then one fails to reject H_0, indicating that the means are very close to the overall mean and/or the distributions overlap.

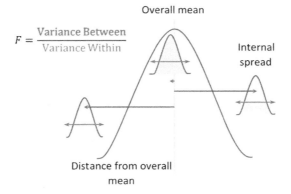

FIGURE 8.4

Schematic of an ANOVA analysis (adapted from Foltz, 2020).

Following Foltz (2020), a hand calculation for data analysis of three sample methods is shown in Example 8.2.9. This analysis is easily duplicated using the ANOVA statistical function in Excel, as demonstrated in Example 8.2.10.

Example 8.2.9. Consider three methods developed for a chemical analysis. Seven observations are made for each method with the following concentrations (in ppm):

(i) Method 1: 8.2, 9.3, 6.1, 7.4, 6.9, 7.0, 5.3; (ii) Method 2: 7.1, 6.2, 8.5, 9.4, 7.8, 6.6, 7.1; and (iii) Method 3: 6.4, 7.3, 8.7, 9.1, 5.6, 7.8, 8.7. Perform an ANOVA analysis to determine if there is any significant difference among the methods.

	A	B	C	D	E	F	G	H
1	Method 1	Method 2	Method 3					PANEL A
2	8.2	7.1	6.4					
3	9.3	6.2	7.3					
4	6.1	8.5	8.7					
5	7.4	9.4	9.1					
6	6.9	7.8	5.6					
7	7.0	6.6	7.8					
8	5.3	7.1	8.7					
9		Measured Concentrations (ppm)						
10	7.17	7.53	7.66	7.45				
11	Mean 1	Mean 2	Mean 3	Overall Mean				
12								
13	Total no. of observations (N):		21					
14	No. of groups/methods (C):		3					
15	No. of observations in each group (n):		7					
16						Ftest$^{(d)}$		
17	Source of Variance	df$^{(a)}$	SS$^{(b)}$	MS$^{(c)}$	F	F$_{crit}$		
18	Between	2	0.887	0.443	0.284	3.55		
19	Within	18	28.1	1.56				
20	Total	20	29.0					
21	(a) Degrees of Freedom (df)		(b) Sum of Squares (SS)		(c) Mean Sum of Squares (MS)		(d) F test	
22	$df_{between} =$	C - 1	SSC=	VAR.S(A10:C10)*(C-1)*n	MSC=	SSC/$df_{between}$	F=	MSC/MSE
23	$df_{within} =$	N - C	SSE=	VAR.S(A2:A8)*(n-1)+ VAR.S(B2:B8)*(n-1)+ VAR.S(C2:C8)*(n-1)	MSE=	SSE/df_{within}	F$_{crit}$=	F.INV.RT(0.05,$df_{between}$,df_{within})
24	$df_{total} =$	$df_{between}+df_{within}$	SST=	SSC+SSE				
25								
26	Anova Excel Analysis: Single Factor							PANEL B
27	SUMMARY							
28	Groups	Count	Sum	Average	Variance			
29	Method 1	7	50.2	7.17	1.73			
30	Method 2	7	52.7	7.53	1.25			
31	Method 3	7	53.6	7.66	1.70			
32	ANOVA							
33	Source of Variation	SS	df	MS	F	P-value	F crit	
34	Between Groups	0.887	2	0.443	0.2837	0.756	3.55	
35	Within Groups	28.1	18	1.56				
36								
37	Total	29.0	20					

FIGURE 8.5

Example calculation of an ANOVA.

Solution:

In panel A of Fig. 8.5, the three separate data sets ($C = 3$) are input into an Excel spreadsheet, where the means are evaluated accordingly. Since each group has the same number of data points ($n = 7$), the "overall" mean can be taken either by averaging the resultant averages for each of the three groups or by averaging all of the observations (with $N = 21$). An ANOVA assessment is performed below the data sets for the two variation sources for the "between" and "within" groups. In this analysis, the number of degrees of freedom (df), the sum of squares (SS), and the mean sum of squares (MS) are evaluated with the formulas given in the footnote entries. Here the "C" and "E" designations (that is, SSC and SSE for the sum of squares and MSC and MSE for the mean sum of squares) refer to a "column (C)" or "error (E)" quantity, respectively, for the two sources of variance groups.

To simplify the calculation for the sum of squares, the Excel variance function (VAR.S) is employed (which is the square of the standard deviation in Eq. (8.4)). To recover the sum of squares, this function must be multiplied by appropriate factors for the degrees of freedom for the *between* and *within* groups. The mean sum of squares (MS) for a given group is simply the sum of squares values (SS) divided by the number of degrees of freedom (df). Finally, the F parameter is a ratio of the sources variances for the *between* and *within* groups for the mean sum of squares (that is, $F = MSC/MSE$). The F value is then tested against a critical value F_{crit}. This latter quantity can be obtained either from Table 8.8 or using the Excel function at a 95% confidence level F.INV.RT(0.05, df for the numerator, df for the denominator). In this example, since $F < F_{crit}$, one fails to reject the null hypothesis, indicating no significant difference in the means for the three methods.

In contrast to a one-way analysis, a two-way ANOVA test can be further performed if subgroups exist in the data sampling. For example, if observations are made at two different temperatures as well in each chemical process, this additional type of test could be employed. An F test simply indicates a significant difference between the groups but it does not tell where the differences lie. One must then perform a post hoc test. Such analyses may include a Bonferroni correction, a Tukey honest significant difference test, or a least significant difference test. The post hoc tests explore the differences between the groups while controlling the experimental error rate.

Example 8.2.10. Perform an analysis for the data in Example 8.2.9 using the Excel ANOVA statistical function.

Solution:
In Excel, select the "Data" tab in the upper ribbon and choose "Data Analysis" at the top right corner of the worksheet. A "Data Analysis" pop-up box appears. In Fig. 8.6(a), select the "Anova: Single Factor" entry for a one-way ANOVA analysis in "Analysis Tools" and click the "OK" button. Another pop-up box as Fig. 8.6(b) appears. Select the entire input range of data including the headings from $A$1:$C$8 (the heading will be reflected accordingly in the Excel output table). Choose "Grouped By: Columns" and check the "Labels in the First Row." Input an "Alpha" value of 0.05 (for a 95% confidence level for the F_{crit} value). Select the cell location for output in the "Output Range" entry and click "OK." The Excel output table in panel B of Fig. 8.5 results. These results match precisely the calculated results in panel A. This latter methodology provides a simple way to perform an analysis with the Excel ANOVA function.

FIGURE 8.6

Sample calculation using the ANOVA Excel function (a) with the Data Analysis dialog for (b) a one-way analysis.

8.3 Regression analysis and software applications

In the derivation of a calibration curve, for example, it may be necessary to derive a "best" straight line through a given set of data points (which only contain random error). This type of regression analyses is handled through scientific software packages such as SigmaPlot and SigmaStat (Systat Software, Inc., 2018), as well as Excel. All packages easily handle linear regression and some nonlinear functions. SigmaPlot also uses a more general model fitting algorithm for nonlinear fittings using a Marquardt–Levenberg algorithm (see Press et al., 1986).

In a *linear regression* analysis, it is assumed that there is a linear relation between the *dependent variable*, y, and the *independent variable*, x:

$$y = a + bx. \tag{8.15}$$

The task at hand is therefore to determine the *intercept*, a, and the *slope*, b, for the best-fit straight line through the scattered measured data. As shown in Fig. 8.7, the *residuals*, which represent the vertical deviations (d) between the experimental data and the least-squares line, must be minimized. This type of fitting is obtained with the following equations:

$$b = \frac{AB}{A^2} \text{ and } a = \bar{y} - b\bar{x}, \tag{8.16}$$

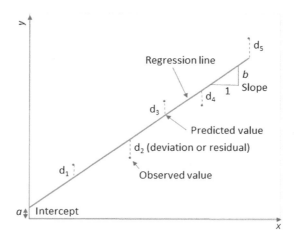

FIGURE 8.7

Linear regression analysis.

where the regression quantities are defined as

$$AB = \sum (x_i - \bar{x})(y_i - \bar{y}) = \sum x_i y_i - \frac{\sum x_i \sum y_i}{n}, \tag{8.17a}$$

$$A^2 = \sum (x_i - \bar{x})^2 = \sum x_i^2 - \frac{\left(\sum x_i\right)^2}{n} = \sum x_i^2 - n\bar{x}, \tag{8.17b}$$

$$B^2 = \sum (y_i - \bar{y})^2 = \sum y_i^2 - \frac{\left(\sum y_i\right)^2}{n} = \sum y_i^2 - n\bar{y}. \tag{8.17c}$$

Here, x_i and y_i are individual pairs of data for the n data points so that for the summations i goes from $i = 1$ to $i = n$. The quantities $\bar{x}$ and $\bar{y}$ are the average values of the variables:

$$\bar{x} = \frac{\sum x_i}{n} \text{ and } \bar{y} = \frac{\sum y_i}{n}. \tag{8.18}$$

The variation of data about the regression line can be quantified in terms of the *standard error of estimate*:

$$s_e = \sqrt{\frac{\sum (y_i - y)^2}{n - 2}} = \sqrt{\frac{B^2 - b^2 A^2}{n - 2}}. \tag{8.19}$$

The quantity $(y_i - y)$ in the first term of Eq. (8.19) is the vertical deviation d_i for point i in Fig. 8.7. The number of degrees of freedom in Eq. (8.19) is two less than the number of points because one degree of freedom is lost in calculating a and one for b. The quantity s_e can be used to predict the confidence intervals for the regression line estimate. In particular, the standard error for the slope b is given by

$$s_b = \sqrt{\left(\frac{s_e}{A}\right)^2} = \frac{s_e}{A}. \tag{8.20}$$

The confidence limit for the slope can be derived using s_b in Eq. (8.20) such that

$$\text{confidence limit for slope} = b \pm t s_b. \tag{8.21}$$

The *prediction interval* for y (at a given value of x^*), using s_e in Eq. (8.19), is given by

$$\text{prediction interval for } y = y \pm t s_e \sqrt{1 + \frac{1}{n} + \frac{(\bar{x} - x^*)^2}{A^2}}. \tag{8.22}$$

Both Eq. (8.21) and Eq. (8.22) are evaluated with a t value for a desired confidence level and $n - 2$ degrees of freedom (Table 8.6).

A further quantity of interest is the *correlation coefficient*, which determines the strength and magnitude of the relationship between the variables x and y. The symbol r is used for the correlation coefficient for the data sample (and ρ for the population). The most popular one is the Pearson product moment correlation coefficient, defined as

$$r = \frac{n \sum x_i y_i - \sum x_i \sum y_i}{\sqrt{\left[n \sum x_i^2 - (\sum x_i)^2\right]\left[n \sum y_i^2 - (\sum y_i)^2\right]}} = \frac{AB}{\sqrt{A^2 B^2}}. \tag{8.23}$$

The range of r is from -1.0 to $+1.0$. A strong negative linear relationship exists when r is near -1, and a strong positive relationship exists when r is close to $+1$. Therefore, the closer $|r|$ (or r^2) is to 1, the more confident one is that a straight line represents the data.

Example 8.3.1. The first two columns of Table 8.9 contain the experimental data for the calibration curve for the peak area (y) versus the concentration of isooctane (in mol%) (x) based on a chromatographic analysis. Perform a linear least squares analysis of the data.

Solution:
Columns 3, 4, and 5 of Table 8.9 contain the computed values for $(x_i)^2$, $(y_i)^2$, and $x_i y_i$ (and their sums). Thus, Eq. (8.17a), Eq. (8.17b), and Eq. (8.17c) yield $AB = 15.81992 - (5.365 \times 12.51 / 5) = 2.39669$; $A^2 = 6.90201 - (5.365)^2 / 5 = 1.14537$; and $B^2 = 36.3775 - (12.51)^2 / 5 = 5.07748$. Substitution of these values into Eq. (8.16) and Eq. (8.18) gives the regression coefficients $b = 2.39669 / 1.14537 = 2.0925 = 2.093$ and $a = (12.51 / 5) - (2.0925 \times 5.365 / 5) = 0.2567 = 0.257$. The equation for the straight fitted line from Eq. (8.15) is therefore $y_{fit} = 0.257 + 2.093x$.

Table 8.9 Calibration data for a chromatographic method.

Mol% isooctane, x_i	Peak area, y_i	$(x_i)^2$	$(y_i)^2$	$x_i y_i$
0.352	1.09	0.12390	1.1881	0.38368
0.803	1.78	0.64481	3.1684	1.42934
1.08	2.60	1.16640	6.7600	2.80800
1.38	3.03	1.90440	9.1809	4.18140
1.75	4.01	3.06250	16.0801	7.01750
5.365	12.51	6.90201	36.3775	15.81992

Using Eq. (8.19) and Eq. (8.20), the standard errors for the estimate and slope are, respectively,

$$s_e = \sqrt{\frac{5.07748 - (2.0925)^2 \times 1.14537}{5 - 2}} = 0.144 \text{ and } s_b = \sqrt{\frac{(0.144)^2}{1.14537}} = 0.135.$$

The confidence limit for the slope follows from Eq. (8.21) and Table 8.6, where for a 95% confidence level and $5 - 2 = 3$ degrees of freedom: 95% confidence limit of slope $= 2.093 \pm (3.18 \times 0.135) = 2.093 \pm 0.429$.

Similarly, using Eq. (8.15) and Eq. (8.22), the prediction interval for y at a 95% confidence level, at the point $x^* = x_1 = 0.352$ (Table 8.9), is

$$(0.2567 + 2.093 \times 0.352) \pm 3.18 \times 0.144 \times \sqrt{1 + \frac{1}{5} + \frac{(5.365 / 5 - 0.352)^2}{1.14537}}$$

$$= 0.993 \pm 0.590.$$

The correlation coefficient is calculated as $r = \dfrac{2.39669}{\sqrt{1.14537 \times 5.07748}} = 0.994$ or ($r^2 = 0.988$). This value of r indicates a strong (positive) linear relationship. These labor-intensive calculations can also be performed easily in Excel (see Fig. 8.8). The 95% confidence intervals for both the population (that is, *prediction interval*) and the *regression line* are shown in Fig. 8.9. These results are obtained with a spreadsheet analysis using the Excel top menu tab "Formulas" → "More Functions" → "Statistical." The two function choices are employed in the drop down menu: "LINEST" (for the regression analysis) and T.INV (for the value of the Student t distribution at a 95% probability, that is, $\alpha = 0.05$).

For the LINEST calculation, one selects the cells for output. The input for the dialog boxes of this function are: y-column values; x-column values; TRUE (indicates a line of the form $y = a + bx$ with a nonzero intercept); and TRUE (to list the estimates). After the dialog boxes are filled, one holds down the three keys "control + shift + enter" and results automatically fill the selected area. The output includes the

⊿	A	B	C	D	E	F
1	Mol%	Peak Area				
2	x	y	y_{fit}			
3	0.352	1.09	0.993	(Calc 1: eq(8.15))		
4	0.803	1.78	1.937			
5	1.08	2.6	2.517			
6	1.38	3.03	3.144			
7	1.75	4.01	3.919			
8	Calc 1: $y_{regression}$ for C3 =$B$11*A3+$C$11					
9						
10	Regression (=LINEST(B3:B7,A3:A7,TRUE,TRUE), then type: cntrl+shift+enter)					
11	b	2.093	0.257	a	t (=TINV(0.05, C14))	3.18
12	s_b	0.135	0.158	s_a	Δb (=$F$11*$B$12)	0.429
13	r^2	0.988	0.144	s_e	Δa (=$F$11*$C$12)	0.504
14	F-statistic	241	3	DoF	F_{crit} (=F.INV.RT(0.05,1,C14))	10.13
15	regression SS	5.015	0.0624	residual SS		
16						
17	Average x	1.073				
18	A^2	1.14537				
19						
20		Confidence Interval regression line				
21		Δy	$y_{fit} + Δy$	$y_{fit} - Δy$		
22	(Calc 2 - see text)	0.371	1.364	0.622		
23		0.236	2.173	1.701		
24		0.205	2.722	2.311		
25		0.244	3.388	2.901		
26		0.356	4.274	3.563		
27	Calc 2: Δy for B22=$F$11*$C$13*SQRT((1/($C$14+2))+(($B$17-A3)^2)/$B$18)					
28						
29		Prediction interval				
30		Δy	$y_{fit} + Δy$	$y_{fit} - Δy$		
31	(Calc 3: eq(8.22))	0.590	1.584	0.403		
32		0.516	2.453	1.421		
33		0.503	3.019	2.014		
34		0.520	3.664	2.625		
35		0.581	4.499	3.338		
36	Calc 3: Dy for B31=$F$11*$C$13*SQRT((1+1/($C$14+2))+(($B$17-A3)^2)/$B$18)					
37						

FIGURE 8.8

Excel spreadsheet analysis for linear regression.

slope (b) and intercept (a) of the fitted regression line and the standard errors of the latter parameters ($Δb$ and $Δa$), as well as the standard error for the estimate (s_e). The number of degrees of freedom, the F-statistic, the regression sum of squares, and the residual sum of squares are further tabulated.

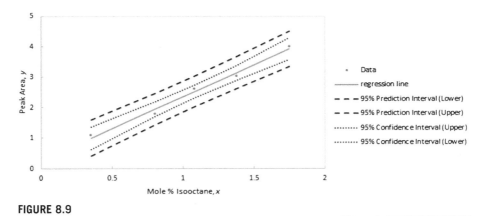

FIGURE 8.9

Graph of calibration curve with confidence and prediction intervals shown.

As mentioned, the r^2 value is generally used as a popular indicator of the goodness of fit. This quantity can be directly estimated from the total sum of squares (SS) and the regression sum of squares (SS):

$$r^2 = \frac{\text{regression SS}}{\text{total SS}} = \frac{\sum\limits_{i}^{n}(y_{i,fit} - \bar{y})^2}{\sum\limits_{i}^{n}(y_i - \bar{y})^2},$$

where $y_{i,fit}$ is the predicted value of y for a given value of x_i using the equation for the fitted regression line. Although a value close to one indicates a good fit, an even better statistical test of the goodness of fit is the Fisher F-statistic. This statistic involves the ratio of the variances (its value is given in Fig. 8.8):

$$F = \frac{\text{variance explained}}{\text{variance unexplained}} = \frac{\text{regression SS} / v_1}{\text{residual SS} / v_2} = \frac{\left(\sum (y_{i,fit} - \bar{y})^2\right) / v_1}{\left(\sum (y_i - y_{i,fit})^2\right) / v_2}.$$

The statistic is used under the null hypothesis that the data are a random scatter of points with a zero slope. The critical value of the F-statistic can also be calculated with Excel. If the F-statistic $> F_{crit}$, the null hypothesis fails and the linear model is significant. For the degrees of freedom $v_1 = 1$ and $v_2 = n - k$, where k is the number of variables in the fitted regression analysis including the intercept and n is the number of data points. The value for v_2 is the number of degrees of freedom. For this example, $v_1 = 1$ and $v_2 = 3$, so that the F_{crit} value is evaluated as 10.13, as shown in Fig. 8.8 using the Excel function "F.INT.RT" (with $\alpha = 0.05$, $v_1 = 1$, and $v_2 =$ number of degrees of freedom). The F-statistic in the example is 241, which is much greater than F_{crit}, so that one is 95% sure that the data are not a random scatter of points and that the linear regression is justified.

The confidence interval for the regression line is calculated with Eq. (8.22) but neglecting the factor of unity in the square-root expression. The intervals in Fig. 8.9 become wider as the value of x is further from $\bar{x}$. In the Excel analysis, the standard errors for the slope and intercept are also available, as listed as s_b and s_a in Fig. 8.8. As expected, the values of the regression coefficients calculated above for a, b, s_e, s_b, r^2, and confidence intervals agree perfectly with the Excel analysis results.

8.4 Propagation of errors

Numerical problems are important in engineering for solving problems on large (or small) computers where errors can occur and propagate.

Significant digits and error

Most digital computers represent numbers in either a fixed system (that is, a fixed number of decimal places such as 62.358, 0.013, 1.00, and so on) or a floating point system (that is, a fixed number of significant digits such as $0.6238 \times 10^3, 0.1714 \times 10^{-3}, -0.2000 \times 10^1$, etc.). The significant digit of a number c is any digit of c, except possibly for zeros to the left of the first nonzero digit that serves only to fix the position of the decimal point. For example, 1360, 1.360, and 0.001360 all have four significant digits.

An error will be caused by chopping (i.e., discarding all decimals from some decimal on) or rounding. A good rule of thumb for rounding is to discard the $(k+1)$th decimal and all subsequent decimals as follows:

(a) If the number thus discarded is less than half a unit in the kth place, then leave the kth decimal unchanged ("rounding down").
(b) If it is greater than half a unit, add one to the kth decimal ("rounding up").
(c) If it is exactly half a unit, round off to the nearest even decimal. For example, rounding 3.45 and 3.55 to one decimal gives 3.4 and 3.6, respectively.

Rule (c) ensures that in discarding exactly half a decimal, rounding up and down typically happen equally on average. For technical simplicity, computers use rounding (by rounding up in case (a)), or in some case chopping.

The final results of computations are approximations because of round-off (or chopping) errors, experimental errors, or truncation errors (for example, truncating an infinite series). If $\tilde{a}$ is an approximate value of a quantity whose exact value is a, the error is simply the difference

$$\epsilon = a - \tilde{a},$$

or $a = \tilde{a} + \epsilon$ (i.e., the true value equals the approximate value plus error). The relative error ϵ_r of $\tilde{a}$ is defined by

$$\epsilon_r = \frac{\epsilon}{a} = \frac{a - \tilde{a}}{a} = \frac{\text{error}}{\text{true value}} \quad (a \neq 0).$$

If $|\epsilon| << |\tilde{a}|$, the relative error can be approximated by $\epsilon_r \approx \dfrac{\epsilon}{\tilde{a}}$. In practice ϵ is unknown; however, one can get an _error band_ for $\tilde{a}$ such that $|\epsilon| \leq \beta$; hence, $|a - \tilde{a}| \leq \beta$. Similarly, for the relative error, $|\epsilon_r| \leq \beta_r$, and hence $\left| \dfrac{a - \tilde{a}}{a} \right| \leq \beta_r$.

8.4.1 Error propagation

Errors in a computation can propagate and affect the accuracy of the result.

(a) In addition and subtraction, an error bound for the results is given by the sum of the error bounds for the terms.
  Derivation: Let $x = \tilde{x} + \epsilon_1$, $y = \tilde{y} + \epsilon_2$, where $|\epsilon_1| \leq \beta_1$ and $|\epsilon_2| \leq \beta_2$. Then the error difference is $|\epsilon| = |x - y - (\tilde{x} - \tilde{y})| = |\epsilon_1 - \epsilon_2| \leq |\epsilon_1| + |\epsilon_2| \leq \beta_1 + \beta_2$. The proof for the sum is similar.

(b) In multiplication and division, an error bound for the relative error of the results is given (approximately) by the sum of the error bounds for the relative errors of the given numbers.
  Derivation: For the relative error ϵ_r of $\tilde{x}\tilde{y}$ one gets from the relative errors ϵ_{r1} and ϵ_{r2} of $\tilde{x}$, $\tilde{y}$ and bounds β_{r1} and β_{r2}:

$$|\epsilon_r| = \left| \frac{xy - \tilde{x}\tilde{y}}{xy} \right| = \left| \frac{xy - (x - \epsilon_1)(y - \epsilon_2)}{xy} \right| = \left| \frac{\epsilon_1 y + \epsilon_2 x - \epsilon_1 \epsilon_2}{xy} \right|$$

$$\simeq \left| \frac{\epsilon_1 y + \epsilon_2 x}{xy} \right| = |\epsilon_{r1} + \epsilon_{r2}| \leq \beta_{r1} + \beta_{r2}.$$

The proof for the quotient is similar.

Accumulation of determinate errors

For the sum or difference $y = a + b - c$, where a, b, and c are measurable quantities, the absolute determinate errors, Δa, Δb, and Δc in these quantities, propagate as $\Delta y = \Delta a + \Delta b - \Delta c$.

Example 8.4.1. Calculate the error in the result of the following calculation: $y = +0.50(+0.02) + 4.10(-0.03) - 1.97(-0.05) = 2.63(\Delta y)$, where the numbers in parentheses are the absolute determinate errors.

Solution:
The absolute error is $\Delta y = 0.02 + (-0.03) - (-0.05) = +0.04$.

For the product or quotient $y = ab/c$, where a, b and c are measurable quantities, the *absolute* determinate errors, Δa, Δb, and Δc in these quantities, propagate in accordance with the following *relative error* expression: $(\Delta y/y) = (\Delta a/a) + (\Delta b/b) - (\Delta c/c)$.

Example 8.4.2. Calculate the error in the result of the following calculation, where the number in parentheses are the absolute determinate errors:

$$y = \frac{4.10(-0.02) \times 0.0050(+0.0001)}{1.97(-0.04)} = 0.01041(\Delta y).$$

Solution:

The error must be based on relative errors such that $\dfrac{\Delta y}{y} = \dfrac{-0.02}{4.10} + \dfrac{0.0001}{0.0050} -$

$\dfrac{-0.04}{1.97} = 0.035$.

To obtain the absolute error in y, $\Delta y = 0.035 \times y = 0.035 \times 0.01041 = 0.0004$. Thus, the final result is $y = 0.0104 \ (+0.0004)$.

Accumulation of indeterminate errors

In contrast to a determinate error no sign can be attached to a standard deviation (since this error is random) and therefore has an equal probability of being either positive or negative. Since these errors are independent (and random), they may cancel one another. In the case of sums or differences, the errors therefore propagate in quadrature (that is, as the sum of individual *absolute variances*). Thus, for the quantity $y = a + b - c$, with standard deviations $\pm s_a$, $\pm s_b$, and $\pm s_c$, the errors propagate as

$$s_y = \sqrt{s_a^2 + s_b^2 + s_c^2}. \tag{8.24}$$

Similarly, for a product or quotient $y = ab/c$, the *relative variance* of the result is equal to the sum of the individual relative variances so that

$$(s_y)_r = \sqrt{(s_a^2)_r + (s_b^2)_r + (s_c^2)_r}, \tag{8.25}$$

where the relative error is the standard deviation normalized by the given quantity (for example, $(s_a)_r = s_a/a$).

Example 8.4.3. Calculate the standard deviation for the following calculation:

$$y = \frac{[14.3(\pm 0.2) - 11.6(\pm 0.2)] \times 0.050(\pm 0.001)}{[820(\pm 10) + 1030(\pm 5)] \times 42.3(\pm 0.4)} = 1.725(\pm s_y) \times 10^{-6}.$$

Solution:

First one must calculate the standard deviation of the sum and difference. For the difference in the numerator, $s_a = \sqrt{(\pm 0.2)^2 + (\pm 0.2)^2} = \pm 0.28$, and for the sum in the denominator, $s_b = \sqrt{(\pm 10)^2 + (\pm 5)^2} = \pm 11$. The equation may be rewritten as $y = \dfrac{[2.7(\pm 0.28)] \times 0.050(\pm 0.001)}{[1850(\pm 11)] \times 42.3(\pm 0.4)}$. Since the equation only contains products and quotients, the relative standard deviations of the individual quantities must be determined:

$$(s_a)_r = \frac{\pm 0.28}{2.7} = \pm 0.104, \qquad (s_b)_r = \frac{\pm 0.001}{0.050} = \pm 0.020,$$

$$(s_c)_r = \frac{\pm 11}{1850} = \pm 0.0060, \quad \text{and} \quad (s_d)_r = \frac{\pm 0.4}{42.3} = \pm 0.0095,$$

so that from Eq. (8.25):

$$(s_y)_r = \sqrt{(\pm 0.104)^2 + (\pm 0.020)^2 + (\pm 0.0060)^2 + (\pm 0.0095)^2} = 0.106.$$

The absolute standard deviation of the result is $s_y = 1.725 \times 10^{-6} \times (\pm 0.106) = \pm 0.18 \times 10^{-6}$. The final result is therefore $1.7(\pm 0.2) \times 10^{-6}$.

If y is a function of one variable, $y(x)$, then

$$\delta y = \left| \frac{dy}{dx} \right| \delta x. \tag{8.26}$$

Substituting in the standard deviations s_y and s_x for δy and δx, respectively, gives

$$s_y = \left| \frac{dy}{dx} \right| s_x. \tag{8.27}$$

Example 8.4.4. Suppose $\theta = (\pi/3) \pm 0.052$ rad. The best estimate of $\cos(\pi/3) = 0.50$. What is the uncertainty in $\cos(\theta)$?

Solution:
Using Eq. (8.27), the uncertainty is $s_{\cos\theta} = |d(\cos\theta)/d\theta| s_\theta = |\sin\theta| s_\theta = |\sin(\pi/3)| \times 0.052 = \pm 0.045$. Therefore, the final result is $\cos\theta = 0.50 \pm 0.05$.

Using the general relation in Eq. (8.27), the indeterminate error for exponential calculations, $y = x^n$, is

$$s_y = \left| nx^{n-1} \right| s_x = \left| \frac{ny}{x} \right| s_x \Rightarrow (s_y)_r = |n|(s_x)_r. \tag{8.28}$$

Similarly, using Eq. (8.27) for the logarithmic expression, $y = \log x = 0.434 \ln x$, the standard deviation is derived as

$$s_y = 0.434(s_x)_r. \tag{8.29}$$

Example 8.4.5. The standard deviation in measuring the diameter d of a sphere is ± 0.02 cm. What is the standard deviation in the volume if $d = 2.15$ cm?

Solution:

We have $V = \frac{4}{3}\pi \left(\frac{d}{2}\right)^3 = \frac{4}{3}\pi \left(\frac{2.15}{2}\right)^3 = 5.20 \text{ cm}^3$. Hence, using Eq. (8.28), $\frac{s_V}{V} = 3 \times \frac{s_d}{d} = 3 \times \frac{0.02}{2.15} = 0.028$. The absolute standard deviation in V is then $s_V = 5.20 \times 0.028 = 0.15$. Therefore, $V = 5.2 \ (\pm 0.2) \text{ cm}^3$.

Example 8.4.6. Calculate the absolute standard deviations in (a) $y = \log[2.00(\pm 0.02) \times 10^{-3}] = -2.6990\pm?$ and (b) $x = \text{alog}[1.200(\pm 0.003)] = 15.849\pm?$.

Solution:

(a) Using Eq. (8.29), $s_y = \pm 0.434 \times \dfrac{0.02 \times 10^{-3}}{2.00 \times 10^{-3}} = \pm 0.004$. Thus,

$$\log[2.00(\pm 0.02) \times 10^{-3}] = -2.699 \ (\pm 0.004).$$

(b) Rearranging Eq. (8.29), $(s_x)_r = \dfrac{s_x}{x} = \dfrac{s_y}{0.434} = \dfrac{\pm 0.003}{0.434} = \pm 0.0069$. Thus,

$$s_x = \pm 0.0069 \times x = \pm 0.0069 \times 15.849 = 0.11.$$

Therefore, $\text{alog}[1.200(\pm 0.003)] = 15.8 \pm 0.1$.

Finally, in general, if y is any function of several variables $x, ..., z$, then for independent random variables,

$$\delta y = \sqrt{\left(\frac{\partial y}{\partial x}\delta x\right)^2 + ... + \left(\frac{\partial y}{\partial z}\delta z\right)^2} \Rightarrow s_y = \sqrt{\left(\frac{\partial y}{\partial x}s_x\right)^2 + ... + \left(\frac{\partial y}{\partial z}s_z\right)^2}. \quad (8.30)$$

It can be seen that Eq. (8.24) and Eq. (8.25) are simply special cases, which follow from Eq. (8.30).

Total uncertainty

For determinate errors (that is, a systematic component, δx_{sys}) and indeterminate errors (that is, a random component, δx_{random}), these quantities can be combined in quadrature (since they are independent) to yield an overall error:

$$\delta y = \sqrt{(\delta x_{random})^2 + (\delta x_{sys})^2}. \quad (8.31)$$

Although δx_{random} can be reduced with a larger number of measurements (see Eq. (8.9) and Eq. (8.12)), the overall uncertainty δx is ultimately limited by δx_{sys}.

Problems

8.1   Given the function $y = ab/c$, where a, b, and c are independent variables with random errors of s_a, s_b, and s_c, show that using the general error propagation formula one obtains the specific result $(s_y)_r = \sqrt{(s_a)_r^2 + (s_b)_r^2 + (s_c)_r^2}$.

8.2   What is the expression for calculating the standard error of the intercept (y_o) of a linear regression line? Hint: Use the confidence interval relation to determine this expression.

8.3   For the logarithmic expression $y = \log x$, show that the standard deviation s_y is given by $s_y = 0.434(s_x)_r$.

8.4   A concentration measurement (in ppm) was made consisting of the following values: 21, 15, 23, 21, and 24. Should the anomalously low value be refused based on a Q critical value at a 90% confidence?

8.5   What is the difference between a t distribution and a normal distribution as used for confidence interval prediction?

Numerical analysis

In engineering, one may be faced with a need to find the zero of a function or for numerical interpolation, spline fitting, or smoothing of acquired data. In addition, there may be a further requirement for numerical integration and differentiation.

One of the most important tasks in engineering is optimization and approximation theory. This may involve the approximation of an equation with a solution at discrete points (see Chapters 6 and 7) or for the solution of a nonlinear system. The determination of roots may arise in the finding of an optimal solution to a real-life problem. It is also important to find the zeros or poles of a transfer function in control theory or for the solution of differential equations using transform methods in Chapter 3. Numerical interpolation and spline fitting provides a means to interpolate numerical listings in mathematical tables or for the interpolation of acquired data obtained in design and testing experiments. Lagrange interpolation polynomials allow one to develop shape functions in finite element methods (Chapter 7) or to assess the error in numerical integration formulas where an analytical solution for the integral does not exist. It can be used for assessment of distorted geometries when loads are applied to deformable bodies. Spline fitting is particularly important to avoid oscillatory behavior in the development of models, which can arise when fitting higher-order polynomials to the data. Data smoothing may also be required for improved data analysis, where there may be a requirement to filter noisy data from measurements that have an inherent uncertainty due to difficult sampling procedures.

Numerical differentiation is at the heart of numerical methods for solution of ordinary and partial differential equations in the modeling of physical behavior (see Chapter 6). For instance, this approach is particularly needed for the solution of complex engineering problems that can arise, where the coefficients representing the thermal and physical properties of the materials in the underlying equations may be a function of the dependent-solution variable itself (e.g., with a dependence of the thermal conductivity on temperature in a heat conduction problem), resulting in a nonlinear problem. Numerical techniques are particularly adaptable to computer programs or algorithms. Computer-aided engineering relies on numerical analysis methods for the modeling of dynamical systems and numerical simulation of coupled differential and algebraic systems for real-time applications.

Techniques for the solution of these various problems are described in this chapter.

9.1 Finding zeros of functions

A solution of the equation

$$\boxed{f(x) = 0} \tag{9.1}$$

involves finding a number $x = s$ such that $f(s) = 0$. Equation (9.1) can consist of an <u>algebraic equation</u> (e.g., a polynomial such as $x^2 - 3x + 2 = 0$) or a <u>transcendental equation</u> (e.g., $\tan x = x$). For a polynomial or transcendental equation, the solution of Eq. (9.1) is called the <u>roots</u> of the equation.

As seen previously, equations of the form of Eq. (9.1) may occur. For example, such problems arise in the solutions of characteristic equations and the zeros of Bessel functions, to name a few. Formulas that exist to give exact numerical values of the solution only exist in simple situations (e.g., solution of quadratic equations), and in most cases, an approximate <u>iterative method</u> is required.

9.1.1 Fixed point iteration

Equation (9.1) can be transformed algebraically into the form

$$x = g(x). \tag{9.2}$$

Thus, choosing a value of x_0, implies that $x_1 = g(x_0), x_2 = g(x_1), ...,$ and in general,

$$\boxed{x_{n+1} = g(x_n)} \quad (n = 0, 1, ...). \tag{9.3}$$

From Eq. (9.1), several different forms of Eq. (9.2) can arise, and the behavior of the corresponding iterative sequences $x_0, x_1, ...$ may differ accordingly. The iterative process in Eq. (9.3) is <u>convergent</u> for x_0 if the sequence $x_0, x_1, ...$ converges. A sufficient condition for convergence for any x_0 in the interval **J** is that if $x = s$ is a solution of $x = g(x)$ and g has a continuous derivative in **J** (containing s), then $|g'(x)| \leq K < 1$ in **J**.

Example 9.1.1. Find a solution of $f(x) = x^2 - 4x + 2 = 0$ by iteration.

Solution:
The roots of a general quadratic equation $ax^2 + bx + c$ can be obtained analytically from

$$x_{1,2} = \frac{-b \pm \sqrt{b^2 - 4ac}}{2a}.$$

Hence, inserting $a = 1, b = -4$, and $c = 2$ into this solution equation, the two roots are evaluated analytically as $x_{1,2} = 2 \pm \sqrt{2}$, yielding the values $x_1 = 0.5858$ and $x_2 = 3.414$.

For the numerical solution, the equation may be written as

$$x = g_1(x) = \frac{1}{4}\left(x^2 + 2\right), \quad \text{thus } x_{n+1} = \frac{1}{4}\left(x_n^2 + 2\right).$$

Then $|g_1'(x)| = \frac{1}{2}|x| < 1$ for any $x < 2$. Choosing $x_0 = 1$, one obtains $x_0 = 1.0000$, $x_1 = 0.7500$, $x_2 = 0.6406$, $x_3 = 0.6026$, $x_4 = 0.5908...$, which is approaching the exact solution for the first root of 0.5858.   [answer]

Note that if one chose $x_0 = 4$, the sequence diverges as expected from the above theorem, that is, $x_0 = 4.000$, $x_1 = 4.500$, $x_2 = 5.563$, $x_3 = 8.235$, $x_4 = 17.455...$. The equation may also be written as

$$x = g_2(x) = 4 - \frac{2}{x}, \quad \text{thus } x_{n+1} = 4 - \frac{2}{x_n}.$$

Then $|g_2'(x)| = \frac{2}{x^2} < 1$ for $x > \sqrt{2}$. Thus, now choosing $x_0 = 4$, one obtains the sequence $x_0 = 4.000$, $x_1 = 3.500$, $x_2 = 3.429$, $x_3 = 3.417$, $x_4 = 3.415...$, which converges to the second root at 3.414.   [answer]

9.1.2 Newton's method

This method is commonly used because of its simplicity and speed. In this method, one approximates the graph of f by suitable tangents (see Fig. 9.1).

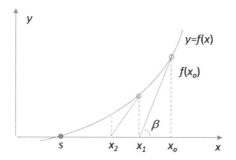

FIGURE 9.1

Evaluation of $f(x) = 0$ using the Newton method.

One starts the iteration process with an approximate value x_0, from which a value is determined on the graph for f. Next using a tangent line at this point to the point of intersection on the x axis, one obtains the next value x_1. The tangent to the curve of f at x_0 is evaluated from

$$\tan \beta = f'(x_0) = \frac{f(x_0)}{x_0 - x_1}, \quad \text{and hence } x_1 = x_0 - \frac{f(x_0)}{f'(x_0)}.$$

This process continues as the improved iterative values for the root approach the value at s using the iterative procedure:

$$x_{n+1} = x_n - \frac{f(x_n)}{f'(x_n)}. \tag{9.4}$$

Example 9.1.2. Find the positive solution of $x^2 - 4x + 2 = 0$ by Newton's method.

Solution.
Setting $f(x) = x^2 - 4x + 2 = 0$, $f'(x) = 2x - 4$ and Eq. (9.4) gives

$$x_{n+1} = x_n - \frac{x_n^2 - 4x_n + 2}{2x_n - 4} = \frac{x_n^2 - 2}{2(x_n - 2)}.$$

The solution is again near $x_0 = 1$. Therefore, successive iterations yield $x_0 = 1.00000$, $x_1 = 0.5000$, $x_2 = 0.5833$, $x_3 = 0.5858$. The value of x_3 is identical to four decimal places to the analytical solution in Example 9.1.1.   [answer]

Order of the method (speed of convergence)

Given an iteration method $x_{n+1} = g(x_n)$, where x_n is an approximate to the solution s, that is, $s = x_n + \epsilon_n$ (or $x_n - s = -\epsilon_n$), if g is differentiable many times, the Taylor formula gives

$$x_{n+1} = g(x_n) = g(s) + g'(s)(x_n - s) + \frac{1}{2}g''(s)(x_n - s)^2 + \cdots$$

$$= g(s) - g'(s)\epsilon_n + \frac{1}{2}g''(s)\epsilon_n^2 + \cdots. \tag{9.5}$$

The <u>order</u> of the iteration process is the exponent of ϵ_n in the first nonvanishing term after $g(s)$ (see the example below for the Newton method). For instance, subtract $g(s) = s$ on both sides of Eq. (9.5):

$$x_{n+1} - s = -\epsilon_{n+1} = -g'(s)\epsilon_n + \frac{1}{2}g''(s)\epsilon_n^2. \tag{9.6}$$

In Newton's method, $g(x) = x - \dfrac{f(x)}{f'(x)}$ and

$$g'(x) = 1 - \frac{f'(x)^2 - f(x)f''(x)}{f'(x)^2} = \frac{f(x)f''(x)}{f'(x)^2}. \tag{9.7}$$

Since $f(s) = 0$, Eq. (9.7) gives $g'(s) = 0$. Differentiating Eq. (9.7) again and setting $x = s$ yields

$$g''(s) = \frac{f''(s)}{f'(s)}. \tag{9.8}$$

Thus, Eq. (9.6) and Eq. (9.8) indicate that the Newton method is of second order, that is,

$$\epsilon_{n+1} = -\frac{f''(s)}{2f'(s)}\epsilon_n^2,$$

if $f(x)$ is three times differentiable and f' and f'' are not zero at a solution s of $f(x) = 0$. This result means that if $\epsilon_n = 10^{-k}$ in some step, then for the second order, $\epsilon_{n+1} = \text{constant} \cdot 10^{-2k}$ so that the number of significant digits is about doubled in each step.

9.1.3 Secant method

Replacing the derivative $f'(x)$ in Eq. (9.4) by the difference quotient

$$f'(x_n) \approx \frac{f(x_n) - f(x_{n-1})}{x_n - x_{n-1}}$$

yields the <u>secant method</u>:

$$x_{n+1} = x_n - f(x_n)\frac{x_n - x_{n-1}}{f(x_n) - f(x_{n-1})}. \tag{9.9}$$

A geometric interpretation is shown in Fig. 9.2.

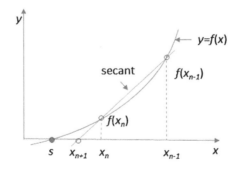

FIGURE 9.2

Evaluation of $f(x) = 0$ using the secant method.

Example 9.1.3. Solve $f(x) = x^2 - 4x + 2 = 0$ by the secant method with $x_0 = 1$ and $x_1 = 0.5$.

Solution.
By Eq. (9.9):

$$x_{n+1} = x_n - \frac{(x_n^2 - 4x_n + 2)(x_n - x_{n-1})}{x_n^2 - 4x_n + 2 - x_{n-1}^2 + 4x_{n-1} - 2} = x_n - \frac{(x_n^2 - 4x_n + 2)(x_n - x_{n-1})}{x_n^2 - x_{n-1}^2 - 4(x_n - x_{n-1})}.$$

The numerical results are $x_0 = 1.0000$, $x_1 = 0.5000$, $x_2 = 0.6000$, $x_3 = 0.5862$, $x_4 = 0.5858$.   [answer]

As in Example 9.1.2, $x_4 = 0.5858$ is exact to four decimal places.

9.2 Interpolation

Interpolation is required to find (approximate) values of a function $f(x)$ for an x between different x values $(x_0, x_1, ..., x_n)$ at which the values of $f(x)$ are known (for example, as in mathematical tables or with recorded data). An interpolation polynomial $p_n(x)$ of degree n can be found that assumes the given values

$$p_n(x_0) = f_0, \; p_n(x_1) = f_1, ..., \; p_n(x_n) = f_n, \qquad (9.10)$$

where $f_n = f(x_n)$.

9.2.1 Lagrange interpolation

In Lagrange's form, the polynomial $p_n(x)$ is given by

$$f(x) \approx p_n(x) = \sum_{k=0}^{n} L_k(x) f_k = \sum_{k=0}^{n} \frac{l_k(x)}{l_k(x_k)} f_k. \qquad (9.11)$$

In Eq. (9.11), $L_k(x_k) = 1$ and $L_k(x_j) = 0$ for $j \neq k$. Thus, $l_k(x)$ is defined by

$$
\begin{aligned}
l_0(x) &= (x - x_1)(x - x_2)...(x - x_n), \\
l_k(x) &= (x - x_0)...(x - x_{k-1})(x - x_{k+1})...(x - x_n), \quad 0 < k < n, \qquad (9.12) \\
l_n(x) &= (x - x_0)(x - x_1)...(x - x_{n-1}).
\end{aligned}
$$

Inspection of Eq. (9.12) shows that $l_k(x_j) = 0$ if $j \neq k$, so that for $x = x_k$ the sum in Eq. (9.11) reduces to the single term $(l_k(x_k))/(l_k(x_k)) f_k = f_k$, as required.

Example 9.2.1. Find the linear Lagrange interpolation polynomial $p_1(x)$, $n = 1$, from Eq. (9.11) and Eq. (9.12).

Solution.
We have

$$p_1(x) = L_0(x) f_0 + L_1(x) f_1 = \frac{x - x_1}{x_0 - x_1} \cdot f_0 + \frac{x - x_0}{x_1 - x_0} \cdot f_1. \qquad \text{[answer]} \quad (9.13)$$

Example 9.2.2. Compute $\sinh(1.5)$ from $\sinh(1.0) = 1.1752$ and $\sinh(2.0) = 3.6269$ by linear Lagrange interpolation.

Solution.

Given $x_0 = 1.0$, $x_1 = 2.0$, $f_0 = \sinh(1.0)$, and $f_1 = \sinh(2.0)$, from Eq. (9.13) in Example 9.2.1 we have

$$\sinh(1.5) \approx p_1(1.5) = \frac{1.5 - 2.0}{1.0 - 2.0}(1.1752) + \frac{1.5 - 1.0}{2.0 - 1.0}(3.6269) = 2.4011. \quad \text{[answer]}$$

Since $\sinh(1.5) = 2.1293$ (exact to four decimal places), the error is $\epsilon = a - \tilde{a} = 2.1293 - 2.4011 = -0.2718$.

The <u>error</u> for any polynomial interpolation method, if $f(x)$ has a continuous $(n + 1)$st derivative, is given by the formula

$$\epsilon_n(x) = f(x) - p_n(x) = (x - x_0)(x - x_1)...(x - x_n)\frac{f^{n+1}(t)}{(n + 1)!}. \qquad (9.14)$$

Example 9.2.3. Estimate the error in Example 9.2.2 using Eq. (9.14).

Solution.

Given $n = 1$, $f(t) = \sinh t$, $f'(t) = \cosh t$, $f''(t) = \sinh t$, one has

$$\epsilon_1(x) = (x - 1.0)(x - 2.0)\frac{1}{2}(\sinh t),$$

which is evaluated as $\epsilon_1(1.5) = -0.1250 \sinh t$ so that:

$t = 1.0$ implies a smaller negative value of $= -0.1469$,
$t = 2.0$ implies a larger negative value of $= -0.4534$.

Hence, one obtains $-0.4534 \le \epsilon_1(1.5) \le -0.1469$. \quad [answer]

This calculated range is in agreement with the reported error in Example 9.2.2 of -0.2718.

9.2.2 Newton's divided difference

With the Lagrange method, the degree of the interpolation polynomial that will give the required accuracy is not known. However, the Newton method provides a means to simply add on another term to improve the accuracy. <u>Newton's divided difference formula</u> is

$$\begin{aligned} f(x) \simeq f_0 + (x - x_0)f[x_0, x_1] + (x - x_0)(x - x_1)f[x_0, x_1, x_2] + \\ ... + (x - x_0)...(x - x_{n-1})f[x_0, ..., x_n], \end{aligned} \qquad (9.15)$$

where

$$f[x_0, x_1] = \frac{f_1 - f_0}{x_1 - x_0},$$

$$f[x_0, x_1, x_2] = \frac{f[x_1, x_2] - f[x_0, x_1]}{x_2 - x_0},$$

$$f[x_0, ..., x_2] = \frac{f[x_1, ..., x_k] - f[x_0, ..., x_{k-1}]}{x_k - x_0}.$$

(9.16)

Example 9.2.4 shows how to use a <u>difference table</u> for the evaluation of Eq. (9.15).

Example 9.2.4. Compute sinh(1.5) from the given values in Table 9.1 using Newton's divided difference interpolation formula.

Table 9.1 Example of a Newton divided difference table for interpolation.

x_j	$f_j = f(x_j)$	$f[x_j, x_{j+1}]$	$f[x_j, x_{j+1}, x_{j+2}]$	$f[x_j, ..., x_{j+3}]$
0.5	0.5211			
		1.3082		
1.0	1.1752		0.7623	
		2.4517		0.4829
2.0	3.6269		1.9697[a]	
		6.3911		
3.0	10.018			

↑

Given
values

Solution.
The divided differences are shown in Table 9.1. For example, a sample calculation for "a" is $(6.3911 - 2.4517)/(3.0 - 1.0) = 1.9697$. The values needed in Eq. (9.15) are placed in boxes, and the calculation is as follows:

$$f(x) \simeq p_3(x) = 0.5211 + 1.3082(x - 0.5) + 0.7623(x - 0.5)(x - 1.0)$$
$$+ 0.4829(x - 0.5)(x - 1.0)(x - 2.0).$$

Therefore at $x = 1.5$, $f(1.5) = 2.0897$.   [answer]

Note that the exact value to four decimal places is $\sinh(1.5) = 2.1293$. With this method, the accuracy increases from term to term:

$$p_1(1.5) = 1.8293, \quad p_2(1.5) = 2.2105, \quad \text{and} \quad p_3(1.5) = 2.0897.$$

Newton's forward difference formula (equal spacing)

Newton's formula in Eq. (9.15) is for arbitrary-spaced nodes. However, if the x_js are regularly spaced (as in function tables) such that

$$x_0, x_1 + h, x_2 = x_0 + 2h, ..., x_n = x_0 + nh,$$

then Eq. (9.15) becomes Newton's forward difference interpolation formula:

$$f(x) \simeq p_n(x) = \sum_{s=0}^{n} \binom{r}{s} \Delta^s f_0 \qquad (x = x_0 + rh, r = (x - x_0)/h)$$
$$= f_0 + r \Delta f_0 + \frac{r(r-1)}{2!} \Delta^2 f_0 + ... + \frac{r(r-1)...(r-n+1)}{n!} \Delta^n f_0,$$
$$\tag{9.17}$$

where the forward differences of f at x_j are defined by

$$\Delta f_j = f_{j+1} - f_j, \quad \Delta^2 f_j = \Delta f_{j+1} - \Delta f_j, \quad ..., \quad \Delta^k f_j = \Delta^{k-1} f_{j+1} - \Delta^{k-1} f_j.$$
$$\tag{9.18}$$

Similarly, the error for this interpolation method is

$$\epsilon_n(x) = f(x) - p_n(x) = \frac{h^{n+1}}{(n+1)!} r(r-1)...(r-n) f^{n+1}(t). \tag{9.19}$$

Example 9.2.5. Compute $\sinh 1.56$ from the given values using Newton's forward difference formula.

Solution.
Using Eq. (9.18) and the given values, the forward differences are computed in Table 9.2. The values needed in Eq. (9.17) are placed in boxes. Also in Eq. (9.17), $r = (1.5 - 0.0)/0.1 = 1.5$. The calculation is as follows:

$$f(1.5) = \sinh(1.5) \simeq p_3(1.5) = 0.0000 + 1.5(1.1752) + \frac{1.5(0.5)}{2}(1.2765)$$
$$+ \frac{1.5(0.5)(-0.5)}{6}(2.6629) = 2.0751.$$

For the error estimate in Eq. (9.19), $\sinh^{(4)} t = \sinh t$ so that

$$\epsilon_3(1.5) = \frac{1.0^4}{4!} \times 1.5(0.5)(-0.5)(-1.5) \sinh t = A \sinh t \quad (A = 0.02344).$$

Here $0.0 \leq t \leq 3.0$ and one obtains an inequality for the largest and smallest $\sinh t$ in the interval

$$A \sinh 0.0 \leq \epsilon_3(1.5) \leq A \sinh 3.0.$$

Table 9.2 Example of a Newton forward difference table for interpolation.

j	x_j	$f_j = \sinh x_j$	Δf_j	$\Delta^2 f_j$	$\Delta^3 f_j$
0	0.0	0.0000			
			1.1752		
1	1.0	1.1752		1.2765 [a]	
			2.4517		2.6629
2	2.0	3.6269		3.9394	
			6.3911		
3	3.0	10.018			

[a] $\Delta^2 f = 2.4517 - 1.1752 = 1.2765.$

Since $f(x) = p_3(x) + \epsilon_3(x)$, the error bands are

$$\underbrace{p_3(1.5) + A \sinh 0.0}_{2.0751} \le \sinh 1.5 \le \underbrace{p_3(1.5) + A \sinh 1.5}_{2.3099}. \qquad \text{[answer]}$$

In fact, the exact answer to four decimal places ($\sinh 1.5 = 2.1293$) lies within these bounds.

9.3 Splines

For various functions $f(x)$, the corresponding interpolation polynomials in Section 9.2 tend to become more numerically unstable as the degree of the polynomial n increases (see Fig. 9.3). Thus, instead of interpolating and approximating by a single

FIGURE 9.3

Various-order interpolation polynomials showing instability for higher-order polynomials.

high-degree polynomial, it is preferable to use a cubic spline $g(x)$ for the function $f(x)$ on the interval $a \le x \le b$, by subdividing it into subintervals with common endpoints (called nodes):

$$a = x_0 < x_1 < ... < x_n = b, \qquad (9.20)$$

where

$$g(x_0) = f(x_0) = f_0, \ g(x_1) = f(x_1) = f_1, \ ...., \ g(x_n) = f(x_n) = f_n. \qquad (9.21)$$

Also if one requires that

$$g'(x_0) = k_0, \ g'(x_n) = k_n \qquad (k_0, k_n \text{ two given numbers}), \qquad (9.22)$$

then the cubic spline can be uniquely determined. For each subinterval $x_j \leq x \leq x_{j+1} = x_j + h$, the spline $g(x)$ is given by a cubic polynomial:

$$p_j(x) = a_{j0} + a_{j1}(x - x_j) + a_{j2}(x - x_j)^2 + a_{j3}(x - x_j)^3. \qquad (9.23)$$

The function $f(x)$ is now approximated by n polynomials. The coefficients for the polynomial $p_j(x)$ in Eq. (9.23) are obtained from a Taylor series formula:

$$
\begin{aligned}
a_{j0} &= p_j(x_j) = f_j, \\
a_{j1} &= p_j'(x_j) = k_j, \\
a_{j2} &= \frac{1}{2}p_j''(x_j) = \frac{3}{h^2}\left(f_{j+1} - f_j\right) - \frac{1}{h}\left(k_{j+1} + 2k_j\right), \\
a_{j3} &= \frac{1}{6}p_j'''(x_j) = \frac{2}{h^3}\left(f_j - f_{j+1}\right) + \frac{1}{h^2}\left(k_{j+1} + k_j\right).
\end{aligned}
\qquad (9.24)
$$

The $k_1, ..., k_{n-1}$ constants in Eq. (9.24) are obtained from the relation

$$k_{j-1} + 4k_j + k_{j+1} = \frac{3}{h}\left(f_{j+1} - f_{j-1}\right), \qquad j = 1, ..., n-1 \qquad (9.25)$$

(noting that k_0 and k_n are defined by Eq. (9.22)).

Example 9.3.1. Interpolate $f(x) = e^{x^2} - 1$ on the interval $-1 \leq x \leq 1$ by the cubic spline $g(x)$ corresponding to the partition $x_0 = -1$, $x_1 = 0$, $x_2 = 1$ and satisfying $g'(x) = f'(-1)$ and $g'(1) = f'(1)$.

Solution.
In standard notation, $f_0 = f(-1) = e - 1$, $f_1 = f(0) = 0$, $f_2 = f(1) = e - 1$. The given interval is partitioned into $n = 2$ parts of length $h = 1$. Hence, the spline g consists of two polynomials of the form of Eq. (9.23):

$$
\begin{aligned}
p_0(x) &= a_{00} + a_{01}(x + 1) + a_{02}(x + 1)^2 + a_{03}(x + 1)^3 & (-1 \leq x \leq 0), \\
p_1(x) &= a_{10} + a_{11}x + a_{12}x^2 + a_{13}x^3 & (0 \leq x \leq 1).
\end{aligned}
$$

As given, $f'(-1) = -2e = g'(-1)$ and $f'(1) = 2e = g'(1)$. Hence, using Eq. (9.22), $k_0 = -2e$ and $k_2 = 2e$. Substitution of these values into Eq. (9.25) yields $k_0 + 4k_1 + k_2 = \frac{3}{1}(f_2 - f_0) = 0$. Thus, $k_1 = 0$.

The coefficients of the two polynomials can now be evaluated from Eq. (9.24):

$$
\begin{aligned}
\text{Coefficients of } p_0: \quad a_{00} &= f_0 = e - 1, \\
a_{01} &= k_0 = -2e,
\end{aligned}
$$

$$a_{02} = 3(f_1 - f_0) - (k_1 + 2k_0) = 3 + e,$$
$$a_{03} = 2(f_0 - f_1) + (k_1 + k_0) = -2,$$

Coefficients of p_1 : $a_{10} = f_1 = 0,$
$$a_{11} = k_1 = 0,$$
$$a_{12} = 3(f_2 - f_1) - (k_2 + 2k_1) = e - 3,$$
$$a_{13} = 2(f_1 - f_2) + (k_2 + k_1) = 2.$$

The cubic splines after using Maple to simplify the resulting expressions are

$$p_0(x) = x^2(e - 3 - 2x),$$
$$p_1(x) = x^2(e - 3 + 2x).$$

The spline $g(x)$ approximating $f(x)$ (see Fig. 9.4) is

$$g(x) = \begin{cases} x^2(e - 3 - 2x) & \text{if} \quad (-1 \leq x \leq 0), \\ x^2(e - 3 + 2x) & \text{if} \quad (0 \leq x \leq 1). \end{cases} \quad \text{[answer]}$$

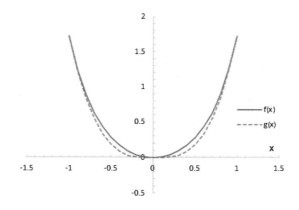

FIGURE 9.4

Example of a cubic spline fit of a function.

9.4 Data smoothing

A practical technique for the smoothing of measured data in engineering applications is the well-known Savitzky–Golay (S–G) filter (Savitzky and Golay, 1964). This technique employs a simple arithmetic calculation for the smoothing of data over an interval with a chosen number of points. For example, a popular choice is a quadratic fit (with a fitting polynomial of order $n = 2$) to a seven-point data interval (where the index i ranges from -3 to $+3$) for the points y_i in the interval. Coefficients for other-order polynomials and data intervals are tabulated in Savitzky and Golay (1964). For the use of this technique, the data must be equally spaced. The measured data are also reduced in size with a loss of the first three and last three data points in this seven-point calculation. As mentioned, the fitting coefficients are evaluated with a simple arithmetic operation providing both a smoothed value of the quantity of interest as well as its derivative as required. The coefficients for the smoothed value and derivative are given, respectively, as

$$b_{20} = \frac{1}{21}[-2y_{-3} + 3y_{-2} + 6y_{-1} + 7y_0 + 6y_1 + 3y_2 - 2y_3],$$
$$b_{21} = \frac{1}{28}[-3y_{-3} - 2y_{-2} - 1y_{-1} + 0y_0 + 1y_1 + 2y_2 + 3y_3].$$

Using a moving calculation with a standard Excel spreadsheet analysis, one simply shifts the center of the interval from the point i to the point $i + 1$ and recalculates for the next point. Thus, the formula is applied for the ith data point after which the corner tab of the cell is pulled down to extend the calculation over the remaining data entries. Thus, the smoothed value (and derivative as needed) can be found at the subsequent points.

Example 9.4.1. Consider a constant rate of release R of material into a closed system, where there is also a loss of material characterized by a first-order rate constant k. The mass balance of material $N(t)$ at time t follows as $\dfrac{dN}{dt} = R - kN$. If there is no material in the system at time zero, $N(0) = 0$. Thus, the solution of this ordinary differential equation using an integrating factor (Example 2.1.8) is $N(t) = \dfrac{R}{k}\left(1 - e^{-kt}\right)$, or equivalently $A(t) \equiv kN(t) = R(1 - e^{-kt})$. At equilibrium as $t \to \infty$, $A \to R$. Fig. 9.5 shows a history plot of $A(t)$ with an associated random error. For this plot, the rate constant is $k = 0.0862$ d^{-1} and the normalized release rate is $R = 1$. This problem is analogous to Problem 7.7 for the coolant activity of fission products in the primary heat transport system of a nuclear reactor.

Given the case of noisy data in Fig. 9.5, evaluate (a) the smoothed quantity $A(t)$ and (b) the smoothed release rate R as a function of time using an S-G filter.

Solution.

(a) The smoothed value $A(t)$ is simply given by applying b_{20} to the individual data points in the figure with an Excel analysis. The equation is first centered on the fourth point (using three points before this point and three points afterwards for the simple arithmetic operation). One then moves to the next point along the curve repeating the calculation with the original data. The smoothed curve for $A(t)$ is shown in Fig. 9.5.

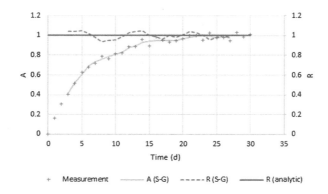

FIGURE 9.5

Mass balance analysis for noisy data using an S–G filter. Note that smoothed values are not available for the first three and last three data points for the seven-point fitting.

(b) The mass balance equation can be rearranged to solve for R as $R = \dfrac{1}{k}\dfrac{dA}{dt} + A = \dfrac{1}{k}\dfrac{b_{21}}{\Delta t} + b_{20}$, where $\Delta t = 1$ d in accordance with the equal spacing of the data. The smoothed curve for $R(t)$ is shown in Fig. 9.5 along with the constant analytic value of R equal to unity.

9.5 Numerical integration and differentiation

If a definite integral cannot be obtained by usual methods of calculus or the integrand is an empirical function given by measured values, numerical integration can be used.

9.5.1 Trapezoidal rule

In this method, the interval of integration $a \le x \le b$ is subdivided into n subintervals of equal length $h = (b - a)/n$ and the function $f(x)$ is approximated by a chord with endpoints $[a, f(a)], [x, f(x)], ..., [b, f(b)]$ on the curve of f (see Fig. 9.6). The area under the curve of f is approximated by n trapezoids of areas $\frac{1}{2}[f(a) + f(x_1)]h, \frac{1}{2}[f(x_1) + f(x_2)]h, ..., \frac{1}{2}[f(x_{n-1}) + f(b)]h$. Hence, the trapezoidal rule is obtained as the following sum:

$$J = \int_a^b f(x)\,dx \simeq h[\frac{1}{2}f(a) + f(x_1) + f(x_2) + ... + f(x_{n-1}) + \frac{1}{2}f(b)].$$

(9.26)

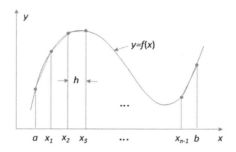

FIGURE 9.6

Calculation of the area under a curve using the trapezoidal rule.

The error in Eq. (9.26) can be derived as follows. Using Eq. (9.14) for the error in the Lagrange interpolation polynomial with $n = 1$ for a single subinterval,

$$f(x) - p_1(x) = (x - x_0)(x - x_1)\frac{f''(t)}{2}.$$

Integrating over x for the first trapezoid (that is, from $a = x_0$ to $x_1 = x_0 + h$) gives

$$\int_{x_0}^{x+h} (f(x) - p_1(x))\, dx = \int_{x_0}^{x+h} (x - x_0)(x - x_0 - h)\frac{f''(t(x))}{2}\, dx.$$

Setting $\xi = x - x_0$ and applying the mean value theorem of integral calculus yields

$$\int_{x_0}^{x+h} (f(x) - p_1(x))\, dx = \frac{f''(\tilde{t})}{2}\int_0^h \xi(\xi - h)\, d\xi = \left(\frac{h^3}{3} - \frac{h^3}{2}\right)\frac{f''(\tilde{t})}{2}$$

$$= -\frac{h^3}{12}f''(\tilde{t}),$$

where $\tilde{t}$ is a suitable value between x_0 and x_1. Hence, the error in ϵ in Eq. (9.26) for n intervals is

$$\epsilon = -\frac{nh^3}{12}f''(\tilde{t}) = -\frac{n\left[\frac{(b-a)}{n}\right]^3}{12}f''(\tilde{t}) = -\frac{(b-a)^3}{12n^2}f''(\tilde{t}),$$

with $\tilde{t}$ being a suitable value between a and b. The error bands are now obtained by taking the largest and smallest values of f'' in the interval of integration (that is, M_2 and M_2^*); hence

$$\boxed{K M_2 \le \epsilon \le K M_2^*, \quad \text{where } K = -\frac{(b-a)^3}{12n^2}.} \tag{9.27}$$

Example 9.5.1. Evaluate $\int_0^1 \sinh(x)\,dx$ by the trapezoid rule and estimate the error with $n = 10$.

Solution.
Using Eq. (9.26) one has

$$J \simeq 0.1 \left[\frac{\sinh(0)}{2} + \sinh(0.1) + \sinh(0.2) + \sinh(0.3) + \sinh(0.4) + \sinh(0.5) \right.$$

$$\left. + \sinh(0.6) + \sinh(0.7) + \sinh(0.8) + \sinh(0.9) + \frac{\sinh(1)}{2} \right] = \underline{0.543533}.$$

The error can be calculated from Eq. (9.27). By differentiation, $f''(x) = \sinh x$. Also, $f'''(x) > 0$ if $0 < x < 1$ so that the minimum and maximum occur at the ends of the interval. Hence, $M_2 = f''(1) = 1.175201$ and $M_2^* = f''(0) = 0$, and $K^{-1} = -1200$. Therefore, Eq. (9.27) gives $-0.000979 \leq \epsilon \leq 0.000000$ so that the exact value must lie between

$$\underline{0.542554} = 0.543533 - 0.000979 \leq J \leq 0.543533 + 0.000000 = \underline{0.543533}.$$

[answer]

Actually, $J = 0.543081$, which is exact to six decimal places, so that J lies within these bounds.

9.5.2 Simpson's rule

A linear piecewise approximation of f gives a trapezoidal rule of integration, whereas a higher accuracy can be obtained with a piecewise quadratic approximation.

Here the interval of integration $a \leq x \leq b$ is divided into an even number of subintervals (that is, $n = 2m$) of length $h = (b - a)/2m$ with endpoints $x_0 = a, x_1, ..., x_{2m-1}, x_{2m} = b$ (see Fig. 9.7). Considering the first two subintervals $x_0 \leq x \leq x_2 = x_0 + 2h$, $f(x)$ can be approximated by the Lagrange interpolation polynomial $p_2(x)$ through the nodes $(x_0, f_0), (x_1, f_1), (x_2, f_2)$, where $f_j = f(x_j)$, using Eq. (9.11),

$$p_2(x) = \frac{(x - x_1)(x - x_2)}{(x_0 - x_1)(x_0 - x_2)} f_0 + \frac{(x - x_0)(x - x_2)}{(x_1 - x_0)(x_1 - x_2)} f_1 + \frac{(x - x_0)(x - x_1)}{(x_2 - x_0)(x_2 - x_1)} f_2.$$

The denominators are $2h^2$, $-h^2$, and $2h^2$, respectively. Setting $s = (x - x_1)/h$ implies $(x - x_0) = (s + 1)h$, $(x - x_1) = sh$, and $(x - x_2) = (s - 1)h$. Hence,

$$p_2(s) = \frac{1}{2} s(s - 1) f_0 - (s + 1)(s - 1) f_1 + \frac{1}{2}(s + 1)s f_2.$$

Integrating from x_0 to x_2 yields

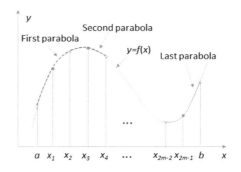

FIGURE 9.7

Calculation of the area under a curve using the Simpson rule.

$$\int_{x_0}^{x_1} f(x)\,dx \approx h \int_{-1}^{1} p_2(s)\,ds = h\left[\frac{f_0}{3} + \frac{4}{3}f_1 + \frac{f_2}{3}\right].$$

A similar formula holds for the next two subintervals from x_2 to x_4, and so on. Therefore, summing these m formulas gives the Simpson rule:

$$J = \int_a^b f(x)\,dx \simeq \frac{h}{3}\left[f_0 + 4f_1 + 2f_2 + 4f_3 + \ldots + 2f_{2m-2} + 4f_{2m-1} + f_{2m}\right].$$

(9.28)

The error bounds for Eq. (9.28) are obtained by a similar method to that of the trapezoidal rule (assuming the fourth derivative of f exists and is continuous):

$$CM_4 \le \epsilon \le CM_4^*, \quad \text{where } C = -\frac{(b-a)^5}{180(2m)^4},$$

(9.29)

where M_4 and M_4^* are the largest and smallest values of the fourth derivative in the interval of integration.

Example 9.5.2. Evaluate $\int_0^1 \sinh(x)\,dx$ by the Simpson rule and estimate the error with $n = 10$.

Solution.
Using Eq. (9.28) one has

$$J \simeq \frac{0.1}{3}\bigg[\sinh(0) + 4(\sinh(0.1)) + 2(\sinh(0.2)) + 4(\sinh(0.3)) + 2(\sinh(0.4))$$

$$+ 4(\sinh(0.5)) + 2(\sinh(0.6)) + 4(\sinh(0.7)) + 2(\sinh(0.8)) + 4(\sinh(0.9)))$$

$$+ (\sinh(1)) \bigg] = \underline{0.54308094}.$$

The error can be calculated from Eq. (9.29). By differentiation, $f^{IV}(x) = \sinh x$. By considering the derivative of f^{IV} (that is, f^{V}), the smallest value of f^{IV} in the interval of integration occurs at $x = 0$ and the largest value at $x = 1$. Therefore, $M_4 = f^{IV}(1) = 1.1752012$ and $M_4^* = f^{IV}(0) = 0$. Since $2m = 10$ and $b - a = 1$, one obtains $C = -1/(180 \times 10^4) = -0.00000056$. Therefore, from Eq. (9.29), $-0.00000065 \leq \epsilon \leq 0.0000000$ so that the exact value must lie between:

$$\underline{0.54308028} = 0.54308094 - 0.00000065 \leq J \leq 0.54308094 + 0.000000$$

$$= \underline{0.54308094}. \qquad \text{[answer]}$$

In fact, $J = 0.54308064$, which is exact to eight decimal places, so that J lies within these bounds. Note that for the same number of nodes, the present result is much better than that obtained in Example 9.5.1 with the trapezoidal rule.

9.5.3 Gaussian integration formula

This method provides a high degree of accuracy; however, it requires an irregular spacing of $x_1, ..., x_n$, where

$$\boxed{\int_{-1}^{1} f(x)\,dx \approx \sum_{j=1}^{n} A_j f_j \qquad [f_j = f(x_j)].} \qquad (9.30)$$

Here the values $x_1, ..., x_n$ are the n zeros of the Legendre polynomial $P_n(x)$ (see Eq. (2.144)), where

$$P_0 = 1, \quad P_1(x) = x, \quad P_2(x) = \frac{1}{2}(3x^2 - 1), \quad P_3(x) = \frac{1}{2}(5x^3 - 3x), \quad ....$$

The numerical value of the zeros of $P_n(x)$, and the corresponding coefficients $A_1, ..., A_n$ are given in Table 9.3.

Table 9.3 Values of zeros and coefficients for the Gaussian integration formula.

n	Zeros of $P_n(x)$	Coefficients, A_j
2	$\pm 1/\sqrt{3}$	1
3	0	8/9
	$\pm\sqrt{3/5}$	5/9
4	$\pm\sqrt{(15 - \sqrt{120})/35}$	0.6521451549
	$\pm\sqrt{(15 + \sqrt{120})/35}$	0.3478548451

The method can be applied to an integral with any constant limits by applying an appropriate transformation so that $\int_a^b f(x)\,dx = \int_{-1}^1 g(x)\,dx$. This transformation is derived by letting $x = mt + c$, where $x = a$ when $t = -1$, and $x = b$ when $t = +1$. These requirements yield the two following equations:

$$a = m(-1) + c,$$
$$b = m(1) + c,$$

with the solution $m = \dfrac{b-a}{2}$ and $c = \dfrac{b+2}{2}$. Hence, the required transformation becomes $x = \left(\dfrac{b-a}{2}\right)t + \left(\dfrac{b+a}{2}\right)$ and $dx = \dfrac{b-a}{2}\,dt$. Finally,

$$\boxed{\int_a^b f(x)\,dx = \left(\frac{b-a}{2}\right)\int_{-1}^1 f\left(\frac{b-a}{2}t + \frac{b+a}{2}\right)dt.} \qquad (9.31)$$

Example 9.5.3. Evaluate $\int_0^1 \sinh x\,dx$ by the Gaussian integration formula with $n = 3$.

Solution.
Using Eq. (9.31) and Table 9.3 one has

$$J = \int_0^1 \sinh x\,dx$$

$$= \frac{1}{2}\int_{-1}^1 \sinh \frac{1}{2}(t+1)\,dt$$

$$\simeq \frac{1}{2}\left\{\frac{5}{9}\sinh\left[\frac{1}{2}\left(-\sqrt{\frac{3}{5}}+1\right)\right] + \frac{8}{9}\sinh\left[\frac{1}{2}\right] + \frac{5}{9}\sinh\left[\frac{1}{2}\left(\sqrt{\frac{3}{5}}+1\right)\right]\right\}$$

$$= \underline{0.54308037.} \qquad \text{[answer]}$$

This result is as accurate as the result in Example 9.5.2 but requires fewer operations.

9.5.4 Numerical differentiation

Numerical differentiation involves the computation of a derivative of a function f from given values of f. Such formulas are basic to the numerical solution of differential equations.

Defining $f_j' = f'(x_j)$, $f_j'' = f''(x_j)$, where $f'(x) = \lim\limits_{h \to 0} \dfrac{f(x+h) - f(x)}{h}$, one obtains the relations

$$f_{1/2}' \approx \frac{\delta f_{1/2}}{h} = \frac{f_1 - f_0}{h}$$

and

$$f_1'' \approx \frac{\delta^2 f_1}{h^2} = \frac{f_2 - 2f_1 + f_0}{h^2}.$$

More accurate approximations can be obtained by differentiating suitable Lagrange interpolation polynomials. For example, given the Lagrange interpolation polynomial as before,

$$p_2(x) = \frac{(x - x_1)(x - x_2)}{2h^2} f_0 - \frac{(x - x_0)(x - x_2)}{h^2} f_1 + \frac{(x - x_0)(x - x_1)}{2h^2} f_2,$$

where $x_1 - x_0 = x_2 - x_1 = h$, one obtains

$$f'(x) \simeq p_2'(x) = \frac{2x - x_1 - x_2}{2h^2} f_0 - \frac{2x - x_0 - x_2}{h^2} f_1 + \frac{2x - x_0 - x_1}{2h^2} f_2.$$

Evaluating this expression at the points x_0, x_1, x_2 yields the "three-point" formula:

$$f_0' \approx \frac{1}{2h} (-3f_0 + 4f_1 - f_2),$$

$$f_1' \approx \frac{1}{2h} (-f_0 + f_2),$$

$$f_2' \approx \frac{1}{2h} (f_0 - 4f_1 + 3f_2).$$

Similarly, using the Lagrange interpolation polynomial $p_4(x)$ yields

$$f_2' \approx \frac{1}{12h} (f_0 - 8f_1 + 8f_3 - f_4).$$

Problems

9.1 Solve $x = \cos x$ by:

(a) fixed point iteration ($x_0 = 1$, 20 steps, six significant figures);

(b) Newton's method ($x_0 = 1$, six decimal places) (sketch the function first);

(c) the secant method ($x_0 = 0.5$, $x_1 = 1$, six decimal places).

9.2 As shown in the moving-boundary Problem 5.7 with a liquid front during the melting of a material, the following transcendental equation arises for the variable γ:
$\gamma e^{\gamma^2} \mathrm{erf}(\gamma) = \dfrac{\sigma (u_w - u_m)}{L\sqrt{\pi}}$. To numerically solve this transcendental equation (if all of the other constants and material property values were known), define the function $f(\gamma)$ to be used and give the iteration scheme for each of the following methods: (i) fixed point iteration, (ii) Newton's method, and (iii) the secant method. For application of the Newton method, what is the expression for $f'(x)$?

9.3 Consider the following tabular data: $\ln 9.0 = 2.1972$, $\ln 9.5 = 2.2513$, and $\ln 11.0 = 2.3979$.

(a) Calculate the Lagrange interpolation polynomial $p_2(x)$ and compute approximations of $\ln 9.4$, $\ln 10$, $\ln 10.5$, $\ln 11.5$, $\ln 12$. Evaluate the errors by using exact values to four decimal places.

(b) Comment on the effect of extrapolation.

(c) Set up Newton's divided difference formula for the same data, and derive from it $p_2(x)$.

9.4 Find the cubic spline to the given data with $k_0 = -2$ and $k_2 = -14$. Data: $f_0 = f(-2) = 1$, $f_1 = f(0) = 5$, $f_2 = f(2) = 17$. What is the advantage of a spline fit compared to a Lagrange polynomial for interpolation?

9.5 Evaluate the definite integral $J = \int_0^1 \dfrac{dx}{1+x^2}$ using:

(a) an exact analytic formula obtained from calculus;

(b) approximate methods: (i) the trapezoidal rule ($n = 4$); (ii) Simpson's rule ($2m = 4$); and (iii) the Gaussian integration formula ($n = 4$).

(c) Compute the error bounds for the trapezoidal rule in part (b).

(d) Compare the results obtained in parts (a) and (b).

9.6 The probability P that a measurement will fall within t standard deviations is given by $P = \displaystyle\int_{\mu-t\sigma}^{\mu+t\sigma} \dfrac{\exp\left(\dfrac{-(x-\mu)^2}{2\sigma^2}\right)}{\sigma\sqrt{2\pi}}\, dx$.

(a) Show how this probability expression reduces to the normal error integral $P = \dfrac{1}{\sqrt{2\pi}}\displaystyle\int_{-t}^{t} e^{-\frac{z^2}{2}}\, dz$.

(b) Using the normal error integral and the Gaussian integration formula (for $n = 2$), calculate the probability that a measurement will fall within one standard deviation (that is, $t = 1$). What is the relative error for this estimation?

9.7 The sine integral is defined as $Si(x) = \displaystyle\int_0^x \dfrac{\sin u}{u}\, du$. Evaluate the quantity $Si(1)$ (that is, for $x = 1$) using a Gaussian integration formula to four significant digits using the zeros of the Legendre polynomial $P_2(x)$. Using the actual value from a mathematical handbook (for example, Spiegel, 1973), estimate the relative and absolute error for the estimate.

9.8 For application of the Gauss integration method, show that using Eq. (9.31) the following integral can be transformed as

$$\int_0^1 e^{-x^2}\, dx = \frac{1}{2}\int_{-1}^1 \exp\left[-\frac{1}{4}(t+1)^2\right]\, dt.$$

Introduction to complex analysis

Complex analysis is important in many fields of engineering. For instance, applications in electrical engineering include control theory (transfer functions), signal analysis in communications, Fourier transform analysis, circuit and electronic theory, electromagnetism (time-harmonic fields), and electrostatics (solution of the Laplace equation for the complex potential). In addition to its application in electrostatics, the complex potential is a powerful methodology that can be further employed to solve analogous problems in fluid and heat flow in chemical engineering, as well as airfoil design in the aerospace industry. Complex analysis can be further used to solve boundary value problems for solution of differential equations, as well as for evaluation of inverse transforms and definite integrals. Complex functions arise in power transmission, and complex number field theory is also an important tool in quantum mechanics.

This chapter introduces the properties of complex functions and integration methods leading to the development of the residue theorem. For example, the residue theorem is used to evaluate definite integrals in Chapter 3, as well as to determine an inverse Laplace transform for solution of a heat conduction problem in Chapter 5. Finally contour mapping is applied in this chapter for the solution of Dirichlet boundary value problems and for various engineering problems in fluid flow, electrostatics, and heat flow.

10.1 Complex functions

A function w of a complex variable z can be written as $w = f(z)$. The function is single-valued if for each value of z there corresponds only one value of w; otherwise it is multivalued.

In general, $w = f(z) = u(x, y) + iv(x, y)$, where u and v are real functions of x and y. For example, $w = z^2 = (x + iy)^2 = x^2 - y^2 + 2ixy = u + iv$. As such $u(x, y) = x^2 - y^2$ and $v(x, y) = 2xy$.

Limits and continuity

The function $f(z)$ is said to have a limit l as z approaches z_0 if, given any $\epsilon > 0$, there exists a $\delta > 0$ such that $|f(z) - l| < \epsilon$ whenever $0 < |z - z_0| < \delta$. Alternatively, $f(z)$ is continuous if $\lim_{z \to z_0} f(z) = f(z_0)$.

Advanced Mathematics for Engineering Students. https://doi.org/10.1016/B978-0-12-823681-9.00018-6

Derivatives

If $f(z)$ is single-valued in some region of the z plane, the derivative of $f(z)$ is defined as

$$f'(z) = \lim_{\Delta z \to 0} \frac{f(z + \Delta z) - f(z)}{\Delta z}, \tag{10.1}$$

provided the limit exists independent of the manner in which $\Delta z \to 0$. If the limit in Eq. (10.1) exists for $z = z_0$, then $f(z)$ is called analytic at z_0. If the limit exists for all z in a region $\Re$, then $f(z)$ is called analytic in $\Re$. In order to be analytic, $f(z)$ must be single-valued and continuous (however, the converse is not necessarily true).

This definition for the derivative is analogous to that of a real function. However, in contrast, complex derivatives and differential functions satisfy a much stronger condition. Complex functions that are differentiable at every point on an open subset of the complex plane are said to be holomorphic. Here, the value of the difference quotient in the limit of Eq. (10.1) must approach the same complex number, regardless of the way that z_0 is approached. Hence, holomorphic functions are infinitely differentiable. They are also analytic at every point in the domain as given by a convergent power series.

Elementary functions of a complex variable are natural extensions of the corresponding real functions, that is, where series expansions for real functions $f(x)$ exists, the complex series is obtained by replacing x with z. For example,

$$e^z = 1 + z + \frac{z^2}{2!} + \frac{z^3}{3!} + ...,$$

$$\sin z = z - \frac{z^3}{3!} + \frac{z^5}{5!} - \frac{z^7}{7!} + ...,$$

$$\cos z = 1 - \frac{z^2}{2!} + \frac{z^4}{4!} - \frac{z^6}{6!} + ...$$

Also from these series it follows that $e^z = e^{x+iy} = e^x(\cos y + i \sin y)$. On letting $z = i\theta$, one obtains the so-called Euler formula: $e^{i\theta} = \cos\theta + i\sin\theta$. The polar form of a complex number is $\rho e^{i\phi}$. Moreover, if p is any real number, de Moivre's theorem states $[\rho(\cos\theta + i\sin\theta)]^p = \rho^p(\cos p\theta + i\sin p\theta)$.

The number w is called the nth root of the complex number z if $w^n = z$ (or equivalently $w = z^{1/n}$). Moreover, if n is a positive integer, de Moivre's theorem yields

$$z^{1/n} = [\rho(\cos\theta + i\sin\theta)]^{1/n}$$

$$= \rho^{1/n}\left[\cos\left(\frac{\theta + 2k\pi}{n}\right) + i\sin\left(\frac{\theta + 2k\pi}{n}\right)\right], \qquad k = 0, 1, 2, ..., n - 1.$$

Thus, there are n different values for $z^{1/n}$ provided that $z \neq 0$.

Example 10.1.1. Show that $\ln z$ is a multivalued function.

Solution.

Define a^b as $e^{b \ln a}$, where a and b are complex numbers. Since $e^{2\pi ki} = 1$, we have $e^{i\phi} = e^{i(\phi + 2k\pi)}$, where k is an integer. One therefore defines $\ln z = \ln(\rho e^{i\phi}) = \ln \rho + i(\phi + 2k\pi)$. Hence, $\ln z$ is a multivalued function (that is, this multi-valued function is composed of various single-valued functions that are called its <u>branches</u>).   [answer]

Rules for differentiating functions of a complex variable are the same as those for real variables. For example, $\dfrac{d}{dz}(z^n) = nz^{n-1}$ and $\dfrac{d}{dz}(\sin z) = \cos z$.

Cauchy–Riemann equations

A necessary condition for $w = f(z) = u(x, y) + iv(x, y)$ to be analytic in a region $\Re$ is that u and v satisfy <u>Cauchy–Riemann equations</u>:

$$\boxed{\frac{\partial u}{\partial x} = \frac{\partial v}{\partial y}, \quad \frac{\partial u}{\partial y} = -\frac{\partial v}{\partial x}.} \tag{10.2}$$

If the partial derivatives in Eq. (10.2) are continuous in $\Re$, the equations are sufficient conditions that $f(z)$ is analytic in $\Re$. If the second derivatives of u and v with respect to x and y exist and are continuous, one finds that differentiating Eq. (10.2),

$$\boxed{\frac{\partial^2 u}{\partial x^2} + \frac{\partial^2 u}{\partial y^2} = 0, \quad \frac{\partial^2 v}{\partial x^2} + \frac{\partial^2 v}{\partial y^2} = 0.} \tag{10.3}$$

Thus, the real and imaginary parts satisfy Laplace's equation in two dimensions. Such functions are called <u>harmonic functions</u>.

10.2 **Complex integration**

If $f(x)$ is defined, single-valued, and continuous in $\Re$, the integral of $f(z)$ can be defined along some path C in $\Re$ from point $z_1(= x_1 + iy_1)$ to point $z_2(= x_2 + iy_2)$ as

$$\int_C f(z)\, dz = \int_{(x_1, y_1)}^{(x_2, y_2)} (u + iv)(dx + i\, dy)$$

$$= \int_{(x_1, y_1)}^{(x_2, y_2)} (u\, dx - v\, dy) + i \int_{(x_1, y_1)}^{(x_2, y_2)} (v\, dx + u\, dy).$$

With this definition, the integral of a function of a complex variable can be made to depend on line integrals of real functions. The rules for complex integration are similar to those for real integrals. For instance,

$$\boxed{\left| \int_C f(z)\, dz \right| \leq \int_C |f(z)|\, |dz| \leq M \int_C ds = ML,} \tag{10.4}$$

where M is an upper bound of $|f(z)|$ on C, that is, $|f(z)| \le M$, and L is the length of the path C.

Cauchy's theorem

Let C be a simple closed curve. If $f(z)$ is analytic within the region bounded by C as well as on C, then Cauchy's theorem states

$$\int_C f(z)\, dz = \oint_C f(z)\, dz = 0, \tag{10.5}$$

where the second integral emphasizes the fact that C is a simple closed curve. Eq. (10.5) is equivalent to the statement that $\int_{z_1}^{z_2} f(z)\, dz$ has a value independent of the path joining z_1 and z_2. Such integrals can be evaluated as $F(z_2) - F(z_1)$, where $F'(z) = f(z)$. For example:

(i) Since $f(z) = 2z$ is analytic everywhere, for any simple closed curve C :
$$\oint_c 2z\, dz = 0.$$

(ii) We have $\int_{2i}^{1+i} 2z\, dz = z^2 \Big|_{2i}^{1+i} = (1+i)^2 - (2i)^2 = 2i + 4.$

Example 10.2.1. Derive Cauchy's theorem from first principles.

Solution.
Eq. (10.5) is a line integral, where $f(z) = u + iv$ is analytic over a region bounded by the closed C and dz is an infinitesimal part of the path around C such that $dz = dx + idy$. Hence, Eq. (10.5) can be written as

$$\oint_C f(z)dz = \oint_C (u+iv)(dx+idy) = \oint_C (u\,dx - v\,dy) + i\oint_C (v\,dx + u\,dy). \tag{10.6}$$

Consider the Green's theorem from Eq. (4.13) in Section 4.4.2:

$$\oint_C (P\,dx + Q\,dy) = \iint_A \left(\frac{\partial Q}{\partial x} - \frac{\partial P}{\partial y} \right) dx\,dy,$$

where $P(x)$ and $Q(x)$ are well-behaved functions and A is the area bounded by the contour C. One can apply this theorem to each of the terms in the last expression of Eq. (10.6). Therefore, for the first term, letting $P = u$ and $Q = -v$ yields

$$\oint_C (u\,dx - v\,dy) = \iint_A \left(-\frac{\partial v}{\partial x} - \frac{\partial u}{\partial y} \right) dx\,dy = 0.$$

This expression equals zero because of the second Cauchy–Riemann condition in Eq. (10.2). Similarly, letting $P = v$ and $Q = u$ in the second term yields

$$\oint_C (v\,dx + u\,dy) = \iint_A \left(\frac{\partial u}{\partial x} - \frac{\partial v}{\partial y} \right) dx\,dy = 0.$$

This expression also equals zero because of the first Cauchy–Riemann condition in Eq. (10.2). Thus, Eq. (10.5) is shown to equal zero.   [answer]

Cauchy's integral formulas

If $f(z)$ is analytic within and on a simple closed curve C and a is any point interior to C, then

$$f(a) = \frac{1}{2\pi i} \oint_C \frac{f(z)}{z - a}\,dz, \tag{10.7}$$

where C is traversed in the positive counterclockwise sense. Also the nth derivative of $f(z)$ at $z = a$ is given by

$$f^{(n)}(a) = \frac{n!}{2\pi i} \oint_C \frac{f(z)}{(z - a)^{n+1}}\,dz. \tag{10.8}$$

Eqs. (10.7) and (10.8) are called Cauchy's integral formulas. These are quite remarkable because they show that if the function $f(z)$ is known on the closed curve C, then it is also known within C, and the various derivatives and points within C can be calculated. Thus, if a function of a complex variable has a first derivative, it has all higher derivatives as well. This result of course is not necessarily true for functions of real variables.

Example 10.2.2. Evaluate $\oint_C \frac{\cos z}{z - \pi}\,dz$, where C is the circle $|z - 1| = 3$.

Solution.
Since $z = \pi$ lies within C, $\dfrac{1}{2\pi i} \oint_C \dfrac{\cos z}{z - \pi}\,dz = \cos \pi = -1$ by Eq. (10.7) with $f(z) = \cos z$ and $z = \pi$. Therefore,

$$\oint_C \frac{\cos z}{z - \pi}\,dz = -2\pi i.   \quad \text{[answer]}$$

Example 10.2.3. Derive the Cauchy integral formula in Eq. (10.7) from first principles.

Solution.
Consider the function $\phi(z) = \dfrac{f(z)}{(z - a)}$, which is analytic inside of C except at the point a, which lies within C, where there is a singularity. Furthermore, let C' be a small circle of radius ρ that is centered at a and make a cut between C and C' along

AB, as shown in Fig. 10.1. Furthermore, consider the path in Fig. 10.1 from A around C to A', along $A'B'$, around C' from B' to B, and back along BA.

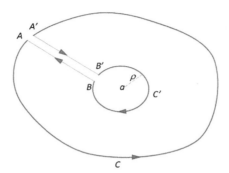

FIGURE 10.1

Schematic of a closed path for evaluation of Cauchy's integral formula.

From Cauchy's theorem in Eq. (10.5),

$$\oint_{AA'B'BA} \phi(z)\,dz = 0$$

for the closed path. Since $\phi(z)$ is analytic in the region between C and C', we have

$$\int_{AA'} \phi(z)\,dz + \int_{A'B'} \phi(z)\,dz + \int_{B'B} \phi(z)\,dz + \int_{BA} \phi(z)\,dz = 0.$$

If now both A and A', and B and B' each come together and coincide, then the integral along the straight line $A'B'$ is equal but opposite to the integral along BA so that these two integrals cancel, leaving the following two integrals:

$$\int_{AA'} \phi(z)\,dz + \int_{B'B} \phi(z)\,dz = 0.$$

Recognizing the direction of the path integrals,

$$\underbrace{\oint_C \phi(z)\,dz}_{counterclockwise} + \underbrace{\oint_{C'} \phi(z)\,dz}_{clockwise} = 0$$

or

$$\oint_C \phi(z)\,dz - \oint_{C'} \phi(z)\,dz = 0, \qquad (10.9)$$

where the integrals in Eq. (10.9) are understood to be counterclockwise. For the circle C' of radius ρ centered at a in Fig. 10.1, one can use polar coordinates so that $z =$

$a + \rho\, e^{i\theta}$ and $dz = i\rho\, e^{i\theta}\, d\theta$, and therefore

$$\int_{C'} \phi(z)\, dz = \oint_{C'} \frac{f(z)}{(z-a)}\, dz = \int_0^{2\pi} \frac{f(z)}{\rho e^{i\theta}} i\rho e^{i\theta}\, d\theta = i \int_0^{2\pi} f(z)\, d\theta.$$

Since this result is valid for any ρ, one can let ρ approach zero for which z approaches a. As $f(z)$ is both continuous and analytic at $z = a$, in the limit $\lim_{z \to a} f(z) = f(a)$. Thus, the integral around C' is evaluated as

$$\oint_{C'} \phi(z)\, dz = i\, f(a) \int_0^{2\pi} d\theta = 2\pi i\, f(a).$$

Hence, substituting this result into Eq. (10.9) yields the final result in Eq. (10.7):

$$f(a) = \frac{1}{2\pi i} \oint_C \frac{f(z)}{(z-a)}\, dz. \qquad \text{[answer]}$$

10.3 Taylor and Laurent series

Let $f(z)$ be analytic inside and on a circle having its center at $z = a$. Then for all points z in the circle, the <u>Taylor series</u> representation of $f(z)$ is

$$f(z) = f(a) + f'(a)(z-a) + \frac{f''(a)}{2!}(z-a)^2 + \frac{f'''(a)}{3!}(z-a)^3 + \dots. \quad (10.10)$$

Example 10.3.1. If $f(z)$ is analytic at all points inside and on a circle of radius R with center at a and if $a + h$ is any point inside C, prove Taylor's theorem that

$$f(a+h) = f(a) + hf'(a) + \frac{h^2}{2!}f''(a) + \frac{h^3}{3!}f'''(a) + \dots.$$

Solution.
By Cauchy's integral formula Eq. (10.7),

$$f(a+h) = \frac{1}{2\pi i} \oint_C \frac{f(z)}{z - a - h}\, dz. \qquad (10.11)$$

However,

$$\frac{1}{z - a - h} = \frac{1}{(z-a)\left[1 - \dfrac{h}{z-a}\right]}$$

$$= \frac{1}{(z-a)} \left\{ 1 + \frac{h}{(z-a)} + \frac{h^2}{(z-a)^2} + \dots + \frac{h^n}{(z-a)^n} + \frac{h^{n+1}}{(z-a)^n(z-a-h)} \right\}.$$

$$(10.12)$$

Substituting Eq. (10.12) into Eq. (10.11) and using the Cauchy integral formulas in Eq. (10.7) and Eq. (10.8) yields

$$f(a+h) = \frac{1}{2\pi i} \oint_C \frac{f(z)dz}{z-a} + \frac{h}{2\pi i} \oint_C \frac{f(z)dz}{(z-a)^2} + \ldots + \frac{h^n}{2\pi i} \oint_C \frac{f(z)dz}{(z-a)^{n+1}} + R_n$$

$$= f(a) + hf'(a) + \frac{h^2}{2!} f''(a) + \ldots + \frac{h^n}{n!} f^{(n)}(a) + R_n,$$

where $R_n = \dfrac{h^{n+1}}{2\pi i} \oint_C \dfrac{f(z)dz}{(z-a)^{n+1}(z-a-h)}.$

Now when z is on C, $\left| \dfrac{f(z)}{z-a-h} \right| \leq M$ and $|z-a| = R$ so that by Eq. (10.4) since $2\pi R$ is the length of C,

$$|R_n| \leq \frac{|h|^{n+1}M}{2\pi R^{n+1}} \cdot 2\pi R.$$

As $n \to \infty$, $|R_n| \to 0$ and the required result follows.   [answer]

Singular points

A singular point of a function $f(z)$ is a value of z at which $f(z)$ fails to be analytic. If $f(z)$ is analytic everywhere in some region except at an interior point $z = a$, the point $z = a$ is called an isolated singularity of $f(z)$. For example, if $f(z) = \dfrac{1}{(z-3)^2}$, the point $z = 3$ is an isolated singularity of $f(z)$.

Poles

If $f(z) = \dfrac{\phi(z)}{(z-a)^n}$, $\phi(a) \neq 0$, where $\phi(z)$ is analytic everywhere in a region including $z = a$, and if n is a positive integer, then $f(z)$ has an isolated singularity at $z = a$ which is called a pole of order n. If $n = 1$, the pole is called a simple pole. If $n = 2$, the pole is a double pole.

For example, $f(z) = \dfrac{z}{(z-3)^2(z+1)}$ has two singularities: a double pole at $z = 3$ and a simple pole at $z = -1$.

As another example, $f(z) = \dfrac{3z-1}{z^2+4} = \dfrac{3z-1}{(z+2i)(z-2i)}$ has two simple poles at $z = \pm 2i$.

A function can have other types of singularities besides poles. For example, $f(z) = \sqrt{z}$ has a branch point at $z = 0$ (which follows since $f(z)$ is a double-valued function). Also the function $\dfrac{\sin z}{z}$ has a singularity at $z = 0$. However, due to the fact that $\lim\limits_{z \to 0} \dfrac{\sin z}{z}$ is finite, this is called a removable singularity.

If $f(z)$ has a pole of order n at $z = a$ but is analytic at every other point and on a circle C with center at a, the function $(z - a)^n f(z)$ is analytic at all points inside and on C and has a Taylor series about $z = a$ so that

$$f(z) = \frac{a_{-n}}{(z - a)^n} + \frac{a_{-n+1}}{(z - a)^{n-1}} + \dots + \frac{a_{-1}}{(z - a)} + a_0 + a_1(z - a) + \dots$$

or

$$f(z) = a_0 + a_1(z - a) + a_2(z - a)^2 + \dots + \frac{a_{-1}}{(z - a)} + \frac{a_{-2}}{(z - a)^2} + \dots.$$

(10.13)

This series is called the <u>Laurent series</u> for $f(z)$. Here $a_0 + a_1(z - a) + a_2(z - a)^2 + \dots$ is called the <u>analytic part</u>, while the remainder (consisting of the inverse powers of $(z - a)$) is the <u>principal part</u>. A Laurent series is more generally defined as

$$\sum_{k=-\infty}^{\infty} a_k(z - a)^k,$$

(10.14)

where the terms with $k < 0$ constitute the principal part. When the principal part of the Laurent series has a finite number of terms and $a_{-n} \neq 0$ while $a_{-n-1}, a_{-n-2}, \dots$ are all zero, then $z = a$ is a pole of order n. If the principal part has infinitely many terms, $z = a$ is called an <u>essential singularity</u>. For example, the function $e^{1/z} = 1 + \frac{1}{z} + \frac{1}{2!z^2} + \dots$ has an essential singularity at $z = 0$.

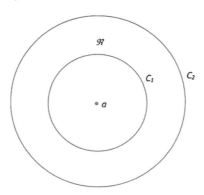

FIGURE 10.2

Schematic of contours for the Laurent series.

Consider C_1 and C_2 as two concentric circles centered at $z = a$, as shown in Fig. 10.2. If $f(z)$ is analytic in the region $\Re$ between the circles, then $f(z)$ can be expanded as a Laurent series that is convergent in $\Re$. In general, for a Laurent series, the power series converges <u>inside</u> a circle C_2, while the inverse power series converges <u>outside</u> the circle C_1. Hence, the radius of C_2 is the <u>radius of convergence</u> for the a_m

series that converges for any value of $|z| \leq \rho_2$, where ρ_2 is the radius of C_2. The a_{m-} series will converge for any value of $|z| \geq \rho_1$, where ρ_1 is the radius of C_1.

Coefficients of a Laurent series

The coefficients $a_0, a_1, a_2...a_{-1}, a_{-2}...$ for the Laurent series expansion of $f(x)$ can be evaluated as follows.

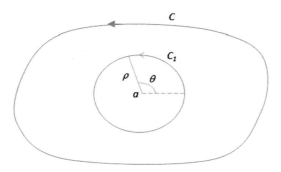

FIGURE 10.3

Schematic of a circle C_1 bounded within a region C.

Let C_1 be a circle of radius ρ centered at a (see Fig. 10.3). Since $(z - a)^{-n}$ is analytic within and on the boundary of the region bounded by C and C_1,

$$I_n = \oint_C \frac{dz}{(z-a)^n} = \oint_{C_1} \frac{dz}{(z-a)^n}.$$

Using polar coordinates on C_1, $|z - a| = \rho$ or $z - a = \rho e^{i\theta}$ and $dz = i\rho \, e^{i\theta} \, d\theta$. The integral therefore equals

$$\int_0^{2\pi} \frac{\rho i e^{i\theta} \, d\theta}{(\rho e^{i\theta})^n} = \frac{i}{\rho^{n-1}} \int_0^{2\pi} e^{(1-n)i\theta} \, d\theta.$$

Consider two cases for the integral:

(i) $(n = 1)$: $I_1 = i \int_0^{2\pi} e^0 d\theta = 2\pi i$;

(ii) $(n \neq 1)$:

$$I_n = \frac{i}{\rho^{n-1}} \frac{1}{i(1-n)} \left[e^{(1-n)i\theta} \, d\theta \right]_0^{2\pi} = \frac{1}{\rho^{n-1}} \frac{1}{(1-n)} \left[e^{(1-n)i2\pi} - 1 \right]$$

$$= \frac{1}{\rho^{n-1}} \frac{1}{(1-n)} \left[\underbrace{\cos[(1-n)2\pi]}_{=+1} + i \underbrace{\sin[(1-n)2\pi]}_{=0} - 1 \right]$$

$$= 0 \quad \text{for any integer } n \text{ positive or negative except } n = 1.$$

Then the integral I_n yields

$$\oint_C \frac{dz}{(z-a)^n} = \begin{cases} 2\pi i, & \text{for } n = 1, \\ 0, & \text{for } n \neq 1, \text{ positive or negative.} \end{cases} \tag{10.15}$$

Now, consider the integral

$$J = \frac{1}{2\pi i} \oint_C \frac{f(z)}{(z-a)^{n+1}} \, dz \tag{10.16}$$

with the Laurent series $f(z) = a_0 + a_1(z-a) + a_2(z-a)^2 + \ldots + \dfrac{a_{-1}}{(z-a)} + \dfrac{a_{-2}}{(z-a)^2} + \ldots$. As mentioned, this series is convergent in a region $\Re$ within two concentric circles C_1 and C_2 centered on the point a (see Fig. 10.2), where C is any simple closed curve surrounding a and lying in $\Re$.

Thus, inserting $f(z)$ into J gives

$$J = \frac{1}{2\pi i} \oint_C \left[\sum_m \frac{a_m(z-a)^m}{(z-a)^{n+1}} + \sum_m \frac{a_{-m}}{(z-a)^m} \frac{1}{(z-a)^{n+1}} \right] dz \tag{10.17}$$

$$= \frac{1}{2\pi i} \oint_C \left[\sum_m \frac{a_m}{(z-a)^{n-m+1}} + \sum_m \frac{a_{-m}}{(z-a)^{n+m+1}} \right] dz. \tag{10.18}$$

From Eq. (10.15), one only gets a nonzero result in the first term when $n - m + 1 = 1$ (that is, $m = n$), which evaluates as a_n for the integral in Eq. (10.16). Hence, the integral J gives the coefficient a_n:

$$a_n = \frac{1}{2\pi i} \oint_C \frac{f(z)}{(z-a)^{n+1}} \, dz.$$

Similarly, with $n + m + 1 = 1$ (that is, $m = -n$), the coefficient a_{-n} is given by

$$a_{-n} = \frac{1}{2\pi i} \oint_C \frac{f(z)}{(z-a)^{-n+1}} \, dz.$$

In summary:

- If all a_{-m} coefficients are zero, then the Laurent expansion reduces to a Taylor series expansion, $f(z)$ is analytic at $z = a$, and a is called a regular point.
- If $a_{-m} = 0$ for all $m > n$, then $f(z)$ is said to have a <u>pole</u> of order n at $z = a$. If $n = 1$, then $f(z)$ is said to have a <u>simple pole</u> at $z = a$.
- If there are an infinite number of nonzero a_{-m} coefficients, then $f(z)$ is said to have an <u>essential singularity</u> at $z = a$.

- The coefficient a_{-1} has a special meaning and is recognized as the <u>residue</u> of $f(z)$ at $z = a$ (see Section 10.4).

Example 10.3.2. Find the Laurent series for the function $\dfrac{e^z}{(z-1)^2}$ about the singularity $z = 1$.

Solution.

Let $z - 1 = u$ so that $z = 1 + u$. Then

$$\frac{e^z}{(z-1)^2} = \frac{e^{1+u}}{u^2} = e \cdot \frac{e^u}{u^2} = \frac{e}{u^2}\left[1 + u + \frac{u^2}{2!} + \frac{u^3}{3!} + \cdots\right]$$

$$= \frac{e}{(z-1)^2} + \frac{e}{(z-1)} + \frac{e}{2!} + \frac{e(z-1)}{3!} + \frac{e(z-1)^2}{4!} + \cdots. \qquad \text{[answer]}$$

Thus, $z = 1$ is a pole of order 2 (that is, a double pole). The series converges for all values of $z \ne 1$.

Example 10.3.3. Given $f(z) = e^z = 1 + z + \dfrac{z^2}{2!} + \dfrac{z^3}{3!} + \cdots$, what is the value of a?

Solution.

On inspection of the Laurent series in Eq. (10.13), $a = 0$. [answer]

Note that there is no series with terms containing a_-. Thus, the function $f(a)$ is analytic at $z = a$. In this example, $f(a) = 1$. Lastly, there is no a_{-1} term, and hence the residue at $z = a = 0$ is zero.

Example 10.3.4. Given $f(z) = \dfrac{e^z}{z^3} = \dfrac{1}{z^3} + \dfrac{1}{z^2} + \dfrac{1}{2z} + \dfrac{1}{3!} + \dfrac{z}{4!} + \cdots$, what is the value of a?

Solution.

Again the value is identified as $a = 0$. [answer]

However, in this example, there is a series containing the a_{-m} terms up to a term in $1/z^3$, thus a is a pole of order three. The residue is the coefficient $a_{-1} = \frac{1}{2}$.

10.4 Residue integration method

The coefficients in Eq. (10.13) can be obtained by writing the coefficient of the Taylor series corresponding to $(z-a)^n f(z)$. The coefficient a_{-1}, called the <u>residue</u> of $f(z)$ at the pole $z = a$, is of considerable importance. It can be evaluated from

$$\boxed{a_{-1} = \lim_{z \to a} \frac{1}{(n-1)!} \frac{d^{n-1}}{dz^{n-1}}\left\{(z-a)^n f(z)\right\},} \qquad (10.19)$$

where n is the order of the pole. For a simple pole

$$a_{-1} = \lim_{z \to a} (z - a) f(z). \qquad (10.20)$$

Example 10.4.1. Find the residues of the function $f(z) = \dfrac{z^2}{(z-2)(z^2+1)}$ at the poles $z = 2, i,$ and $-i$.

Solution.
   These are simple poles and Eq. (10.20) applies so that:

Residue at $z = 2$: $\lim_{z \to 2} (z \cancel{-} 2) \dfrac{z^2}{(z \cancel{-} 2)(z^2+1)} = \dfrac{4}{5}$;

Residue at $z = i$: $\lim_{z \to i} (z \cancel{-} i) \dfrac{z^2}{(z-2)(z+i)(z \cancel{-} i)} = \dfrac{i^2}{(i-2)(2i)} = \dfrac{1-2i}{10}$;

Residue at $z = -i$: $\lim_{z \to -i} (z \cancel{+} i) \dfrac{z^2}{(z-2)(z \cancel{+} i)(z-i)} = \dfrac{i^2}{(-i-2)(-2i)} = \dfrac{1+2i}{10}$.

[answer]

10.5 **Residue theorem**

If $f(z)$ is analytic in a region $\Re$ except for a pole of order n at $z = a$ and C is a simple curve in $\Re$ containing $z = a$, $f(z)$ has the form of Eq. (10.13). Integrating the series and using the previous result from Eq. (10.15),

$$\oint_C \frac{dz}{(z-a)^n} = \begin{cases} 0 & \text{if } n \neq 1, \\ 2\pi i & \text{if } n = 1, \end{cases} \qquad (10.21)$$

it follows that

$$\oint_C f(z)\, dz = 2\pi i a_{-1}. \qquad (10.22)$$

Thus, Eq. (10.22) states that the integral of $f(z)$ around a closed path enclosing a single pole of $f(z)$ is $2\pi i$ times the residue at the pole.

Example 10.5.1. Prove the residue theorem in Eq. (10.22).

Solution.
Inserting the Laurent series in Eq. (10.13) into the integral $\oint_C f(z)\, dz$ yields

$$\oint_C f(z)\, dz = \oint_C \left(a_0 + a_1(z-a) + a_2(z-a)^2 \ldots + \frac{a_{-1}}{(z-a)} + \frac{a_{-2}}{(z-a)^2} + \ldots \right) dz$$
$$= a_0 I_0 + a_1 I_1 + a_2 I_2 + \ldots + a_{-1} I_{-1} + a_{-2} I_{-2} + \ldots.$$

Since I_{-1} is the only nonzero integral in accordance with Eq. (10.21), the final result of Eq. (10.22) follows.   [answer]

More generally, the <u>residue theorem</u> states that if $f(z)$ is analytic within and on the boundary C of a region $\Re$ except at a finite number of poles $a, b, c...$ within $\Re$, having residues $a_{-1}, b_{-1}, c_{-1}, ...$, respectively, then

$$\oint_C f(z)\,dz = 2\pi i\,(a_{-1} + b_{-1} + c_{-1} + ...). \tag{10.23}$$

Cauchy's theorem and integral formulas are special cases of this result.

Example 10.5.2. Prove the more general residue theorem in Eq. (10.23).

Solution.
Let C_1, C_2, C_3, and so on be contours, which each enclose one singularity, that is, C_1 encloses a, C_2 encloses b, C_3 encloses c, and so on. Furthermore, let C be a contour that makes cuts in the contours C_1, C_2, C_3, etc., as depicted in Fig. 10.4. Moreover, since the straight-line paths in this figure are traversed in opposite directions, these straight-line sections cancel in pairs so that

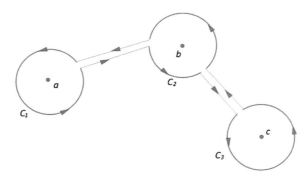

FIGURE 10.4

Multiple contours, each containing a singularity for evaluation of the residue theorem.

$$\oint_C f(z)\,dz = \oint_{C_1} f(z)\,dz + \oint_{C_2} f(z)\,dz + \oint_{C_3} f(z)\,dz...$$
$$= 2\pi i a_{-1} + 2\pi i b_{-1} + 2\pi i c_{-1} + ...,$$

where a_{-1}, b_{-1}, and c_{-1} are the residues of $f(z)$ at the singularities $z = a$, $z = b$, and $z = c$, respectively. Thus, the general result is

$$\oint_C f(z)\,dz = 2\pi i \cdot \text{sum of the residues of } f(z) \text{ enclosed by } C. \quad \text{[answer]}$$

The evaluation of various definite integrals can often be achieved by using the residue theorem together with suitable functions $f(z)$ and a suitable path or contour C.

Example 10.5.3. Show that $\displaystyle\int_0^{2\pi} \frac{\cos 3\theta}{5 - 4\cos\theta}\, d\theta = \frac{\pi}{12}$.

Solution.

If z in polar form is $z = 1 \cdot e^{i\theta}$, then $\cos\theta = \dfrac{z + z^{-1}}{2}$, $\cos 3\theta = \dfrac{e^{3i\theta} + e^{-3i\theta}}{2} = \dfrac{z^3 + z^{-3}}{2}$, and $dz = iz\, d\theta$. The integral therefore becomes

$$\int_0^{2\pi} \frac{\cos 3\theta}{5 - 4\cos\theta}\, d\theta = \oint_C \frac{(z^3 + z^{-3})/2}{5 - 4\left(\dfrac{z + z^{-1}}{2}\right)}\frac{dz}{iz} = -\frac{1}{2i}\oint_C \frac{z^6 + 1}{z^3(2z - 1)(z - 2)}\, dz,$$

where C is the circle of unit radius with center at the origin (see Fig. 10.5).

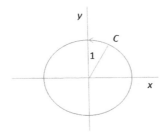

FIGURE 10.5

Schematic of a circle C of unit radius.

The integrand has a pole of order 3 at $z = 0$ and a simple pole $z = 1/2$ within C:

Residue at $z = 0$: $\displaystyle\lim_{z \to 0} \frac{1}{2!}\frac{d^2}{dz^2}\left\{ z^3 \cdot \frac{z^6 + 1}{z^3(2z - 1)(z - 2)}\right\} = \frac{21}{8}$;

Residue at $z = 1/2$: $\displaystyle\lim_{z \to 1/2}\left\{ \frac{(2z - 1)}{2} \cdot \frac{z^6 + 1}{z^3(2z - 1)(z - 2)}\right\} = -\frac{65}{24}$.

Thus, $-\dfrac{1}{2i}\oint_C \dfrac{z^6 + 1}{z^3(2z - 1)(z - 2)}\, dz = -\dfrac{1}{2i}(2\pi i)\left\{\dfrac{21}{8} - \dfrac{65}{24}\right\} = \dfrac{\pi}{12}$ by the residue theorem.   [answer]

10.6 Applications of conformal mapping

Consider the analytic function $w = f(z) = u(x, y) + iv(x, y)$. This function provides a transformation (or mapping) that establishes a correspondence between points on the uv and xy planes, where $u = u(x, y)$ and $v = v(x, y)$. If there is a mapping of curves C_1 and C_2 on the xy plane into curves C_1' and C_2' on the uv plane, where the angle between C_1 and C_2 remains unchanged in both magnitude and sense with the mapping to the curves C_1' and C_2', then the mapping is said to be <u>conformal</u> (Spiegel, 1964). Moreover, if $f(z)$ is analytic with $f'(z) \neq 0$ in a region $\mathfrak{R}$, it follows that the mapping $w = f(z)$ is conformal at all points of $\mathfrak{R}$. <u>Riemann's mapping theorem</u> explicitly states that there exists a function $w = f(z)$ in the region $\mathfrak{R}$ that maps each point of $\mathfrak{R}$ into a corresponding point of $\mathfrak{R}'$, as well as each point of the simple closed boundary curve in the z plane of $\mathfrak{R}$, into a so-called <u>unit circle</u> in the w plane (which forms the boundary of the region $\mathfrak{R}'$) (see Fig. 10.6). The correspondence for this mapping is one-to-one.

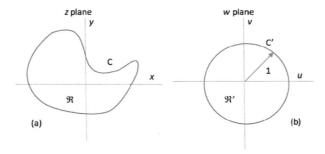

FIGURE 10.6

Schematic for the Riemann mapping from (a) the z plane to (b) the w plane.

Application to Dirichlet boundary value problems

Boundary value problems in Chapter 5 require the solution of a partial differential equation with associated boundary conditions. Consider a <u>Dirichlet problem</u>, for instance, that requires a solution of a harmonic function ϕ of the Laplace equation in a simply connected region $\mathfrak{R}'$ bounded by a simple closed boundary curve C (see Fig. 10.6(a)). The solution function is subject to prescribed values on this boundary C. In fact, if C is a unit circle $|z| = 1$, a function that satisfies Laplace's equation at each point (r, θ) with prescribed values $F(\theta)$ on C' (where $\phi(r, \theta) = F(\theta)$) in $\mathfrak{R}$ is given by the <u>Poisson formula for a circle</u> (Spiegel, 1964):

$$\phi(r, \theta) = \frac{1}{2\pi} \int_0^{2\pi} \frac{(1 - r^2)\, F(\phi)\, d\phi}{1 - 2r \cos(\theta - \phi) + r^2}.$$

Similarly, for a function in the half-plane $y > 0$ (that is, $\text{Im}\{z\} > 0$), with prescribed values $G(x)$ on the x axis (that is, $\phi(x, 0) = G(x)$, $-\infty < x < \infty$), is given by the <u>Poisson formula for a half-plane</u> (Spiegel, 1964):

$$\phi(x, y) = \frac{1}{\pi} \int_{-\infty}^{\infty} \frac{y\, G(\eta)\, d\eta}{y^2 + (x - \eta)^2}.$$

As an application of conformal mapping, the Dirichlet problem for any simply connected region $\Re$ can be solved with a mapping function onto either the unit circle or the half-plane as follows:

- **(i)** Use the mapping function to transform the boundary value problem for the region $\Re$ into a unit circle or half-plane.
- **(ii)** Solve the problem for the unit circle or half-plane.
- **(iii)** Employ an inverse conformal mapping function to obtain the solution. For example, if $w = f(z)$, then there exists a unique inverse mapping function $z = g(w)$ in $\Re$ provided $f'(z) \neq 0$.

Example 10.6.1. Determine the harmonic function $\phi(x, y)$ using the Poison formula on the upper half z plane, where $G(x) = \begin{cases} 1, & x > 0, \\ 0, & x < 0 \end{cases}$ on the x axis.

Solution.
We have

$$\phi(x, y) = \frac{1}{\pi} \int_{-\infty}^{\infty} \frac{y \, G(\eta) \, d\eta}{y^2 + (x - \eta)^2} = \frac{1}{\pi} \int_{-\infty}^{0} \frac{y \, [0] \, d\eta}{y^2 + (x - \eta)^2} + \frac{1}{\pi} \int_{0}^{\infty} \frac{y \, [1] \, d\eta}{y^2 + (x - \eta)^2}$$

$$= \frac{1}{\pi} \tan^{-1} \frac{\eta - x}{y} \Big|_{0}^{\infty} = 1 - \frac{1}{\pi} \tan^{-1} \left(\frac{y}{x} \right).$$

Application of conformal mapping to engineering problems

If $\phi(x, y)$ satisfies the Laplace equation

$$\frac{\partial^2 \phi}{\partial x^2} + \frac{\partial^2 \phi}{\partial y^2} = 0, \tag{10.24}$$

then the function ϕ is <u>harmonic</u>. Moreover, there must be a <u>conjugate harmonic function</u> $\psi(x, y)$ which is analytic, such that

$$\boxed{\Omega(z) = \phi(x, y) + i \psi(x, y).} \tag{10.25}$$

The function Ω is called the <u>complex potential</u>.

Example 10.6.2. Using the Cauchy–Riemann equations for the given harmonic functions in Eq. (10.25), show that

$$\frac{d\Omega}{dz} = \Omega'(z) = \frac{\partial \phi}{\partial x} + i \frac{\partial \psi}{\partial x} = \frac{\partial \phi}{\partial x} - i \frac{\partial \phi}{\partial y}.$$

Solution.
The total derivatives of ϕ and ψ are $d\phi = \frac{\partial \phi}{\partial x} dx + \frac{\partial \phi}{\partial y} dy$ and $d\psi = \frac{\partial \psi}{\partial x} dx + \frac{\partial \psi}{\partial y} dy$.

Hence,

$$d\Omega = d\phi + i \, d\psi = \left(\frac{\partial \phi}{\partial x} + i \frac{\partial \psi}{\partial x} \right) dx + \left(\frac{\partial \phi}{\partial y} + i \frac{\partial \psi}{\partial y} \right) dy.$$

Using the Cauchy–Riemann equations in Eq. (10.2) yields

$$d\Omega = \left(\frac{\partial \phi}{\partial x} + i \frac{\partial \psi}{\partial x} \right) dx + \left(-\frac{\partial \psi}{\partial x} + i \frac{\partial \phi}{\partial x} \right) dy$$

$$= \left(\frac{\partial \phi}{\partial x} + i \frac{\partial \psi}{\partial x} \right) (dx + i dy).$$

Therefore, the result follows:

$$\frac{d\Omega}{dz} = \Omega'(z) = \left(\frac{\partial \phi}{\partial x} + i \frac{\partial \psi}{\partial x} \right) = \left(\frac{\partial \phi}{\partial x} - i \frac{\partial \phi}{\partial y} \right). \qquad \text{[answer]}$$

If a mapping such that $w = f(z) = u(x, y) + iv(x, y)$ is analytic, the one-parameter families of curves in Eq. (10.25) are <u>orthogonal</u> if

$$\phi(x, y) = \alpha \text{ and } \psi(x, y) = \beta,$$

where α and β are constants. As shown in Fig. 10.7(a), each member of one family of curves is perpendicular to the other family of curves at the point of intersection. Moreover, the corresponding image curves in the w plane are parallel to the u and v axes, which also form orthogonal families in Fig. 10.7(b). Thus, when the mapping of $f(z)$ is analytic, the angles between the intersecting curves in the z plane are equal in both magnitude and sense to the intersecting image curves in the w plane. This result is in fact the underlying basis of conformal mapping.

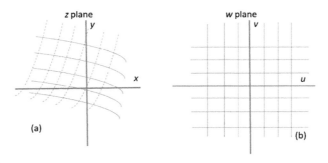

(a)

(b)

FIGURE 10.7

One-parameter families of intersecting curves in (a) the z plane and (b) the w plane.

Applications of conformal mappings to various fields of engineering are briefly described in the following sections.

Fluid flow

The function $\Omega(z)$ is important to characterize the flow pattern. If V_x and V_y are the components of the fluid velocity in the x and y directions, these quantities relate to the velocity potential as

$$V_x = \frac{\partial \phi}{\partial x} \text{ and } V_y = \frac{\partial \phi}{\partial y}. \tag{10.26}$$

Moreover, if a fluid is incompressible so that its density ρ remains constant, the mass flux in the x and y directions is $F_x = \rho V_x$ and $F_y = \rho V_y$, respectively. In the steady state, if there is no accumulation of the fluid in a region C, then the amount of material entering the region is equal to the quantity leaving it, and from Eq. (5.8),

$$\frac{\partial V_x}{\partial x} + \frac{\partial V_y}{\partial y} = 0. \tag{10.27}$$

This latter relation is simply a statement of conservation of continuity. Hence, Eq. (10.26) and Eq. (10.27) yield a velocity potential ϕ that is harmonic and satisfies the Laplace equation in Eq. (10.24). Moreover, using the result of Example 10.6.2 with Eq. (10.26),

$$\Omega'(z) = \frac{\partial \phi}{\partial x} - i \frac{\partial \phi}{\partial y} = V_x - i V_y.$$

Consequently, the complex velocity is the complex conjugate such that $\bar{V} = V_x + i V_y$ with the magnitude $V = |\bar{V}| = \sqrt{V_x^2 + V_y^2}$.

Conformal mapping is useful in obtaining the complex potential $\Omega(z)$ in order to characterize the flow pattern. For instance, to account for flow around an obstacle, one can map a uniform flow in the w plane given by $V_0 w$ (see Fig. 10.8(a)) using the mapping function $w = z + a^2/z$. Here, the upper half w plane is mapped into the upper half z plane that is exterior to the circle C, as depicted in Fig. 10.8(b) (Spiegel, 1964).

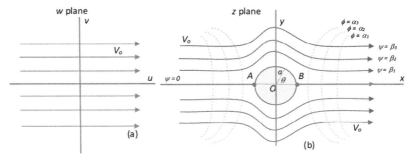

FIGURE 10.8

Mapping of (a) a uniform flow in the w plane to (b) a flow with an obstacle in the z plane (adapted from Spiegel, 1964).

Other types of mapping functions are given in Spiegel (1964) for different applications and problems.

Example 10.6.3. The complex potential for a flowing fluid with a solid body immersed into it is $\Omega(z) = V_o\left(z + \dfrac{a^2}{z}\right)$. The constant V_o is equal to the magnitude of the free stream velocity away from the immersed object and a is the radius of the object (see Fig. 10.8(b)).

(a) Determine the equations for the streamlines and equipotential lines in Fig. 10.8 letting $z = re^{i\theta}$.
(b) Describe the flow around the circular object of radius a.
(c) Find the velocity at any point and away from the object.
(d) Determine the stagnation points.

Solution.

(a) We have

$$\Omega(z) = V_o\left(re^{i\theta} + \frac{a^2}{r}e^{-i\theta}\right) = \underbrace{V_o\left(r + \frac{a^2}{r}\right)\cos\theta}_{\phi} + \underbrace{i\,V_o\left(r - \frac{a^2}{r}\right)\sin\theta}_{\psi}.$$

The stream lines are given by $\psi = V_o\left(r - \dfrac{a^2}{r}\right)\sin\theta = \beta$, which are indicated by the curves showing the actual path of the fluid in Fig. 10.8(b). Similarly, the equipotential lines are given by $\phi = V_o\left(r + \dfrac{a^2}{r}\right)\cos\theta = \alpha$, which are indicated by the dashed lines that are orthogonal to the streamline family of curves in Fig. 10.8.

(b) The circle $r = a$ represents a streamline. Since there is no flow across a streamline, it represents the circular obstacle.

(c) We have

$$\Omega'(z) = V_o\left(1 - \frac{a^2}{z^2}\right) = V_o\left(1 - \frac{a^2}{r^2}e^{-2i\theta}\right)$$

$$= V_o\left(1 - \frac{a^2}{r^2}\cos 2\theta\right) + i\left(\frac{V_o a^2}{r^2}\right)\sin 2\theta.$$

Thus, the complex velocity is $\bar{V} = V_o\left(1 - \dfrac{a^2}{r^2}\cos 2\theta\right) - i\left(\dfrac{V_o a^2}{r^2}\right)\sin 2\theta$, with magnitude

$$V = \sqrt{\left\{V_o\left(1 - \frac{a^2}{r^2}\cos 2\theta\right)\right\}^2 + i\left\{\left(\frac{V_o a^2}{r^2}\right)\sin 2\theta\right\}^2}$$

$$= V_o\sqrt{1 - \frac{2a^2\cos 2\theta}{r^2} + \frac{a^4}{r^4}}.$$

Thus, far from the obstacle, as $r \to \infty$, $V = V_o$.

(d) For a stagnation point, $\Omega'(z) = 0$ so that $V_o \left(1 - \dfrac{a^2}{z^2} \right) = 0$ or $z = a$ and $z = -a$, corresponding to points A and B in Fig. 10.8(b).

Electrostatics

A charge distribution will establish an electric field. The intensity of this field $\mathscr{E}$ can be related to an electrostatic potential ϕ as

$$\mathscr{E} = -\nabla \phi.$$

Hence, for a two-dimensional charge distribution,

$$E_x = -\frac{\partial \phi}{\partial x} \text{ and } E_y = -\frac{\partial \phi}{\partial y}. \tag{10.28}$$

In the z plane, using Eq. (10.28), the electric field intensity is

$$\mathscr{E} = E_x + i E_y = -\frac{\partial \phi}{\partial x} - i \frac{\partial \phi}{\partial y}. \tag{10.29}$$

Using Stokes and Green's theorems from Section 4.4.2, along with Maxwell's equation $\nabla \times E = 0$ (Jackson, 1975) with the definition of the vector cross-product, it can be shown that the tangential component of the electric field intensity for a simple curve C in the z plane is given by

$$\oint_C E_t \, ds = \oint_C E_x \, dx + E_y \, dy = 0. \tag{10.30}$$

Similarly, using the divergence and Green's theorems, with Maxwell's equation $\nabla \cdot E = 0$ (Jackson, 1975) (that is, for the case that there is no charge in any region) with the definition of the vector dot product, the normal component of the electric field intensity is given by

$$\oint_C E_n \, ds = \oint_C E_x \, dy - E_y \, dx = 0. \tag{10.31}$$

It also follows directly from Maxwell's equation that with no charge

$$\frac{\partial E_x}{\partial x} + \frac{\partial E_y}{\partial y} = 0. \tag{10.32}$$

Using Eq. (10.28) and Eq. (10.32), one obtains the Laplace equation in Eq. (10.24) for the electrostatic potential. As such, ϕ is harmonic for all points not occupied by any charge. Consequently, there must be a harmonic conjugate function ψ, where Ω is the complex electrostatic potential such that

$$\Omega(z) = \phi(x, y) + i \psi(x, y).$$

Thus, from Eq. (10.29)

$$\mathscr{E} = -\frac{\partial \phi}{\partial x} - i\frac{\partial \phi}{\partial y} = -\frac{\partial \phi}{\partial x} + i\frac{\partial \psi}{\partial y} = -\overline{\Omega'(z)}. \tag{10.33}$$

Again, equipotential lines and <u>flux lines</u> are given, respectively, by

$$\phi(x, y) = \alpha \text{ and } \psi(x, y) = \beta.$$

Heat flow

The heat flux across a surface is given by Fourier's law in Example 5.1.1:

$$\mathscr{Q} = -K\nabla\phi,$$

where ϕ is the temperature and K is the thermal conductivity of the solid material. One has a similar relationship for the temperature as for the potential in electrostatics, where in two dimensions,

$$\mathscr{Q} = Q_x + iQ_y = -K\left(\frac{\partial \phi}{\partial x} + i\frac{\partial \phi}{\partial y}\right), \tag{10.34}$$

where for the second relation one has used

$$Q_x = -K\frac{\partial \phi}{\partial x} \text{ and } Q_y = -\frac{\partial \phi}{\partial y}. \tag{10.35}$$

Analogously, for a closed simple curve C in the z plane (that represents the cross-section of a cylinder), the tangential and normal components of the heat flux for steady-state conditions, in which there is no accumulation of heat inside of C and no sinks or sources, are

$$\oint_C Q_n \, ds = \oint_C Q_x \, dy - Q_y \, dx = 0 \text{ and } \oint_C Q_t \, ds = \oint_C Q_x \, dx + Q_y \, dy = 0. \tag{10.36}$$

The first equation of Eq. (10.36) gives the similar result as for electrostatics:

$$\frac{\partial Q_x}{\partial x} + \frac{\partial Q_y}{\partial y} = 0. \tag{10.37}$$

Hence, using Eq. (10.35) and Eq. (10.37), one obtains the Laplace equation for the temperature where ϕ is harmonic. Again one has the harmonic conjugate function ψ:

$$\Omega(z) = \phi(x, y) + i\psi(x, y).$$

Similarly, the same families of curves are identified as <u>isothermal lines</u> and <u>flux lines</u> for the complex temperature $\Omega(z)$:

$$\phi(x, y) = \alpha \text{ and } \psi(x, y) = \beta.$$

Thus, the methodology for solving temperature problems is analogous to that of electrostatics.

Summary

As a summary of results, the functions $\phi(x, y) = \alpha$ and $\psi(x, y) = \beta$ yield the analogous quantities in fluid flow, electrostatics, and heat flow, where, respectively:

(i) For steady *fluid flow*, these relations give rise to equipotential curves and stream curves. The stream curves represent the actual path fluid particles take in a flow pattern.

(ii) For *electrostatics*, they result in equipotential lines and flux lines.

(iii) For *heat flow*, they give isothermal lines and flux lines.

Problems

10.1 Expand the function $f(z) = \dfrac{1}{(z+1)(z+3)}$ in a Laurent series, which is valid for $1 < |z| < 3$.

Hint: Use partial fractions and a binomial series expansion, $(1+x)^{-1} = 1 - x + x^2 - x^3...$, for x values in the interval of convergence of $-1 < x < 1$.

10.2 Consider the integral $\oint_C \dfrac{2\,dz}{z^2 - 1}$, where C is a circle of radius 1/2 and centered at 1, and positively oriented.

(a) Using a partial fraction method, show that $\dfrac{2}{z^2 - 1} = \dfrac{1}{z - 1} - \dfrac{1}{z + 1}$.

(b) Evaluate the integral using Cauchy's integral formula and Cauchy's theorem with these partial fractions, respectively.

10.3 The complex inversion formula in Eq. (3.9) with a branch point $s = 0$ yields (Spiegel, 1971): $f(t) = \mathscr{L}^{-1}\left\{\dfrac{e^{-a\sqrt{s}}}{s}\right\} = 1 - \text{erf}\left\{\dfrac{a}{2\sqrt{t}}\right\} = \text{erfc}\left\{\dfrac{a}{2\sqrt{t}}\right\}$. This transform is used in Example 5.2.11.

(a) Show that $\mathscr{L}^{-1}\left\{e^{-a\sqrt{s}}\right\} = \dfrac{a}{2\sqrt{\pi}} t^{-3/2} e^{-a^2/(4t)}$. To find this transform, use the Laplace transform of a derivative $\mathscr{L}\{f'(t)\} = s F(s)$ (that is, for $f(0) = 0$), with the previous result $f(t) = \mathscr{L}^{-1}\{F(s)\} = \text{erfc}\left\{\dfrac{a}{2\sqrt{t}}\right\}$, where $F(s) = \dfrac{e^{-a\sqrt{s}}}{s}$.

10.4 Given the function $u(x, y) = x^2 - y^2$.

(a) Prove that this equation satisfies the Laplace equation.

(b) Find the function $v(x, y)$ such that $f(z) = u + iv$ is an analytic function.

10.5 Show that $f(z) = \sin z$ is analytic in the following steps:

(a)   Find the real and imaginary parts of $f(z)$.

(b)   Show that $u(x, y)$ and $v(x, y)$ from part (a) satisfy the Cauchy–Riemann conditions.

10.6   Evaluate the integral $\oint_C z^2 \, dz$ along the path shown in Fig. 10.9.

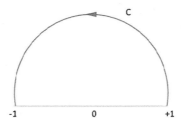

FIGURE 10.9

Schematic of a closed semicircle contour C.

10.7   Evaluate the integral $\oint_C \dfrac{dz}{(z^2 - a^2)} \, dz$, where C is a circle of radius b centered at the origin with $b > a$.

10.8   Evaluate the definite integral $\displaystyle\int_{-\infty}^{+\infty} \dfrac{dx}{(1 + x^2)}$, where the contour runs along the x axis from $-\rho$ to $+\rho$, and then closes by a semicircle in the upper half-plane of radius ρ centered at the origin.

10.9   Show that using contour integration and the residue theorem, the definite integral $\displaystyle\int_{-\infty}^{+\infty} \dfrac{-e^{iwx} dw}{(w^2 - iw + 2)}$ is evaluated as $g(x) = \begin{cases} \dfrac{-\sqrt{2\pi}}{3} e^x & \text{for } x < 0, \\ \dfrac{-\sqrt{2\pi}}{3} e^{-2x} & \text{for } x > 0. \end{cases}$ This result is used in the solution of Problem 3.10.

Nondimensionalization

11

A mathematical model or an equation describes the behavior of a real-life system. A model is described by dependent and independent variables, and parameters. When a system is defined in terms of position as a function of time, the position acts as the dependent variable and time as the independent variable. On the other hand, parameters, in general, can be constants or vary in a model. A variable instead of a parameter can be used if the value changes continuously (e.g., over time). A parameter can be used also if the value does not change or only changes at particular moments of time. A solution is sought to determine the relation between dependent and independent variables in a model.

Variables and parameters used in modeling the behavior of a real-life system represent physical properties and hence have physical dimensions. Both sides of an equation must have the same dimension (sometimes called the dimensional homogeneity). Wherever a sum of quantities appears, all terms in the sum must have the same dimension.

Modeling a real-life system may become difficult with too many parameters and variables. Consequently, techniques are applied to reduce the number of variables/parameters while simplifying equations. A dimensional analysis is applied to relate parameters/variables with each other while developing a nondimensional expression between independent and dependent variables/parameters. Conversely, a nondimensionalization technique is applied to reduce the number of parameters in a system. In both techniques (i.e., dimensional and nondimensionalization analyses), the physical dimensions of variables and parameters can be fruitfully exploited, resulting in a simplified nondimensional equation or expression.

Dimensional analysis is used to check if dimensional homogeneity is achieved. In addition, with the application of Buckingham's π theorem (which is a tool for dimensional analysis), nondimensional expressions are developed to relate dependent and independent variables/parameters. In Buckingham's π theorem, no definite basic equation is defined but rather relationships.

Nondimensionalization is a technique that is applied to remove, either fully or partially, dimensions from an equation by substituting in relevant variables. Nondimensionalization finds application, in particular, for systems that can be described by differential equations. It is employed to: (i) generalize a problem; (ii) simplify an equation by reducing the number of variables; (iii) rescale parameters and variables so that quantities are of similar order/magnitude; and (iv) analyze the system regardless of units used to measure variables.

Advanced Mathematics for Engineering Students. https://doi.org/10.1016/B978-0-12-823681-9.00019-8

Dimensionless numbers, such as those used in fluid mechanics (for example, the Reynolds number, defined as the ratio of inertia forces to viscous forces) and heat transfer (e.g., the Prandtl number, defined as the ratio of momentum diffusivity to thermal diffusivity), are outcomes of simplifications, such as nondimensionalization of differential equations. These numbers can be used to identify the relative importance of terms. For example, the Reynolds number will indicate which force (i.e., inertia forces or viscous forces) becomes dominant for a given condition of fluid velocity, kinematic viscosity, and characteristic length.

Experiments cannot be always performed on one-to-one scale prototypes; experimental systems need to be scaled down. In addition, sometimes, different fluids are used in the modeling of a system for cost saving purposes. Nondimensionalization or dimensional analysis techniques need to be applied to facilitate a scale-up of laboratory results to the actual system (for example, the scaling up of a modeled wind tunnel experiment). The use of dimensionless numbers facilitates this scale-up.

In the following sections, dimensional analysis and Buckingham's π theorem will be discussed, followed by similarity laws and the nondimensionalization technique.

11.1 Dimensional analysis

The dimension of a physical quantity can be expressed by the product of basic physical dimensions such as mass (M), length (L), time (T), absolute temperature (Θ), amount of a substance (N), electric current (I), and luminous intensity (J). These physical dimensions are basic as they can be easily measured in experiments. Dimensions are not the same as units. For example, the physical quantity of speed can be measured in units of kilometers per hour or miles per second, but regardless of its unit, speed is always defined as a length divided by a time, or simply L/T. It should be noted that some quantities do not have dimensions. These quantities include: trigonometric and exponential functions, logarithms, and counted quantities. Despite the fact that angles are dimensionless, they have units, such as a degree.

Example 11.1.1. Determine the dimensions of the following quantities: area, $(\text{area})^{1/2}$, $(\text{volume})^3$, density, acceleration, force, pressure, work, kinetic energy, charge, and electric field.

Solution:

Area: L^2 $(\text{area})^{1/2}$: L $(\text{volume})^3$: L^9

Density ($=$ mass/volume): M/L^3

Acceleration ($=$ velocity/time): L/T^2

Force ($=$ mass $\times$ acceleration): $M{\cdot}L/T^2$

Pressure ($=$ force/area): $M/L{\cdot}T^2$

Work ($=$ force $\times$ distance): $M{\cdot}L^2/T^2$

Kinetic energy ($=$ mass $\times$ (velocity)^2/2): $M{\cdot}L^2/T^2$

Charge (= current × time): I·T

Electric field (= force/charge): M·L/I·T^3

Example 11.1.2. Determine the SI units of the following quantities: density, force, pressure, (kinetic) energy, and electric field.

Solution:

Density: kg/m^3

Force: kg·m/s^2

Pressure: kg/m·s^2

Energy: kg·m^2/s^2

Electric field: kg·m/A·s^3

Example 11.1.3. Determine the dimensions of k in the expressions $\cos(\sqrt{k/m} \cdot t)$, e^{ikt}, and 10^{kx}, where m is mass, t is time, i is the imaginary unit, and x is length.

Solution:

$k = M/T^2$ in expression $\cos(\sqrt{k/m} \cdot t)$

$k = T^{-1}$ in expression e^{ikt}

$k = L^{-1}$ in expression 10^{kx}

Dimensional analysis is also applied to ensure the dimensional homogeneity of an equation in which both sides must have the same dimension.

Example 11.1.4. Check the dimensional homogeneity of the equation written for pumping power:

$$Q + W = \dot{m}(H_{outlet} - H_{inlet}) + \frac{1}{2}\dot{m}(v_{outlet}^2 - v_{inlet}^2) + \dot{m}g(z_{outlet} - z_{inlet}),$$

where Q is the rate of heat input to the coolant in $J \cdot s^{-1}$, W is the rate of work input to the coolant in $J \cdot s^{-1}$, $\dot{m}$ is the mass flow rate in $kg \cdot s^{-1}$, H is the enthalpy in $J \cdot kg^{-1}$, v is the velocity in $m \cdot s^{-1}$, g is the gravitational acceleration, which is $9.81 \ m \cdot s^{-2}$ and z is the height in m.

Solution:

We have

$$\underbrace{Q}_{\frac{J}{s}} + \underbrace{W}_{\frac{J}{s}} = \underbrace{\dot{m}}_{\frac{kg}{s}} \underbrace{(H_{outlet} - H_{inlet})}_{\frac{J}{kg}} + \frac{1}{2} \underbrace{\dot{m}}_{\frac{kg}{s}} \underbrace{(v_{outlet}^2 - v_{inlet}^2)}_{\frac{m^2}{s^2}}$$

$$+ \underbrace{\dot{m}}_{\frac{kg}{s}} \underbrace{g}_{\frac{m}{s^2}} \underbrace{(z_{outlet} - z_{inlet})}_{m}.$$

The Joule (J) is defined as the energy transfer to (or work done on) an object when a force of 1 Newton ($= kg \cdot m \cdot s^{-2}$ in SI units) acts on that object in the direction of its motion through a distance of 1 meter. It can be further expressed as $J = kg \cdot m^2 \cdot s^{-2}$. Therefore, the units on both sides of the equation are equal and given by $kg \cdot m^2 \cdot s^{-3}$. Using dimensions of the terms, instead of units, would generalize the application of the dimensional homogeneity. Therefore, it can be demonstrated that both sides have the dimensions of $M \cdot L^2 \cdot T^{-3}$.

Differentiation of a quantity with respect to another adds the dimension of the variable in the denominator. For example, the derivative of position with respect to time, (dx/dt), is the velocity and has the dimension of $L \cdot T^{-1}$. The second derivative with respect to time, (d^2x/dt^2), is the acceleration and has the dimension of $L \cdot T^{-2}$.

Polynomials of mixed degree with dimensionless coefficients should satisfy dimensional homogeneity. For example, a projectile motion can be expressed as

$$y = ax + bx^2, \tag{11.1}$$

where a and b are constants, y is the vertical displacement with dimension L, and x is the horizontal displacement with dimension L. Dimensional homogeneity can be satisfied only if the a and b coefficients have the correct units; otherwise if $x = 2$ meters, the right-hand side of the equation becomes $a(2 \text{ m}) + b(4 \text{ m}^2)$ and the left-hand side, y, is given in meters. A closer look into the projectile motion (see Fig. 11.1) sheds light on how to choose these coefficients.

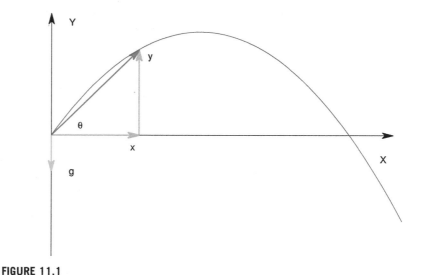

FIGURE 11.1

Projectile motion.

At any time, t, with the known initial launch angle θ and initial velocity v_0, the projectile's horizontal and vertical displacement are given by

$$x = v_0 t \cos(\theta), \tag{11.2}$$

$$y = v_0 t \sin(\theta) - \frac{1}{2} g t^2. \tag{11.3}$$

If time, t, is eliminated from Eq. (11.2) and Eq. (11.3), then the vertical displacement can be expressed in terms of the horizontal displacement as

$$y = \tan(\theta) \cdot x - \frac{g}{2 v_0^2 \cos^2(\theta)} \cdot x^2. \tag{11.4}$$

Since g, θ, and v_0 are constant, the coefficients a and b in Eq. (11.1) become

$$a = \tan(\theta), \qquad b = -\frac{g}{2 v_0^2 \cos^2(\theta)}. \tag{11.5}$$

The coefficient a does not have any dimension, whereas the coefficient b has the dimension of L^{-1}. In this case, the dimensional homogeneity is sustained with both sides of the equation having a dimension of L.

Another application of dimensional analysis is the conversion from one unit to another. For example, the inch and cm are both units of length, and 1 inch $= 2.54$ cm. By dividing both sides by the same expression, one gets $1 = 2.54$ cm/1 inch. Given that any quantity can be multiplied by 1 without changing it, the expression of "2.54 cm/1 inch" can be used to convert it from inch to cm by multiplying it with the quantity to be converted. Therefore, 10 inch will be 25.4 cm ($= 10$ inch $\times$ 2.54 cm/1 inch).

Example 11.1.5. Convert 50 miles per hour into meters per second (note that 1 mile $= 1609.344$ meters).

Solution.
We have

$$\frac{50 \text{ mile}}{1 \text{ hour}} \times \frac{1609.344 \text{ meter}}{1 \text{ mile}} \times \frac{1 \text{ hour}}{3600 \text{ second}} = 22.352 \frac{\text{meter}}{\text{second}}.$$

11.2 Buckingham's π theorem – dimensional analysis

Buckingham's π theorem is key in dimensional analysis. It provides a method for computing sets of dimensionless parameters from given (dimensional) variables (and parameters), even if the form of the equation is unknown.

Nondimensionalization is useful only when the equation (for example, a conservation equation) is known. In many real-world phenomena, the equations are either unknown or too difficult to solve. Experimentation is the only method of obtaining reliable information. In most experiments, scaled models[1] are used due to time and cost saving purposes. Experimental conditions and results need to be scaled properly so that results are meaningful for the full-scale prototype.

Unlike the nondimensionalization technique, which is applied to basic (fundamental) equations, the analysis of Buckingham's π theorem is based on defining relationships between variables. The concept is based on the assumption that the space of fundamental and derived physical units forms a vector space over the rational numbers. The fundamental units are basis vectors, with multiplication of physical units as the "vector addition" operation, and raising to powers as the scalar multiplication operation. Making the physical units match across sets of physical equations can then be regarded as imposing linear constraints in the vector space of the physical units. Therefore, a dimensionless variable is a quantity with fundamental dimensions raised to the zeroth power (the zero vector of the vector space over the fundamental dimensions).

The significance of Buckingham's π theorem is that it:

- allows analysis of a system when little is known about the physics acting on it;
- allows measurements to be generalized from a particular set-up of the system;
- describes the relation between the number of variables and fundamental dimensions;
- generates nondimensional parameters that help in the design of experiments (physical and/or numerical) and in reporting of results;
- predicts trends in the relationship between parameters;
- reduces the number of variables that must be specified to describe an event, which often leads to an enormous simplification;
- provides a similarity law index for the phenomenon under consideration so that prototype behavior can be predicted from model behavior (discussed in the next section).

Buckingham's π theorem states: "If there are n variables (and parameters) in a problem and these variables (and parameters) contain m basic dimensions (for example M, L, T), the equation relating all the variables will have $(n\text{-}m)$ dimensionless groups." Buckingham referred to these groups as Π groups.

The dimensional complete set of variables and parameters is expressed as

$$f(x_1, \cdots, x_n) = 0. \tag{11.6}$$

This expression can be replaced by the following set of nondimensional groups:

$$F(\Pi_1, \cdots, \Pi_{(n-m)}) = 0, \tag{11.7}$$

---

[1] See the next section for similarity.

where the Π_i are dimensionless parameters constructed from the dimensional x_i by $(n$-$m)$ dimensionless equations—the so-called Π groups—of the form

$$\Pi_i = x_1^{a_1} x_2^{a_2} \cdots x_n^{a_n}. \tag{11.8}$$

The exponents a_i are rational numbers; however, they can always be taken to be integers by redefining Π_i as being raised to a power that clears all denominators.

A dimensional matrix $\mathbf{D}$ whose rows are the basic dimensions and whose columns are the dimensions of the variables can be formed for a system of n dimensional variables (with physical dimensions) in m basic dimensions. For dimensionless $\Pi_i = x_1^{a_1} x_2^{a_2} \cdots x_n^{a_n}$, a vector $\mathbf{a} = [a_1, a_2, \cdots a_n]$ is sought such that the matrix-vector product $\mathbf{Da}$ equals the zero vector.

Nondimensional Π parameters can be generated by several methods, but the "method of repeating variables" is discussed here. While using the theorem, the following steps are applied:

- List relevant variables/parameters. Clearly define the problem, identify all the variables that are important, and determine which is the main variable of interest (that is, the dependent variable). Indicate the total number of independent and dependent variables and parameters, n.
- Indicate basic dimensions of each parameter. Express fundamental dimensions of each of n variables/parameters, and hence determine the number of basic dimensions, m.
- Determine the number of Π groups, which is expressed by $(n$-$m)$.
- Choose repeating variables: While choosing the repeating variables, the following considerations need to be taken into account:
  - never pick the dependent variable;
  - never pick variables that are already dimensionless;
  - never pick two variables with the same dimensions;
  - chosen variables must represent all the primary dimensions;
  - choose common variables as they may appear in each of the Π groups;
  - choose simple variables over complex ones.
- Form dimensionless Π groups and check that they are actually all dimensionless. Combine repeating parameters into products with each of the remaining parameters, one at a time, to create the Π groups.

Example 11.2.1. Investigate the pressure drop in a pipe using Buckingham's π theorem.

Solution:
Independent variables and parameters, with their dimensions in capital letters, are

- pressure drop, $[\Delta P]$, $\text{M·L}^{-1}\text{·T}^{-2}$;
- viscosity, $[\mu]$, $\text{M·L}^{-1}\text{·T}^{-1}$;

- density, $[\rho]$, $M \cdot L^{-3}$;
- length, $[l]$, L;
- diameter, $[D]$, L;
- velocity, $[v]$, $L \cdot T^{-1}$,

where pressure drop, $[\Delta P]$, is the dependent variable.

The model is of the form

$$f(\Delta P, \mu, \rho, l, D, v) = 0. \tag{11.9}$$

There are $m = 3$ basic physical dimensions in this equation, i.e., time (T), mass (M), and length (L), and $n = 6$ dimensional variables, i.e., ΔP, μ, ρ, l, D, and v. Consequently, there will be $6 - 3 = 3$ dimensionless Π parameters. The variables l and D cannot be chosen as repeating variables at the same time. Both of these variables have the same dimension (repeating parameters by themselves should not be able to form a dimensionless group) and D is chosen as the repeating variable. Repeating variables should be less complex than the prime quantities. In this case, μ can also be chosen as prime quantity, as it is more complex than the others. Therefore, ΔP, which is the dependent variable of interest, and l and μ can be chosen as prime quantities. The rest, D, ρ, and v, are chosen as the repeating variables. Based on this analysis, the form of the model is

$$\Pi_0 = f(\Pi_1, \Pi_2), \tag{11.10}$$

where

$$\Pi_0 = \Delta P^{a_1} \rho^{a_2} v^{a_3} D^{a_4}, \tag{11.11}$$

$$\Pi_1 = l^{b_1} \rho^{b_2} v^{b_3} D^{b_4}, \tag{11.12}$$

$$\Pi_2 = \mu^{c_1} \rho^{c_2} v^{c_3} D^{c_4}. \tag{11.13}$$

For the dimensionless Π_0 parameter, the dimensional **D** matrix, where the rows correspond to the basis dimensions T, M, and L and the columns to the dimensional variables ΔP, ρ, v and D, is given by

	ΔP	ρ	v	D
T	-2	0	-1	0
M	1	1	0	0
L	-1	-3	1	1

or

$$\mathbf{D} = \begin{bmatrix} -2 & 0 & -1 & 0 \\ 1 & 1 & 0 & 0 \\ -1 & -3 & 1 & 1 \end{bmatrix} \tag{11.14}$$

The matrix-vector multiplication is shown below:

$$\begin{bmatrix} -2 & 0 & -1 & 0 \\ 1 & 1 & 0 & 0 \\ -1 & -3 & 1 & 1 \end{bmatrix} \begin{bmatrix} a_1 \\ a_2 \\ a_3 \\ a_4 \end{bmatrix} = [0, \ 0, \ 0], \tag{11.15}$$

which leads to the following linear equations:

$$\begin{aligned} -2a_1 - a_3 &= 0, \\ a_1 + a_2 &= 0, \\ -a_1 - 3a_2 + a_3 + a_4 &= 0. \end{aligned} \tag{11.16}$$

Taking the index of prime quantity $a_1 = 1$, the other indices become $a_2 = -1$, $a_3 = -2$, and $a_4 = 0$. The vector, $\mathbf{a}$, becomes

$$\mathbf{a} = \begin{bmatrix} 1 \\ -1 \\ -2 \\ 0 \end{bmatrix}. \tag{11.17}$$

Substituting the values of the vector $\mathbf{a}$ as exponents in Eq. (11.11),

$$\Pi_0 = \Delta P^1 \rho^{-1} v^{-2} D^0. \tag{11.18}$$

Rearranging Eq. (11.18),

$$\Pi_0 = \frac{\Delta P}{\rho v^2}. \tag{11.19}$$

For the dimensionless Π_1 parameter, the dimensional $\mathbf{D}'$ matrix, where the rows correspond to the basis dimensions T, M, and L and the columns to the dimensional variables l, ρ, v, and D, is given by

	l	ρ	v	D
T	0	0	-1	0
M	0	1	0	0
L	1	-3	1	1

or

$$\mathbf{D}' = \begin{bmatrix} 0 & 0 & -1 & 0 \\ 0 & 1 & 0 & 0 \\ 1 & -3 & 1 & 1 \end{bmatrix}. \tag{11.20}$$

The matrix-vector multiplication is shown below:

$$\begin{bmatrix} 0 & 0 & -1 & 0 \\ 0 & 1 & 0 & 0 \\ 1 & -3 & 1 & 1 \end{bmatrix} \begin{bmatrix} b_1 \\ b_2 \\ b_3 \\ b_4 \end{bmatrix} = [0, \ 0, \ 0], \tag{11.21}$$

which leads to the following linear equations:

$$\begin{aligned} -b_3 &= 0, \\ b_2 &= 0, \\ b_1 - 3b_2 + b_3 + b_4 &= 0. \end{aligned} \tag{11.22}$$

Taking the index of the prime quantity, $b_1 = 1$, and with $b_2 = 0$ and $b_3 = 0$, b_4 becomes equal to -1. The vector, $\mathbf{b}$, can therefore be expressed as

$$\mathbf{b} = \begin{bmatrix} 1 \\ 0 \\ 0 \\ -1 \end{bmatrix}. \tag{11.23}$$

Substituting the values of the vector $\mathbf{b}$ as exponents in Eq. (11.12),

$$\Pi_1 = l^1 \rho^0 v^0 D^{-1}. \tag{11.24}$$

Rearranging Eq. (11.24),

$$\Pi_1 = \frac{l}{D}. \tag{11.25}$$

Similarly, the matrix-vector multiplication for c indices (for Π_2) is written as

$$\begin{bmatrix} -1 & 0 & -1 & 0 \\ 1 & 1 & 0 & 0 \\ -1 & -3 & 1 & 1 \end{bmatrix} \begin{bmatrix} c_1 \\ c_2 \\ c_3 \\ c_4 \end{bmatrix} = [0, \ 0, \ 0], \tag{11.26}$$

leading to the following linear equations:

$$\begin{aligned} -c_1 - c_3 &= 0, \\ c_1 + c_2 &= 0, \\ -c_1 - 3c_2 + c_3 + c_4 &= 0. \end{aligned} \tag{11.27}$$

Taking the index of the prime quantity, $c_1 = 1$, the other indices become $c_2 = -1$ and $c_3 = -1$, and c_4 becomes equal to -1. The vector, $\mathbf{c}$, can therefore be expressed as

$$\mathbf{c} = \begin{bmatrix} 1 \\ -1 \\ -1 \\ -1 \end{bmatrix}. \tag{11.28}$$

Substituting the values of the vector $\mathbf{c}$ as exponents in Eq. (11.13),

$$\Pi_2 = \frac{\mu}{\rho v D}. \tag{11.29}$$

It should be noted that the term $\rho v D / \mu$, which appeared in Eq. (11.29), is the Reynolds number (Re), and the Darcy friction factor, f_D, in a pipe flow can be expressed in terms of the Reynolds number as $f_D \propto 1/\text{Re} = 1/(\rho v D/\mu)$.

Inserting Π groups into Eq. (11.10), the relationship between dimensionless pressure drop and the independent variables, in terms of dimensionless groups, is expressed as

$$\frac{\Delta P}{\rho v^2} = f\left(\frac{l}{D}, \frac{\mu}{\rho v D}\right)$$

or

$$\frac{\Delta P}{\rho v^2} = f\left(\frac{l}{D}, \frac{1}{\text{Re}}\right). \tag{11.30}$$

Eq. (11.30) can also be written as a relationship, rather than as a functionality, as

$$\frac{\Delta P}{\rho v^2} = C\left(\frac{l}{D}\right)\left(\frac{1}{\text{Re}}\right)^k, \tag{11.31}$$

where C is the constant of proportionality and k is any number which is a constant.

11.3 Similarity laws

In hydraulic and aerospace engineering, fluid flow conditions are tested with scaled models to study complex (fluid dynamics) problems where calculations and computer simulations are not reliable. Scale models are usually smaller than the final design to allow testing of a design prior to building, which in many cases is a critical step in the development process.

While constructing a scale model, an analysis needs to be applied as some parameters need to be altered to match the conditions of interest to the prototype. The geometry may be simply scaled, but other parameters, such as velocity, pressure, temperature, and type of fluid may need to be altered. Similarity is achieved when testing conditions are created such that the test results are applicable to the real design. Similarity implies a certain equivalence between two physical phenomena.

Under particular conditions where certain relationships (in the form of dimensionless groups) are established and satisfied, the two cases are considered similar. For experimental purposes, different systems that share the same description in terms of the dimensionless numbers are equivalent. Two systems with similar dimensionless numbers are called similar. Therefore, similarity is achieved between the model tested and the prototype to be designed, if all relevant dimensionless parameters have the same corresponding values for the model and prototype.

Similarity between the model and prototype is achieved if the following conditions are satisfied.

Geometric similarity: The length dimension L must be ensured. In general, geometric similarity is established when a model and prototype are geometrically similar if and only if all body dimensions in all three coordinates have the same linear scale ratio: each dimension must be scaled by the same factor. Therefore, all of the angles and orientations of the model and prototype with respect to the surroundings must be identical.

Kinematic similarity: The velocity at any point in the model must be proportional, i.e., the model and prototype have the same length and time scale ratio. Thus, the velocity scale ratio will be the same for both. Fluid flow of both the model and the real application must undergo similar time rates of change in motion (fluid streamlines are similar). Geometric similarity should be met before achieving kinematic similarity.

Dynamic similarity: The same ratio should be established between resultant forces at corresponding locations on boundaries of the model and the prototype system. Dynamic similarity is achieved when the model and prototype have the same length scale ratio, time scale ratio, and force scale (mass scale) ratio. Therefore, the other two similarities should be met before achieving dynamic similarity.

As shown in Example 11.2.1, certain dimensionless numbers appear as a result of simple analyses and are associated with a particular type of force, such as the association of the Reynolds number, Re, with viscous and inertia forces. Therefore, to ensure that inertia and viscous forces are in the same ratio in the model and the prototype, the Reynolds number must be the same on both. Most of the problems require geometric and dynamic similarity. Dynamic similarity exists simultaneously with kinematic similarity if the model and prototype force scale ratios are identical. These similarities are, in fact, ratios (independent dimensionless groups), based on the π theorem. Since Eq. (11.8) is entirely general, it applies to any system, which is a function of the same variables. Therefore, a true model can be designed and operated with a matching dimensionless *dependent* Π parameter to the prototype,

$$(\Pi_0)_{prototype} = (\Pi_0)_{model}, \tag{11.32}$$

if the dimensionless *independent* Π parameters are equal,

$$(\Pi_1)_{prototype} = (\Pi_1)_{model},$$

$$\vdots \tag{11.33}$$

$$(\Pi_{(n-m)})_{prototype} = (\Pi_{(n-m)})_{model}.$$

Example 11.3.1. Consider the following engineering problem: Sea water at 10°C close to the surface (atmospheric pressure) will be pumped from location A to B using a pipe system, which is horizontal without elevation and form losses. In order to pump the water at a speed of 0.5 m/s, the pressure drop needs to be known to determine the pump power. Consequently, a model of 1/10 scale with a similar l/D ratio is built to determine the pressure drop. The water used in the model system is fresh water at 25°C. In order to ensure the dynamic similarity between the model and prototype, what would be the velocity of the water in the model system and what would be the pressure drop in the seawater pumping system if it is found that the pressure drop in the model is 50 kPa when the model is built and tested for the pressure drop?

Solution:
The Buckingham π theorem as described in Example 11.2.1 shows that the pressure drop in a pipe can be described (see Eq. (11.30)) with two dimensionless numbers (that is, l/D and the Reynolds number) and one dependent variable ($\Delta P/(\rho v^2)$). Since the dimensionless parameters will stay constant for both the model and the prototype, they will be used to formulate scaling laws for the model. Geometric similarity, which is the length-to-diameter ratio (l/D), should be satisfied while building the model. Dynamic similarity can be satisfied by equating the Reynolds numbers:

$$\left(\frac{\rho v D}{\mu}\right)_{prototype} = \left(\frac{\rho v D}{\mu}\right)_{model}.$$

The properties of seawater at atmospheric pressure and 10°C are $\rho = 1025.0$ kg/m³ and $\mu = 0.00148$ kg/m·s, while for fresh water at 25°C they are $\rho = 997.1$ kg/m³ and $\mu = 0.0008899$ kg/m·s. Hence,

$$\left(\frac{1025.0\,v\,D}{0.00148}\right)_{prototype} = \left(\frac{997.1\,v\,(D/10)}{0.0008899}\right)_{model}. \tag{11.34}$$

Rearranging Eq. (11.34), the velocity ratio is

$$\frac{v_{model}}{v_{prototype}} = \left(\frac{1025.0}{997.1}\right)\left(\frac{0.0008899}{0.00148}\right) \times 10 = 6.18.$$

For the seawater flowing at a speed of 0.5 m/s in the prototype (pipe), the speed of fresh water in the model will be 3.09 m/s.

Given that l/D and the Reynolds number (Re) are the same for both the prototype and the model, the nondimensional pressure drop term should be the same, which is given as

$$\left(\frac{\Delta P}{\rho v^2}\right)_{prototype} = \left(\frac{\Delta P}{\rho v^2}\right)_{model}. \tag{11.35}$$

Rearranging Eq. (11.35),

$$\frac{\Delta P_{prototype}}{\Delta P_{model}} = \frac{\left(\rho v^2\right)_{prototype}}{\left(\rho v^2\right)_{model}} = \frac{256.2}{9523.8} = 0.027. \tag{11.36}$$

For the tested pressure drop of 50 kPa in the model, the pressure drop in the prototype will be 1.35 kPa.

11.4 **Nondimensionalization technique**

Unlike the Buckingham π theorem in the dimensional analysis, the form of the equation should be known in nondimensionalization. The sizes of certain dimensionless parameters indicate the importance of certain terms in an equation, providing possibilities to neglect terms. Nondimensional equations can be applied to similar systems where the only changes are those of the basic dimensions of the system. Proper scaling parameters, with a suitable combination of parameters and constants of the equations, need to be selected so that the resulting equation can be dimensionless and represent the characteristics of the physical domain. While nondimensionalizing an equation, nondimensional numbers/parameters (e.g., the Reynolds number) often appear.

One can nondimensionalize a system of equations by taking the following steps:

1. Identify all the independent and dependent variables.
2. List primary dimensions of all dimensional variables and parameters/constants.
3. Select characteristic units (that is, scaling parameters) to nondimensionalize variables: If x is a variable, then x_c is the characteristic unit used to scale the variable x, where x_c has the same dimension as x.
4. Replace each of the dimensional variables with a quantity that is scaled relative to a characteristic unit.
5. Divide through by the coefficient of the highest-order polynomial or derivative term.
6. Choose the characteristic unit for each variable to make as many coefficients as possible unity.

Example 11.4.1. Apply the nondimensionalization technique to the following first-order ordinary differential equation with constant coefficients:

$$\frac{dx}{dt} = a + bx + cx^2. \tag{11.37}$$

Solution.

- Identify all the independent and dependent variables: x is a dependent variable and t is an independent variable.
- List primary dimensions of all dimensional variables and parameters/constants: $[x] = L$; $[t] = T$.

  Note that in order to satisfy the dimensional homogeneity, the units of all the constant coefficients a, b, and c must be different in order to be added in Eq. (11.37): $[a] = L \cdot T^{-1}$; $[b] = T^{-1}$; $[c] = L^{-1} \cdot T^{-1}$.
- Select characteristic units (that is, scaling parameters) to nondimensionalize variables: A constant value can be selected for the characteristic unit (x_c) as x_0 and for t_c as t_0.
- Replace each of dimensional variables with a quantity scaled relative to a characteristic unit. Given that the objective is to replace the dimensional quantities or variables with nondimensional variables, ξ is chosen as the nondimensional variable for x, and τ is chosen for t. Therefore, the dimensional variable in terms of nondimensional variables and characteristic units are expressed as

$$x = \xi x_0, \qquad t = \tau t_0. \tag{11.38}$$

Inserting Eq. (11.38) into Eq. (11.37),

$$\frac{x_0}{t_0}\frac{d\xi}{d\tau} = a + (bx_0)\xi + (cx_0^2)\xi^2. \tag{11.39}$$

- Divide through by the coefficient of the highest-order polynomial or derivative term. The coefficient of the highest-order term is in front of the first derivative term. Dividing by this coefficient gives

$$\frac{d\xi}{d\tau} = \frac{at_0}{x_0} + (bt_0)\xi + (cx_0t_0)\xi^2. \tag{11.40}$$

- Choose the characteristic unit for each variable to make the coefficients of as many terms as possible unity. By choosing the constant term (i.e., the first term in the right-hand side of Eq. (11.40)) and the coefficient in front of ξ (i.e., the second term in the right-hand side of Eq. (11.40)) equal to 1,

$$bt_0 = 1 \Rightarrow t_0 = \frac{1}{b}, \qquad \frac{at_0}{x_0} = 1 \Rightarrow x_0 = \frac{a}{b}. \tag{11.41}$$

The coefficient of ξ^2 (i.e., the last term in Eq. (11.40)) becomes $\kappa = ca/b^2$.

Finally, the nondimensional equation can be expressed as

$$\frac{d\xi}{d\tau} = 1 + \xi + \kappa\xi^2. \tag{11.42}$$

Instead of three undetermined parameters (a, b, c) in the dimensional version (that is, Eq. (11.37)), the nondimensional version has only one parameter, κ.

Problems

11.1 Determine the variables on which a pendulum's amplitude (or sometimes called the pendulum position) depends (see Fig. 11.2). Hint: Use the Buckingham π theorem; the relevant dimensional variables can be listed as pendulum's mass (m), length of the rigid rod (l), pendulum's period (τ), time (t), gravitational acceleration (g), and initial position of the pendulum $(\phi_0 = \phi(0))$.

FIGURE 11.2

Motion of a pendulum.

11.2 Apply the nondimensionalization technique to the Navier–Stokes equation for the x-component; the incompressible Navier–Stokes momentum equation is given by

$$\rho[\frac{\partial u}{\partial t} + (u \cdot \nabla)u] = -\nabla P + \mu\nabla^2 u + \rho g,$$

where ρ is the density, P is the pressure, μ is the dynamic viscosity, u is the flow velocity, t is time, and g is the gravitational acceleration.

11.3 Consider a simplification of the problem in Example 11.2.1 for the pressure drop Δp in a straight pipe of diameter D. Here, in an experiment, it is observed that with laminar flow conditions in the pipe, the pressure drop is observed to depend on three parameters: the distance l along the pipe between the pressure sampling points, the velocity v of the fluid, and the fluid viscosity μ. For this problem, using Buckingham's π theorem, determine how the pressure drop varies with the pipe diameter.

11.4 As shown in Lewis et al. (2017), fuel swelling in nuclear fuel is caused by the presence of fission gases in small bubbles. Consider a gas bubble of radius R embedded in a solid medium with no hydrostatic stress, where the pressure p acting to expand the bubble is balanced by the surface tension γ of the solid (see Fig. 11.3). Using Buckingham's π theorem, determine the dependence of the pressure on the bubble radius and solid surface tension. Hint: For this analysis, use the basic physical dimensions for the system of force (F), length (L), and time (T).

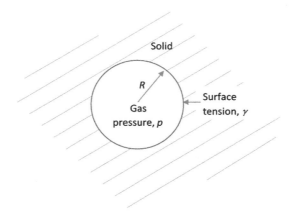

FIGURE 11.3

Gas-filled bubble in mechanical equilibrium with a solid (adapted from Fig. 4.61 in Lewis et al., 2017).

Nonlinear differential equations

Nonlinear phenomena vary from extremely short subfemtosecond processes of laser–atom interactions to large-scale gravitational clustering of galaxies. Nonlinear physics covers a wide range of topics, such as nonlinear processes in fluids, optics and high-power lasers, acoustics, geophysics, astrophysics and cosmology, and quantum systems. In nonlinear systems, the change in the output is not proportional to the change in the input. For example, it is not possible to predict accurately long-term weather forecasts due to the nonlinearity of the system. Most systems, such as atmospheric conditions, economics, quantum mechanics, optics, and fluid dynamics, are nonlinear in nature, where it is difficult to find a general solution. Nonlinear dynamical systems may appear chaotic or unpredictable.

Unlike linear problems where a family of linearly independent solutions can be used to construct a general solution through a superposition principle, it is not generally possible to combine known solutions into a new solution with nonlinear systems. It is possible sometimes to find very specific solutions to nonlinear equations, which makes methods of solution or analysis dependent on the problem itself.

Examples of nonlinear differential equations include: (i) the Navier–Stokes equations in fluid dynamics, (ii) the nonlinear Schrödinger equation in optics and quantum theory, (iii) the Boltzmann equation in thermodynamics for a nonequilibrium state in Hamiltonian mechanics and galactic dynamics, (iv) the Korteweg–de Vries (KdV) equation that arises in the theory of shallow water waves, (v) the Vlasov equation in plasma dynamics with a long-range Coulomb interaction, (vi) the Landau–Ginzburg theory in superconductivity phase transitions, and (vii) the Lotka–Volterra equations in biology.

Nonlinear differential equations are formed by the products of the unknown function and its derivatives. Unlike a linear equation, where the variable and its derivatives appear only with a power of one, in the nonlinear equation any power different from one appears. In addition, the appearance of any function (e.g., log, sin, except for a first-order polynomial), which depends on the dependent variable, results in a nonlinear equation.

Example 12.0.1. Determine if the equations given below are linear or nonlinear.

$$\frac{\partial^2 x}{\partial t^2} + \frac{\partial x}{\partial t} + 5x = 0; \qquad \frac{\partial x}{\partial t} + \frac{1}{x} = 0; \qquad \frac{\partial^2 x}{\partial t^2} + x^3 = 0;$$

$$\frac{\partial x}{\partial t} + \frac{\partial y}{\partial t} = 0; \qquad \frac{\partial^2 x}{\partial t^2} + \sin(t) = 0; \qquad \frac{\partial x}{\partial t} + \sin(x) = 0;$$

Advanced Mathematics for Engineering Students. https://doi.org/10.1016/B978-0-12-823681-9.00020-4

$$\frac{\partial^3 x}{\partial t^3} + t\frac{\partial x}{\partial t} = 0.$$

Solution:
We have:

$$\frac{\partial^2 x}{\partial t^2} + \frac{\partial x}{\partial t} + 5x = 0, \qquad \text{linear;}$$

$$\frac{\partial x}{\partial t} + \frac{1}{x} = 0, \qquad \text{nonlinear because } 1/x \text{ is not of the first power;}$$

$$\frac{\partial^2 x}{\partial t^2} + x^3 = 0, \qquad \text{nonlinear because } x^3 \text{ is not of the first power;}$$

$$\frac{\partial x}{\partial t} + \frac{\partial y}{\partial t} = 0, \qquad \text{linear;}$$

$$\frac{\partial^2 x}{\partial t^2} + \sin(t) = 0, \qquad \text{linear;}$$

$$\frac{\partial x}{\partial t} + \sin(x) = 0, \qquad \text{nonlinear because sin is a function of the dependent variable;}$$

$$\frac{\partial^3 x}{\partial t^3} + t\frac{\partial x}{\partial t} = 0, \qquad \text{linear.}$$

In general, nonlinear equations are difficult to solve analytically, and numerical methods are employed to approximate solutions. In some cases, equations can be solved analytically, but due to the availability of powerful computing tools, nonlinear equations may be chosen to be solved numerically. Nevertheless, it is desirable to compare the approximate solution with the exact one if an exact solution exists. Analytical as well as numerical methods are applied to find roots of a nonlinear function or to find the solution of an initial and boundary value problem of a nonlinear equation. Except for some very special functions, it is not possible to find an analytical expression for the root from which the solution can be exactly determined. This result is true for even some polynomial functions.

An exact solution for a nonlinear system may exist for a specific case. Some techniques, such as inverse scattering and Bäcklund transforms, have been proposed and applied to solve analytically a particular type of equation, such as the KdV and sine-Gordon equations. Besides analytical solutions, numerical solutions, such as the finite difference method, have been applied. In some cases, nonlinear dynamical equations are approximated by linear equations (that is, linearization). Linearization may work up to some accuracy within some range of input parameters, but some features can be lost with linearization. Analytical and numerical solutions to initial and boundary value problems, specific to nonlinear equations, are discussed in the following sections.

Nonlinear waves are fundamental covering a wide range of phenomena from aerodynamics and hydrodynamics, solid state physics and plasma physics, to optics and field theory, population dynamics, nuclear physics, and gravity. In particular, solitons are the solutions of a widespread class of weakly nonlinear dispersive partial differential equations. They find application in fiber optics, quantum mechanics, biology, and magnets. Solitons result from the cancelation of nonlinear and dispersive effects in the medium (that is, where the speed of the waves varies according to the frequency).

 Since solutions to nonlinear problems are case-specific, it is not possible to discuss each and every case and solution. Some well-known nonlinear partial differential equations include:

- The nonlinear Schrödinger equation:

$$i\psi_t = -\frac{1}{2}\psi_{xx} + \kappa|\psi|^2\psi,$$

where ψ is the complex field and κ is the wave number. The principal applications of the nonlinear Schrödinger equation are: (i) propagation of light in nonlinear optical fibers, (ii) Bose–Einstein condensates for special cases, (iii) Langmuir waves in hot plasma, (iv) propagation of plane-diffracted wave beams in the focusing regions of the ionosphere, (v) propagation of Davydov's alpha-helix solitons responsible for energy transport along molecular chains, and (vi) small-amplitude gravity waves on the surface of deep inviscid water.

- The Navier–Stokes equations,

$$\rho\left(u_t + u \cdot \nabla u\right) = -\nabla P + \mu\nabla^2 u + \frac{1}{3}\mu\nabla\left(\nabla \cdot u\right) + \rho g,$$

where u is the flow velocity, P is the pressure, μ is the dynamic viscosity, ρ is density, and g is gravitational acceleration. The Navier–Stokes equations may be used to model air flow around a wing, the weather, ocean currents, and water flow in a pipe. These equations find application in the study of blood flow and in the design of aircraft and cars. Coupled with Maxwell's equations, they can be further used to model and study magnetohydrodynamic phenomena.

- Reaction–diffusion equations,

$$u_t = Du_{xx} + R(u),$$

where u is the concentration variable, D is the (diagonal) diffusion coefficient (scalar matrix) independent from coordinates, and $R(u)$ is the local reaction kinetics. Reaction–diffusion equations are employed in defining systems in chemistry, biology, geology, physics (neutron diffusion theory), and ecology. If the reaction term vanishes, the equation represents a pure diffusion process known as Fick's second law of diffusion in Eq. (5.9). With $R(u) = u(1 - u)$, which is the Fisher equation, the reaction–diffusion equation is used to describe the spreading of biological populations. With $R(u) = u(1 - u^2)$, which is the Newell–Whitehead–Segel equation, the reaction–diffusion equation is used to describe Rayleigh–Bénard convection. The reaction–diffusion equation with the reaction term defined as $R(u) = u(1 - u)(u - \alpha)$ with $0 < \alpha < 1$ is the general Zeldovich equation used in combustion theory.

In the following nomenclature, u_t and ψ_t denote partial derivatives with respect to time, t, and u_x, u_{xx} ψ_x denote partial derivatives with respect to the space coordinate, x.

In this chapter, solutions are sought for waves which have wide application to optics, acoustics, and hydrodynamics. Several analytical methods, followed by some numerical solutions, are discussed in the following sections. Of particular importance is the KdV equation, which is a hyperbolic partial differential equation (see, for example, Section 6.2.3), where solutions of hyperbolic equations are "wave-like." If a disturbance is introduced in the initial data of a hyperbolic differential equation, the disturbance is not felt at every point of space at once. Relative to a fixed time coordinate, disturbances have a finite propagation speed and travel along the characteristics of the equation. Unlike a hyperbolic equation, a perturbation of the initial or boundary data of an elliptic or parabolic equation is felt at once by essentially all points in the domain.

The KdV equation is particularly important as a prototypical example of an exactly solvable nonlinear system (that is, completely integrable infinite dimensional system). The KdV equation describes shallow water waves that are weakly and nonlinearly interacting[1], ion acoustic waves in a plasma, long internal waves in a density-stratified ocean, and acoustic waves on a crystal lattice. In addition, it is the governing equation for a string in the Fermi–Pasta–Ulam–Tsingou problem in the continuum limit. The KdV equation is one of the most familiar models for solitons and it is an important foundation for the study of other equations. The KdV equation can be expressed in different forms, some of which are in the nondimensional form, as shown below:

$$
\begin{aligned}
(a) \quad & u_t - 6uu_x + u_{xxx} = 0, \\
(b) \quad & u_t + 6uu_x + u_{xxx} = 0, \qquad\qquad (12.1) \\
(c) \quad & u_t + (1 + u)u_x + u_{xxx} = 0,
\end{aligned}
$$

where u is the average nondimensional velocity. Due to the dispersive term (that is, of third order), waves decay while waves will steepen due to the nonlinear term.

---

[1] Assume compound waves with two peaks, the taller to the left, both propagating to the right. The taller wave moves faster than, catches up on, and interacts with the shorter wave. It then moves ahead of it. In this case, the interaction is not linear. Unlike the linear interaction where the two waves satisfy linear superposition, in the nonlinear interaction, the waves are phase-shifted.

12.1 Analytical solution

In this section, analytical solutions to nonlinear initial and boundary value problems are discussed. The most common analytical solution that is applied to many exactly solvable nonlinear systems is the "inverse scattering transform." Another method is the "traveling wave solution" for waves maintaining a fixed shape and advancing in a particular direction.

12.1.1 Inverse scattering transform

The inverse scattering transform is a nonlinear analog of the Fourier transform (see Section 3.2). In general, this method can be applied to solve many linear partial differential equations. Unlike the direct scattering method, where the scattering matrix is constructed from the potential, in the inverse scattering method, the potential is recovered from its scattering matrix. To be more precise, the time evolution of the potential is recovered from the time evolution of its scattering data.

For example, in the case of the KdV equation, the following steps need to be applied in order to determine the time evolution of the flow velocity $u(x, t)$ from the initial condition given by $u(x, 0)$:

1. Find the Lax pair consisting of two linear operators L and M for the nonlinear equation of interest.
2. Determine the initial scattering matrix $S(\lambda, 0)$ at $t = 0$ from the initial potential $u(x, 0)$ by solving the direct scattering problem, using the first operator L.
3. Determine the time evolution of the scattering matrix $S(\lambda, t)$ at time t, using the second operator M.
4. Determine the potential $u(x, t)$, which is the sought-after solution to the nonlinear equation of interest from the scattering data $S(\lambda, t)$ by solving the inverse scattering problem.

The concept of the inverse scattering transform is demonstrated in Fig. 12.1.

The application of the method for the four steps depicted in Fig. 12.1 is detailed below.

1. *Lax pair for the nonlinear equation.*
   A derivation of the Lax pair, which recovers the nonlinear equation (12.1)(a), is developed as follows. Thus, linear operators L and M are defined such that they satisfy the original equation, which in this case is the KdV equation. The operators L and M depend on an unknown function, $u(x, t)$. The first linear differential operator, L, may depend on x, but not explicitly on t; it describes the spectral (scattering) problem to establish the scattering matrix at $t = 0$ as

$$L\phi = \lambda\phi, \tag{12.2}$$

where ϕ is the eigenfunction and λ is the time-independent eigenvalue.

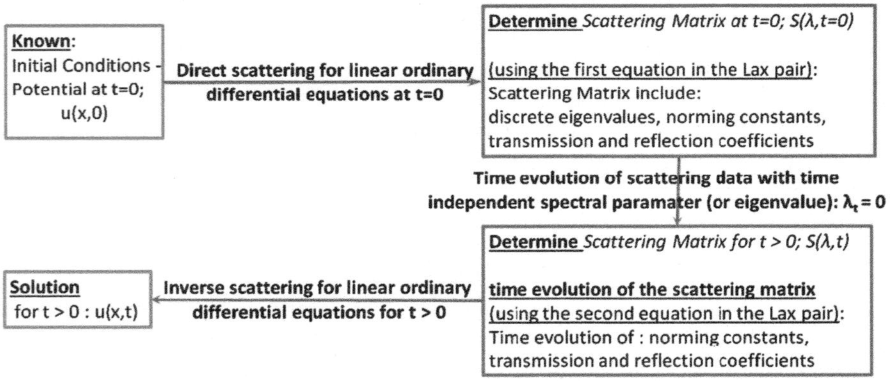

FIGURE 12.1

Diagram showing the steps involved in the inverse scattering transform.

The operator M describes how the eigenfunction evolves in time, satisfying the following equation:

$$\frac{\partial \phi}{\partial t} = M\phi. \tag{12.3}$$

For consistency, these equations can be solved for $L\partial\phi/\partial t$ by taking the time derivative of Eq. (12.2). The resultant equation becomes

$$\frac{\partial L}{\partial t}\phi + L\frac{\partial \phi}{\partial t} = \frac{\partial \lambda}{\partial t}\phi + \lambda\frac{\partial \phi}{\partial t}. \tag{12.4}$$

Rearranging Eq. (12.4) using Eq. (12.3),

$$\frac{\partial L}{\partial t}\phi + LM\phi - \frac{\partial \lambda}{\partial t}\phi - ML\phi = 0. \tag{12.5}$$

Since the spectral parameter (or eigenvalue) λ is independent of time (that is, $\partial \lambda/\partial t = 0$), the Lax equation is expressed as

$$\frac{\partial L}{\partial t} + LM - ML = 0, \tag{12.6}$$

which is a nonlinear partial differential equation for $u(x, t)$ due to the nonzero commutator $LM - ML$ of the two nonconstant operators. In addition, $[L, M] = LM - ML$ is a multiplicative operator.

Let L be the Schrödinger operator,

$$L = \frac{\partial^2}{\partial x^2} - u. \tag{12.7}$$

The M operator becomes

$$M = -4\frac{\partial^3}{\partial x^3} + 6u\frac{\partial}{\partial x} + 3\frac{\partial u}{\partial x}. \tag{12.8}$$

The M operator, along with the Schrödinger operator, Eq. (12.7), satisfies Eq. (12.6), and hence u satisfies the KdV equation, in the form of

$$\frac{\partial u}{\partial t} - 6u\frac{\partial u}{\partial x} + \frac{\partial^3 u}{\partial x^3} = 0. \tag{12.9}$$

Inserting Eq. (12.7) and Eq. (12.8) into Eq. (12.2) and Eq. (12.3), respectively, the Lax pair can be rewritten as

$$L\phi = -\frac{\partial^2 \phi}{\partial x^2} + u(x,t)\phi = \lambda\phi, \tag{12.10}$$

$$\frac{\partial \phi}{\partial t} = M\phi = -4\frac{\partial^3 \phi}{\partial x^3} + 6u\frac{\partial \phi}{\partial x} + 3\frac{\partial u}{\partial x}\phi. \tag{12.11}$$

The Lax equation, Eq. (12.6), with its pair, Eq. (12.7) and Eq. (12.8), is the compatibility condition for the equation of interest, that is, the KdV equation. In the Lax equation, $L_t = \partial L/\partial t$ is the time derivative of L, where it explicitly depends on t. With the combination of the appropriate Lax pair, the Lax equation recovers the original nonlinear partial differential equation. The suitability of the Lax pair for the KdV equation is demonstrated below.

Inserting Eq. (12.7) and Eq. (12.8) into Eq. (12.6),

$$\frac{\partial L}{\partial t} = -\frac{\partial u}{\partial t} = \left[-4\frac{\partial^3}{\partial x^3} + 6u\frac{\partial}{\partial x} + 3\frac{\partial u}{\partial x}, \frac{\partial^2}{\partial x^2} - u \right], \tag{12.12}$$

where the right-hand side of the equation is equal to $-(LM - ML) = (ML - LM)$, with

$$LM = -4\frac{\partial^5}{\partial x^5} + 10u\frac{\partial^3}{\partial x^3} + 15\frac{\partial u}{\partial x}\frac{\partial^2}{\partial x^2} + \left(12\frac{\partial^2 u}{\partial x^2} - 6u^2 \right)\frac{\partial}{\partial x}$$
$$+ \left(3\frac{\partial^3 u}{\partial x^3} - 3u\frac{\partial u}{\partial x} \right), \tag{12.13}$$

$$ML = -4\frac{\partial^5}{\partial x^5} + 10u\frac{\partial^3}{\partial x^3} + 15\frac{\partial u}{\partial x}\frac{\partial^2}{\partial x^2} + \left(12\frac{\partial^2 u}{\partial x^2} - 6u^2 \right)\frac{\partial}{\partial x}$$
$$+ \left(4\frac{\partial^3 u}{\partial x^3} - 9u\frac{\partial u}{\partial x} \right). \tag{12.14}$$

Finally,

$$[L, M] = (LM - ML) = \left(-\frac{\partial^3 u}{\partial x^3} + 6u \frac{\partial u}{\partial x} \right). \tag{12.15}$$

Using the commutator operator $[L, M]$, one recovers the original KdV equation in Eq. (12.9), where $\frac{\partial u}{\partial t} = [L, M]$.

2. *Direct scattering – determine the initial scattering matrix* $S(\lambda, 0)$ *at* $t = 0$.

For a given potential, $u(x)$, the problem is to find the spectrum of the linear operator L (that is, a set of admissible values for λ) and to construct the corresponding functions $\phi(x; \lambda)$. The first Lax equation represents the time-independent Schrödinger equation:

$$\frac{\partial^2 \phi}{\partial x^2} - u\phi = -\lambda\phi. \tag{12.16}$$

Here, λ is a constant eigenvalue, ϕ is an unknown eigenfunction of t and x associated with the eigenvalue λ, and $u(x, t)$ is the solution of the KdV equation that is unknown except at the time $t = 0$.

It is assumed that $u \to 0$ sufficiently rapidly as $x \to \pm\infty$ so that the term $(u\phi)$ vanishes. Hence, as follows from Eq. (12.16), $\phi_{xx} \sim -\lambda\phi$. This latter equation provides an asymptotic solution for ϕ, which is a linear combination of $\exp(\pm i\sqrt{\lambda}x)$ as derived from Section 2.2.2. Consequently, ϕ decays exponentially at infinity if $\lambda < 0$, and oscillates sinusoidally at infinity if $\lambda > 0$. Therefore, solutions are sought for two different spectra: (a) a discrete spectrum (bound states), where $\lambda < 0$, to obtain discrete eigenvalues (λ or κ) and the norming coefficients of the eigenfunctions, C_n, and (b) a continuous spectrum, where $\lambda > 0$ for determination of reflection and transmission coefficients.

(a) Discrete spectrum (bound states) at $t = 0$: With $\lambda < 0$, $\kappa = \sqrt{-\lambda} > 0$.

If the potential $u(x, 0)$ is negative near the origin of the x axis, the spectral problem implies the existence of a finite number of eigenfunctions $\phi = \phi_n(x, \lambda)$, $n = 1, \cdots, N$, where N is the number of bound states. The discrete spectrum has the following characteristics:

- $\lambda_n = -\kappa_n^2 < 0$, $\kappa_1 > \kappa_2 \cdots \kappa_n$, $n = 1, \cdots, N$;
- the spectrum is nondegenerate with a one-to-one correspondence of λ_n with ϕ_n;
- $\min\{u(x, 0)\} < \lambda_1 < \lambda_2 < \cdots < 0$;
- eigenfunctions ϕ_n that are square integrable are used for the normalization, $\|\phi_n\|^2 = \int_{-\infty}^{\infty} \phi_n^2 dx = 1$.

The asymptotic behavior will be consistent with the Schrödinger equation, $\phi_{xx} \sim -\lambda\phi$, at $|x| \to \infty$, where u vanishes, if

$$\phi_n(x, 0) \sim C_n(0) e^{-\kappa_n x} \qquad \text{as} \quad x \to \infty. \tag{12.17}$$

Here $C_n(0)$ are the norming coefficients at $t = 0$.

(b) Continuous spectrum at $t = 0$: With $\lambda > 0$, $\kappa = \sqrt{\lambda} > 0$, all solutions of ϕ in Eq. (12.16) are sinusoidal as $x \to \pm\infty$. Therefore, for all $\lambda > 0$, eigenfunctions are bounded but not square integrable. The solution can be expressed as a super-position of an incident wave, with reflected and transmitted waves, as shown in Fig. 12.2.

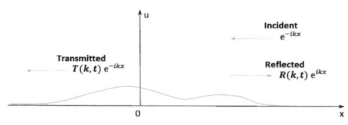

FIGURE 12.2

Scattering by a potential.

For $k = \sqrt{\lambda} > 0$, a solution is sought for

$$\phi_k(x, 0) \sim \begin{cases} T(k, 0)\, e^{-ikx} & \text{as} \quad x \to -\infty, \\ e^{-ikx} + R(k, 0)\, e^{ikx} & \text{as} \quad x \to \infty, \end{cases} \tag{12.18}$$

where $T(k, 0)$ is a transmission coefficient, $R(k, 0)$ is a reflection coefficient, and k is the wave number. Note that the functions $R(k, 0)$ and $T(k, 0)$ are not inde-pendent, which can be expressed by the total probability relationship

$$|T|^2 + |R|^2 = 1.$$

(c) Scattering data at $t = 0$:
Finally, from the analysis of discrete and continuous spectra, the scattering data are expressed in terms of discrete eigenvalues, κ_n, norming coefficients, $C_n(0)$, transmission coefficients $(T(k, 0))$, and reflection coefficients $(R(k, 0))$:

$$S(\lambda, 0) = \left(\{\kappa_n, C_n(0)\}_{n=1}^{N}, R(k, 0), T(k, 0) \right). \tag{12.19}$$

3. *Time evolution of the scattering data $S(\lambda, t)$.*

(a) Discrete spectrum at $t > 0$:
With the potential $u(x, t)$ vanishing at $x \to \infty$, the second operator, M (see Eq. (12.8)) in the Lax pair reduces to

$$M = -4 \frac{\partial^3}{\partial x^3}. \tag{12.20}$$

Inserting Eq. (12.17) into Eq. (12.3) with M defined in Eq. (12.20),

$$\frac{\partial}{\partial t}\phi_n(x,t) = M\phi_n(x,t) = -4\frac{\partial^3}{\partial x^3}C_n(t)\,\mathrm{e}^{-\kappa_n x} = 4\kappa_n^3 C_n(t)\,\mathrm{e}^{-\kappa_n x}. \qquad (12.21)$$

Given that

$$\frac{dC_n(t)}{dt} = 4\kappa_n^3 C_n(t),$$

the time evolution of the norming coefficients, $C_n(t)$, is

$$C_n(t) = C_n(0)\,\mathrm{e}^{4\kappa_n^3 t}. \qquad (12.22)$$

The norming coefficients used in the Marchenko kernel for the inverse scattering transform are

$$C_n'(t) = (C_n(t))^2 \quad \text{so that} \quad C_n'(t) = C_n'(0)\,\mathrm{e}^{8\kappa_n^3 t}, \qquad (12.23)$$

as follows from Eq. (12.22).

(b) Continuous spectrum at $t > 0$:
In this section, the derivation of the time evolution of the continuous spectrum is discussed briefly. The details can be found in Drazin (1985) and Koelink (2006).

The eigenfunctions defined in Eq. (12.18) for $x \to \infty$ can be rewritten as

$$\phi_k(x,t) = A(t)\,\mathrm{e}^{-ikx} + B(t)\,\mathrm{e}^{ikx}, \qquad (12.24)$$

with $A(0) = 1$ and $B(0) = R(k,0)$. Inserting Eq. (12.24) into Eq. (12.3) with M as defined in Eq. (12.20) for $x \to \infty$ yields

$$\frac{\partial}{\partial t}\phi_k(x,t) = M\phi_k(x,t) = -4(-ik)^3 A(t)\,\mathrm{e}^{-ikx} - 4(ik)^3 B(t)\,\mathrm{e}^{ikx}. \qquad (12.25)$$

Here $A(t) = \mathrm{e}^{-4ik^3 t}$ and $B(t) = R(k,0)\,\mathrm{e}^{4ik^3 t}$, where

$$R(k,t) = R(k,0)\,\mathrm{e}^{8ik^3 t}. \qquad (12.26)$$

Similarly, rewriting the eigenfunction defined in Eq. (12.18) for $x \to -\infty$,

$$\phi_k(x,t) = C(t)\,\mathrm{e}^{ikx} + D(t)\,\mathrm{e}^{-ikx}, \qquad (12.27)$$

where $C(0) = 0$ and $D(0) = T(k,0)$. Inserting Eq. (12.27) into Eq. (12.3) with M defined in Eq. (12.20) for $x \to -\infty$ gives

$$\frac{\partial}{\partial t}\phi_k(x,t) = M\phi_k(x,t) = -4(ik)^3 C(t)\,\mathrm{e}^{ikx} - 4(-ik)^3 D(t)\,\mathrm{e}^{-ikx}. \qquad (12.28)$$

Here $C(t) = 0$ and $D(t) = T(k,0) e^{-4ik^3 t}$, where

$$T(k,t) = T(k,0). \tag{12.29}$$

Eq. (12.29) indicates that the transmission coefficient is time-independent.

(c) Scattering data at $t > 0$:
Finally, from the analysis of discrete and continuous spectra at $t > 0$, the scattering data are expressed in terms of discrete eigenvalues (κ_n), norming coefficients ($C_n(t)$), transmission coefficients ($T(k,t)$), and reflection coefficients ($R(k,t)$):

$$S(\lambda, t) = \left(\{\kappa_n, C_n(t)\}_{n=1}^N, R(k,t), T(k,t) \right). \tag{12.30}$$

4. *Inverse scattering transform.*
   The following steps are followed to recover the potential, $u(x,t)$, for $t > 0$ from the scattering data, $S(\lambda, t)$, in Eq. (12.30):

   • The scattering data are used in the Marchenko kernel:

   $$F(y,t) = \frac{1}{2\pi} \int_{-\infty}^{\infty} dk\, R(k,t) e^{iky} + \sum_{n=1}^N C_n(t) e^{-\kappa_n y}. \tag{12.31}$$

   • The Marchenko integral,

   $$K(x,y,t) + F(x+y,t) + \int_x^{\infty} dz\, K(x,z,t) F(y+z,t) = 0, \qquad x < y < \infty, \tag{12.32}$$

   is solved to obtain a solution for $K(x,y,t)$.
   • The potential, $u(x,t)$, is then obtained by employing the relation

   $$u(x,t) = -2 \frac{\partial K(x,x,t)}{\partial x}. \tag{12.33}$$

A special case is the reflectionless potential, where $R(k,t) = 0$ with $N = 1$ in Eq. (12.31). This case results in a soliton with an amplitude $2\kappa^2$ that propagates to the right with the velocity $4\kappa^2$:

$$u(x,t) = -2\kappa^2 \text{sech}^2 \left[\kappa \left(x - 4\kappa^2 t - x_0 \right) \right]. \tag{12.34}$$

Moreover, the solution of the KdV equation corresponding to a reflectionless potential with N-solitons can be represented by a superposition of N single-soliton solutions propagating to the right (see Drazin, 1985),

$$u(x,t) \sim - \sum_{n=1}^N 2\kappa_n^2 \text{sech}^2 \left[\kappa_n \left(x - 4\kappa_n^2 t - x_n \right) \right], \qquad t \to \infty, \tag{12.35}$$

where the phase lag, x_n, is given by

$$x_n = \frac{1}{2\kappa_n} \ln \left[\frac{c_n(0)}{2\kappa_n} \prod_{m=1}^{n-1} \left(\frac{\kappa_n - \kappa_m}{\kappa_n + \kappa_m} \right)^2 \right]. \tag{12.36}$$

In the KdV equation, the discrete spectrum gives N solitary waves, whose amplitudes, $-2\kappa_n^2$, are proportional to their velocities, such that the bigger waves move away faster to the right. The continuous spectrum gives the dispersive wave components of the solution. Dispersive waves travel to the left and spread out where the amplitudes die away.

12.1.2 Traveling wave solution

A traveling wave moves in space, where the traveling wave variable can be defined as $\xi = x - ct$, where c is the constant wave velocity. Substituting this solution form into the partial differential equations gives a system of ordinary differential equations known as a traveling wave equation. The permanent wave solution in the form of $u(x, t) = f(\xi)$ is sought when applied to the KdV equation in the form

$$\frac{\partial u}{\partial t} - 6u \frac{\partial u}{\partial x} + \frac{\partial^3 u}{\partial x^3} = 0. \tag{12.37}$$

The function $f(\xi)$ and c should be determined by substituting the assumed form of solution into the KdV equation, yielding an ordinary differential equation for $f(\xi)$ as

$$-cf' - 6ff' + f''' = 0, \tag{12.38}$$

where f' is defined as

$$f' = \frac{d}{d\xi} f(\xi). \tag{12.39}$$

Note

$$\frac{\partial f}{\partial t} = \frac{\partial f}{\partial \xi} \frac{\partial \xi}{\partial t} = -c \frac{\partial f}{\partial \xi}, \qquad \frac{\partial}{\partial t} = -c \frac{\partial}{\partial \xi}$$

and

$$\frac{\partial f}{\partial x} = \frac{\partial f}{\partial \xi} \frac{\partial \xi}{\partial x} = \frac{\partial f}{\partial \xi}, \qquad \frac{\partial}{\partial x} = \frac{\partial}{\partial \xi}.$$

The integration of Eq. (12.38) gives

$$-cf - 3f^2 + f'' - A = 0, \tag{12.40}$$

where A is the integration constant. Multiplying Eq. (12.40) by f' and integrating the resultant equation yields

$$-\frac{1}{2}cf^2 - f^3 + \frac{1}{2}(f')^2 - Af = B, \tag{12.41}$$

where B is the integration constant. This equation can also be interpreted as the "conservation of energy," where $1/2 \left(f'\right)^2$ represents the kinetic energy and $-1/2cf^2 - f^3 - Af$ represents the potential energy. For the potential, critical points occur if $-cf - 3f^2 - A = 0$ in Eq. (12.40). The discriminant (that is, $b^2 - 4ac$) is $c^2 + 12A$. If $c^2 + 12A \leq 0$, then the potential energy monotonically increases, and the traveling waves are not bounded. In the case of $c^2 + 12A > 0$, there are two equilibria.

This equation can also be written as

$$\frac{1}{2} \left(f'\right)^2 = f^3 + \frac{1}{2}cf^2 + Af + B \equiv F(f). \tag{12.42}$$

This is a first-order ordinary nonlinear differential equation, and the roots of the cubic $F(f)$ equation (see the right-hand side of Eq. (12.42)) are important and can be identified (see Fig. 12.3) as:

1. three distinct real roots (see Fig. 12.3 (C));
2. two real roots, one of which is a double root and the other one is a simple root (see Fig. 12.3 (B) and (D));
3. one real and two imaginary roots (see Fig. 12.3 (A) and (E));
4. one root of order three (see Fig. 12.3 (F)).

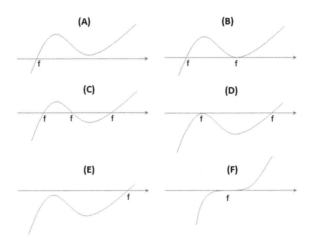

FIGURE 12.3

Roots of the function $F(f)$ for different cases (reproduced from Drazin, 1985).

Depending on the cases of interest (e.g., cnoidal or solitary waves) that define the roots of the equation, some solutions to Eq. (12.42) arise as discussed below.

1. In the case where the function $F(f)$ has one root of order three (see Fig. 12.3 (F)), $f = -c/6$ and the exact solution of Eq. (12.42) is

$$f(\xi) = -\frac{c}{2} + \frac{2}{(\xi - \xi_0)^2}.$$ (12.43)

Note that, in this case, the solution is not bounded.

2. In the case where the function $F(f)$ has three distinct real roots (see Fig. 12.3 (C)), defining cnoidal waves, the solution is obtained using the Jacobian elliptic function. Conversely, the frequency of the cnoidal wave is described as

$$\begin{aligned} \omega &= kc \\ &= -2k(f_1 + f_2 + f_3), \end{aligned}$$

where f_1, f_2, and f_3 are the roots of the function $F(f)$. The solution is obtained as

$$u(x, t) = f(\xi) = f\left(\frac{kx - \omega t}{k}\right).$$ (12.44)

3. In the case where the function $F(f)$ has two real roots, one of which is double and the other one is simple (see Fig. 12.3 (B)), this is an example of a solidary wave, where $f(\xi)$ decays rapidly. In the case of solidary waves, the boundary conditions can be defined as f, f', $f'' \to 0$ as $\xi \to \infty$, resulting in $A = 0$, $B = 0$, and $[1/2(f')^2 = f^2(f + 1/2c)]$. Consequently,

$$\xi = \int \frac{df}{f'} = \int \frac{df}{f\sqrt{2f + c}},$$

and the solution is

$$f(\xi) = -\frac{c}{2}\text{sech}^2\left[\frac{\sqrt{c}}{2}(\xi - \xi_0)\right],$$

or

$$u(x, t) = -\frac{c}{2}\text{sech}^2\left[\frac{\sqrt{c}}{2}(x - ct - x_0)\right].$$ (12.45)

The maximum of u lies at $x = ct$, leading to an amplitude of $c/2$. The width is inversely proportional to $\sqrt{c}$.

12.2 Numerical solution

Numerical methods can be applied to:

- find the roots of a nonlinear equation (these methods include, but are not limited to, (i) interval halving (bisection method), (ii) linear interpolation, (iii) Newton's method, (iv) quotient difference, and (v) the secant method (see Section 9.1);
- solve nonlinear initial and boundary value problems.

12.2.1 **Roots of a nonlinear function**

Newton's method, as discussed in Section 9.1.2, is one of the most widely used techniques to find roots of a nonlinear equation. Other techniques can be found in books specialized in numerical methods.

12.2.2 **Nonlinear initial and boundary value problems**

One of the common methods that is applied to solve a nonlinear differential equation is the finite difference method (see Chapter 6). In this section, the application of the finite difference method is discussed for the KdV equation. The first and third derivatives in the KdV equation can be approximated by the Taylor series as presented in Chapter 6. The scheme that is described below was proposed by Zabusky and Kruskal (1965). It uses the central difference technique. The subscript i represents the space variable, and the subscript j represents the time variable, with $u_{i,j} = u(i\Delta x, j\Delta t)$.

Using the central difference, the time derivative of u can be approximated by

$$u_t = \frac{u_{i,j+1} - u_{i,j-1}}{2\Delta t}. \tag{12.46}$$

Using the central difference, the first space derivative of u can be approximated by

$$u_x = \frac{u_{i+1,j} - u_{i-1,j}}{2\Delta x}. \tag{12.47}$$

The nonlinear term in the KdV equation can be approximated by taking a local spatial average of u (instead of simply u_i) as

$$u = \frac{u_{i+1,j} + u_{i,j} + u_{i-1,j}}{3}. \tag{12.48}$$

Using the central difference, the dispersive term, which is the third derivative of u, can be approximated by

$$u_{xxx} = \frac{u_{i+2,j} - 2u_{i+1,j} + 2u_{i-1,j} - u_{i-2,j}}{2\Delta x^3}. \tag{12.49}$$

Inserting the discretized forms into the KdV equation given by Eq. (12.1)(a), the explicit leapfrog finite difference scheme (Zabusky and Kruskal, 1965) is expressed as

$$
\begin{aligned}
u_{i,j+1} = u_{i,j-1} &+ 2\frac{\Delta t}{\Delta x}\left(u_{i+1,j} + u_{i,j} + u_{i-1,j}\right)\left(u_{i+1,j} - u_{i-1,j}\right) \\
&- \frac{\Delta t}{\Delta x^3}\left(u_{i+2,j} - 2u_{i+1,j} + 2u_{i-1,j} - u_{i-2,j}\right).
\end{aligned}
\tag{12.50}
$$

The truncation error is of order $(O(\Delta t)^2 + O(\Delta x)^2)$. With $u \to -u$, the linear stability requirement for this scheme (Taha and Ablowitz, 1984) is

$$\frac{\Delta t}{\Delta x}\left| -2u_0 + \frac{1}{(\Delta x)^2} \right| \leq \frac{2}{3\sqrt{3}}.$$

Other schemes and methods (that is, implicit schemes and finite Fourier transform or pseudospectral methods) that were proposed to solve the KdV equation are also discussed in Taha and Ablowitz (1984).

The KdV equation defines the waves traveling in one direction (i.e., right-moving, $+x$ direction). In the derivation of the KdV equation, mass (i.e., continuity) and momentum conservation equations as well as appropriate boundary conditions are used. Similarly, waves moving in both directions (i.e., left and right) are described by the Boussinesq equation. Adopting the formulation given by Whitham (1974), the mass and momentum equations, which are discussed next, can be solved separately for a more robust numerical stability.

The continuity equation is given by

$$\frac{\partial \eta}{\partial t} + \frac{\partial}{\partial x}[(1+\eta)u] = 0 \tag{12.51}$$

and the momentum equation is given by

$$\frac{\partial u}{\partial t} + u\frac{\partial u}{\partial x} + \frac{\partial \eta}{\partial x} = \frac{1}{3}\frac{\partial^3 u}{\partial x^2 \partial t}, \tag{12.52}$$

where η is the wave amplitude, x is in the units of h_0, which is defined as the undisturbed depth of water, and t is in the units of h_0/c_0. Here c_0 is the speed of gravity waves as defined as $\sqrt{gh_0}$. Also, g is the acceleration constant of gravity and u is the average velocity over a plane normal to the wave propagation direction.

Based on the numerical scheme suggested by Peregrine (1966), the following steps are applied: (1) solve the continuity equation to calculate the amplitude for an advanced time step, (2) using the amplitude from the previous step, solve the momentum equation to determine the average horizontal velocity, and (3) substitute the average horizontal velocity from the previous step into the continuity equation, to obtain a new corrected amplitude. With the time steps denoted by j, the space steps by i, and the uncorrected amplitude by *, the discretization scheme is defined as follows:

1. Discretize the continuity equation to obtain the uncorrected amplitude for an advanced time step,

$$\frac{\eta^*_{i,j+1} - \eta_{i,j}}{\Delta t} + \left[1 + \frac{1}{2}\left(\eta^*_{i,j+1} + \eta_{i,j}\right)\right] + \frac{u_{i+1,j} + u_{i-1,j}}{2\Delta x}$$
$$+ u_{i,j}\frac{\eta_{i+1,j} - \eta_{i-1,j}}{2\Delta x} = 0. \tag{12.53}$$

2. Discretize the momentum equation to obtain the average horizontal velocity for an advanced time step,

$$
\frac{u_{i,j+1} - u_{i,j}}{\Delta t} + u_{i,j} \frac{u_{i+1,j+1} - u_{i-1,j} + u_{i+1,j} - u_{i-1,j}}{4\Delta x}
$$
$$
+ \frac{\eta^*_{i+1,j+1} - \eta^*_{i-1,j+1} + \eta_{i+1,j} - \eta_{i-1,j}}{4\Delta x}
$$
$$
= \frac{1}{3} \frac{u_{i+1,j+1} - 2u_{i,j+1} + u_{i-1,j+1} - u_{i+1,j} + 2u_{i,j} - u_{i-1,j}}{\Delta x^2 \Delta t}. \tag{12.54}
$$

3. Discretize the continuity equation to obtain the corrected amplitude for an advanced time step,

$$
\frac{\eta_{i,j+1} - \eta_{i,j}}{\Delta t} + \left(1 + \eta_{i,j}\right) + \frac{u_{i+1,j+1} - u_{i-1,j+1} + u_{i+1,j} - u_{i-1,j}}{4\Delta x}
$$
$$
+ \frac{1}{2} \left(u_{i,j+1} + u_{i,j}\right) \frac{\eta_{i+1,j} - \eta_{i-1,j}}{2\Delta x} = 0. \tag{12.55}
$$

This scheme, with discretized initial and boundary conditions, was applied to predict the height of waves in a tubular test section (see Onder, 2004).

The analytical and numerical methods discussed in this section can be used as a basis to seek solutions to other nonlinear problems.

Problems

12.1 Discuss the finite difference solution(s) to the Navier–Stokes equations.

12.2 Obtain the solution for the KdV equation given in Eq. (12.45) by assuming a traveling wave of the form $u(x,t) = f(\xi)$, where $\xi = (x - ct - x_o)$. Solve this problem with direct integration by explicitly separating the differentials and using antiderivatives. Use Maple to solve the resulting integral that arises in this derivation,

$$
\int \frac{df}{f\sqrt{c + 2f}}.
$$

12.3 Show that the classical *advection equation* $\dfrac{\partial u}{\partial t} + c\dfrac{\partial u}{\partial x} = 0$, with $u = u(x,t)$, can be reproduced with the Lax operator pair

$$
L = \frac{\partial^2}{\partial x^2} - uI,
$$
$$
M = -c\frac{\partial}{\partial x},
$$

where I is the identity operator.

12.4 Classify the following equations as linear or nonlinear:

(a) $u_{tt} = c^2 u_{xx}$,

(b) $\sigma \rho u_t = \nabla \cdot (K(u) \nabla u)$, where σ and ρ are constants,

(c) $u' = f(u, t)$,

(d) $\nabla^2 u = \lambda \sin(u)$, where λ is a constant,

(e) $\dfrac{\partial u}{\partial t} + u \dfrac{\partial u}{\partial x} = 0$.

12.5 Consider the nonlinear ordinary differential equation

$$u'(t) = au(t) + b(u(t))^2,$$

where a and b are constants. The time derivative for this equation can be discretized as

$$u'(t) = \frac{u_{n+1} - u_n}{\Delta t},$$

where u_n and u_{n+1} are the values of the function $u(t)$ at the respective nth and $n+1$st mesh points.

(a) What is the iteration scheme for this nonlinear equation using the forward Euler discretization method in Eq. (6.4)?

(b) Consider a Crank–Nicolson discretization method for this ordinary differential equation with a central difference approach given by

$$\frac{u_{n+1} - u_n}{\Delta t} = a\, u_{n+\frac{1}{2}} + b\left[u_{n+\frac{1}{2}} \right]^2.$$

What is the resulting iteration scheme if $u_{n+\frac{1}{2}}$ is approximated by a standard *arithmetic mean* $u_{n+\frac{1}{2}} = \dfrac{1}{2}[u_n + u_{n+1}]$ but linearized with a *geometric mean* for the nonlinear term $\left[u_{n+\frac{1}{2}} \right]^2 \sim u_n \cdot u_{n+1}$?

12.6 A well-known model for population growth involves the nonlinear ordinary differential equation

$$u' = a(u)\, u, \quad u(0) = 1,$$

where the growth term is given by

$$a(u) = \rho \left(1 - \frac{u}{M} \right)^r$$

and r is a constant. Here the growth rate factor $a(u) \sim \rho$ allows the population to grow initially with unlimited access while the population u is small but tails off where $a(t)$ approaches zero as the population u approaches a maximum M that the environment can sustain.

The standard Crank–Nicolson approach for the iteration scheme for this differential equation from Section 6.2.2 would be

$$\frac{u_{n+1} - u_n}{\Delta t} = \frac{1}{2}\left[a(u_n)u_n + a(u_{n+1})u_{n+1}\right].$$

However, this equation is nonlinear. Thus, linearize the iteration scheme using a geometric mean as presented in Problem 12.5.

Integral equations

13

In previous chapters—for example, Chapter 2 and Chapter 5—differential equations arise that involve an unknown function with one or more of its derivatives. An integral equation can also occur involving an unknown function that is contained within an integral. Differential equations provide a means to describe physical laws in engineering that arise in chemical kinetics, heat and mass transfer, and electric circuits, to name a few. The differential equations can be transformed into equivalent integral equations. Moreover, integral equations also arise in the modeling of complex behavior of phenomena such as radiation and diffraction in optics, seismic responses of dams, harmonic vibrations in material, unsteady heat transfer, and eddy currents in turbulent fluid flow (Rahman, 2007). These problems in many cases require a numerical solution. Important representations can also arise as an integro-differential equation, which occur with radiative transfer of energy in absorbing, emitting, and scattering media, as well as in neutron transport theory.

Various solution methods for these equations are presented in Section 13.1.1. There is a close connection between differential and integral equations. Integral equations may in some cases offer a more powerful method of solution as compared to differential equations since the boundary conditions are built into the integral equation rather than imposed as the last step of the solution method. As shown, for example, in Section 13.1.2, values on the boundaries in integral equations are contained in a kernel. This methodology naturally leads to the Green's function in Section 13.2. Here the Green's function is in fact an integral kernel.

The Green's function can be used to solve a large family of problems such as ordinary differential equations with initial or boundary value conditions and inhomogeneous partial differential equations with boundary conditions. The Green's function method has wide applications in physics and engineering, including quantum field theory, areoacoustics, seismology, statistical field theory, electrodynamics, and mechanical oscillators, to name a few.

13.1 Integral equations

The integral equations can be broadly classified as <u>Fredholm</u> and <u>Volterra</u> integral equations, depending on the nature of the limits of integration. Following Arfken et al. (2013) and Press et al. (1986), integral equations can be classified in two ways:

- The *type* of integral equation depends on whether the limits of integration are fixed or variable. The <u>Fredholm</u> equation has constants as integration limits. The <u>Volterra</u> equation has one limit that is variable.
- The integral equation can be further classified as a "first kind" if the unknown function only appears under the integral sign or as a "second kind" if the function appears both inside and outside of the integral. In addition, they can be classified as homogeneous or nonhomogeneous integral equations, leading to several cases as follows (BYU, 2020; Rahman, 2007):

(a) *Fredholm's integral equation* of the *first* kind:

$$\int_a^b K(x, y) u(y) \, dy = f(x), \qquad Ku = f \quad \text{(nonhomogeneous)},$$

$$\int_a^b K(x, y) u(y) \, dy = 0, \qquad Ku = 0 \quad \text{(homogeneous)};$$

Fredholm's integral equation of the *second* kind:

$$u(x) = \lambda \int_a^b K(x, y) u(y) \, dy + f(x), \qquad u = \lambda Ku + f \quad \text{(nonhomogeneous)},$$

$$u(x) = \lambda \int_a^b K(x, y) u(y) \, dy, \qquad u = \lambda Ku \quad \text{(homogeneous)}.$$

(b) *Volterra's integral equation* of the *first* kind:

$$\int_0^x K(x, y) u(y) \, dy = f(x);$$

Volterra's integral equation of the *second* kind:

$$u(x) = \lambda \int_0^x K(x, y) u(y) \, dy + f(x).$$

(c) *Integro-differential equations* include an unknown function under the integral sign and also any derivative of the unknown function:

$$\frac{du}{dx} = u(x) + \int_G K(x, y) u(y) \, dy + f(x),$$

which is applicable over the domain space $G = (a, b)$.

(d) *Singular integral equations* where the limits of the integral are infinite or the kernel K becomes unbounded such that

$$u(x) = f(x) + \lambda \int_{-\infty}^{\infty} u(t) \, dt,$$

$$f(x) = \int_0^x \frac{u(t)}{(x - t)^\beta} \, dt, \quad 0 < \beta < 1.$$

In these integral equations, the dependent variable, u, is an unknown function, and the kernel, $K(x, y)$, $f(x)$, and the parameter λ are assumed to be known. The equation is homogeneous if $f(x) = 0$.

The integral equations include an integral term in the form of an integral operator with the kernel $K(x, y)$, such that

$$Ku = \int_a^{b(x)} K(x, y) u(y) \, dy.$$

The Fredholm equations above are shown along with a simplified form using an integral operator.

Existence and uniqueness of solution

As detailed in BYU (2020), the existence and uniqueness of a solution for a Fredholm equation of the second kind,

$$u = \lambda Ku + f, \tag{13.1}$$

can be determined as follows. Here, K is a bounded operator that satisfies the *Lipschitz condition* $\|Ku_1 - Ku_2\| \le k\|u_1 - u_2\|$ for $k \ge 0$, where $\|f\|_C = max|f(x)|$ and C is a vector space contained in the complete space $\mathfrak{R}^n$. One can further rewrite the integral as

$$u = Tu, \tag{13.2}$$

where T is the operator: $Tu = \lambda Ku + f$. The Banach fixed point theorem can be used to guarantee the existence and uniqueness of fixed points of self-maps, where, in this application, u is a particular fixed point of T. This theorem in fact provides an abstract formulation of Picard's iteration method. This latter method is described as a solution method for integral equations below. Thus, the fixed point theorem can be used to ensure that the integral operator has a unique fixed point. Using the Lipschitz condition, one can show that

$$\begin{aligned}
\|Tu_1 - Tu_2\| &= \|\lambda Ku_1 + f - (\lambda Ku_2 + f)\| \\
&= \|\lambda Ku_1 - \lambda Ku_2\| = |\lambda|\|K(u_1 - u_2)\| \\
&\le |\lambda|k\|\lambda u_1 - u_2\|.
\end{aligned}$$

Moreover, if $|\lambda|k < 1$, then the operator T is a *contraction*. As such, from the Banach fixed point theorem, there exists a unique fixed point for Eq. (13.2). This fixed point is also a solution of the Fredholm integral equation in Eq. (13.1). Thus, the Fredholm equation of the second kind with a bounded kernel has a solution for a value of $\lambda < 1/k$.

Typical applications of integral equations and integro-differential equations that arise in engineering are presented as follows.

Example 13.1.1. As a generalization of Hooke's law (see Example 2.2.5), viscoelasticity is an important phenomena that arises in materials engineering to describe the behavior of rubber, plastics, and glass, to name a few. The viscoelastic constitutive relations for the stress $\sigma(t)$ as a function of the strain rate $\dot{\epsilon}(t)$ can be expressed as a Volterra type integral equation:

$$\sigma(t) = \sigma_o + \int E(t - \eta)\dot{\epsilon}(\eta)\,d\eta,$$

where σ_o is a constant and E is a relaxation modulus.

Example 13.1.2. Several models have been proposed over the many years to describe population growth. A more general model that describes the rate of change of the population with time dP/dt is given by the following overall balance equation. Here the population is assumed to grow proportionally to the current population $c_1 P$. Unlimited growth is restricted with a loss term $-c_2 P^2$ and an additional term for accumulated toxicity since time zero, $-c_3 P \int_0^t P(x)dx$. Thus, c_1, c_2, and c_3 are constants where c_1 is a birth rate coefficient, c_2 is a crowding coefficient, and c_3 is a toxicity coefficient. The resultant Volterra type integral equation for this model is

$$\frac{dP}{dt} = c_1 P - c_2 P^2 - c_3 P \int_0^t P(x)\,dx.$$

This integral equation is subject to the initial condition $P(t = 0) = P_o$.

Example 13.1.3. In nuclear engineering in the design of nuclear reactors, complex neutron transport phenomena can be modeled with reactor physics considerations employing the general Boltzmann transport equation. The integro-differential form of the time-dependent neutron transport equation is (Lewis et al., 2017)

$$\left[\frac{1}{v_{el}(E)}\frac{d}{dt} + \nabla \cdot \mathbf{\Omega} + \Sigma_s(\mathbf{r}, E, t) + \Sigma_a(\mathbf{r}, E, t) \right]\varphi(\mathbf{r}, \mathbf{\Omega}, E, t) = S(\mathbf{r}, E, \mathbf{\Omega}, t)$$
$$+ \int_0^\infty \int_{4\pi} \left[\Sigma_s(\mathbf{r}, E', t)f_s(\mathbf{\Omega}' \cdot \mathbf{\Omega}, E' \to E) + \frac{\chi(E)}{4\pi}v(E')\Sigma_f(\mathbf{r}, E', t) \right]$$
$$\times \varphi(\mathbf{r}, \mathbf{\Omega}', E', t)d\mathbf{\Omega}'dE'.$$

Here a neutron of energy E' and direction $\mathbf{\Omega}'$ will be scattered into energy E and direction $\mathbf{\Omega}$, where the differential scattering kernel f_s accounts for the fractional probability of scattering. The function φ is the neutron angular flux of the neutron at the position $\mathbf{r}$, direction $\mathbf{\Omega}$, energy E, and time t. The speed of the neutron is $v_{el}(E)$. The quantities Σ_s, Σ_a, and Σ_f are the macroscopic cross-sections for scattering, absorption, and fission, respectively. The parameter $\chi(E)$ is the probability of a neutron released from fission having energy E and $v(E')$ is the number of neutrons emitted per fission initiated with a neutron of energy E'. The overall source term S also allows for a production of neutrons by radioactive decay of certain isotopes and nuclear reactions.

Example 13.1.4. Solve the following Volterra integro-differential equation: $u''(x) = x \cosh x - \int_0^x t\, u(t)\, dt$, subject to the initial conditions $u(0) = 0$ and $u'(0) = 1$.

Solution.

Substituting a power series (see Section 2.5) $u(x) = \sum_{n=0}^{\infty} a_n x^n$ and a Taylor series expansion for $\cosh x = \sum_{k=0}^{\infty} \frac{x^{2k}}{(2k)!}$ into the integro-differential equation yields

$$\sum_{n=2}^{\infty} n(n-1)a_n x^{n-2} = x\left(\sum_{k=0}^{\infty} \frac{x^{2k}}{(2k)!}\right) - \int_0^x t \sum_{n=0}^{\infty} a_n t^n \, dt.$$

Applying the initial condition $u(0) = 0$ in the power series expansion gives $u(0) = a_0 = 0$. Similarly, $u'(x) = \sum_{n=1}^{\infty} n a_n x^{n-1}$ implies $u'(0) = a_1 = 1$. Expanding the given series produces the following terms with powers of x:

$$2\cdot 1 a_2 + 3\cdot 2 a_3 x + 4\cdot 3 a_4 x^2 + 5\cdot 4 a_5 x^3$$
$$= x\left(1 + \frac{x^2}{2!} + \frac{x^4}{4!} + \dots\right) - \left(\frac{x^3}{3} + \frac{1}{4}a_2 x^4 + \dots\right).$$

Comparing each of the terms with various powers of x^n yields $a_2 = 0$, $a_3 = \frac{1}{3!}$, $a_4 = 0$, $a_5 = \frac{1}{5!}$, and so on. More generally, $a_{2n} = 0$ and $a_{2n+1} = \frac{1}{(2n+1)!}$ for $n \geq 0$. Thus, the solution is $u(x) = x + \frac{x^3}{3!} + \frac{x^5}{5!} + \dots = \sinh x$.   [answer]

13.1.1 Solution methods of integral equations
A solution of an integral equation is any function $u(x)$ satisfying

$$u = \lambda K u + f, \qquad \text{nonhomogeneous equation,}$$
$$u = \lambda K u, \qquad \text{homogeneous equation.}$$

The value of the parameter λ for which the homogeneous integral equation has a nontrivial solution is called an <u>eigenvalue</u> of the kernel $K(x, y)$, and the corresponding solution is called an <u>eigenfunction</u> of this kernel. One can distinguish eigenvalue problems for the integral kernel (integral equation)

$$u = \lambda K u$$

and for the integral operator

$$Ku = \frac{1}{\lambda}u.$$

Here the eigenvalues of the integral operator are reciprocal to eigenvalues of the integral kernel, whereas the eigenfunctions are the same in both cases.

General methods of solution are summarized below from BYU (2020) and Rahman (2007). Solution methods used for the Fredholm equation include techniques such as successive approximation, <u>Neumann series</u>, and successive substitution using the <u>resolvent method</u>. Successive approximation can also be employed for the Volterra equation.

Successive approximation (Fredholm equation)

The nonhomogeneous integral equation $u = \lambda Ku + f$ can be solved iteratively using a method of successive approximations, such that

$$u_0(x) = f(x),$$
$$u_n(x) = \lambda Ku_{n-1} + f \qquad (n = 1, 2...).$$

This methodology yields the solution

$$u_n(x) = \sum_{k=0}^{n} \lambda^k K^k f, \qquad K^k = \underbrace{K(K(...K))}_{k \text{ times}}.$$

The resultant series $\sum_{k=0}^{n} \lambda^k K^k f$ is called a <u>Neumann series</u> and is convergent for $|\lambda| < \dfrac{1}{M(b-a)}$, where M is a bounded number, that is, $M > 0$, such that for (x, y) belonging to the interval $[a, b]$, $M = \underset{x,y\in[a,b]}{max} |K(x, y)|$. The nonhomogeneous equation $u = \lambda Ku + f$ can be equivalently written in the form $(I - \lambda K)u = f$, where I is an identity operator. Hence, a solution can be obtained using an inversion operator, such that $u = (I - \lambda K)^{-1} f$. The inverse operator exists if $|\lambda| < \dfrac{1}{M(b-a)}$.

Example 13.1.5. Find the solution to the integral equation

$$u(x) = e^x + \frac{1}{e} \int_0^1 u(y)\,dy.$$

Solution:

By inspection, $K(x, y) = 1$, $f(x) = e^x$, $b - a = 1$, $\lambda = \dfrac{1}{e}$, and $M = 1$, where the convergence criterion is satisfied: $|\lambda| < \dfrac{1}{M(b-a)}$, that is, $\dfrac{1}{e} < \dfrac{1}{1 \cdot 1}$.

This problem can be solved either (i) with an iterative approximation method or (ii) by a Neumann series.

(i) *Iteration method*: We have

$$u_0(x) = e^x,$$

$$u_1(x) = e^x + \frac{1}{e} \int_0^1 u_0(y)\,dy = e^x + \frac{1}{e} \int_0^1 e^y\,dy = e^x + 1 - \frac{1}{e},$$

$$u_2(x) = e^x + \frac{1}{e} \int_0^1 u_1(y)\,dy = e^x + 1 - \frac{1}{e} \int_0^1 \left(e^y + 1 - \frac{1}{e}\right)dy = e^x + 1 - \frac{1}{e^2},$$

$$\cdots$$

$$u_n(x) = e^x + \frac{1}{e} \int_0^1 u_{n-1}(y)\,dy = e^x + 1 - \frac{1}{e^n}.$$

The solution is finally obtained by taking the limit of the nth iteration:

$$u(x) = \lim_{n\to\infty} u_n(x) = e^x + 1. \qquad \text{[answer]}$$

The solution is confirmed with substitution into the integral equation.

(ii) *Neumann series*:

Expanding the Neumann series

$$\sum_{k=0}^{\infty} \lambda^k K^k f = f(x) + \lambda^1 K^1 f + \lambda^2 K^2 f + \dots.$$

The individual terms of the series are evaluated as

$$f(x) = e^x,$$

$$Kf = \int_0^1 e^y\,dy = e - 1,$$

$$K^2 f = \int_0^1 (e-1)\,dy = e - 1,$$

$$\cdots$$

$$K^n f = e - 1.$$

Inserting these terms into the Neumann series yields

$$u(x) = e^x + \frac{1}{e}(e-1) + \frac{1}{e^2}(e-1) + \dots + \frac{1}{e^n}(e-1) + \dots$$

$$= e^x - (e-1) + (e-1) + \frac{1}{e}(e-1) + \frac{1}{e^2}(e-1) + \dots + \frac{1}{e^n}(e-1) + \dots$$

$$= e^x - (e-1) + (e-1)\sum_{n=0}^{\infty}\frac{1}{e^n} = e^x - e + 1 + \frac{(e-1)}{1-\frac{1}{e}} = e^x - e + 1 + e$$

$$= e^x + 1. \qquad \text{[answer]}$$

This result is the same as that obtained by the iteration method in item (i).

Successive substitution – resolvent method (Fredholm equation)

If the integral operator K has a continuous kernel, then one can have an <u>iterated kernel</u> $K^n = K\ (K^{n-1}) = (K^{n-1})\ K$ (with $n = 2, 3...$). Now consider the given terms for the repeated operator:

$$(K^1 f)(x) = \int_G \underbrace{K(x, y)}_{K^1(x,y)} f(y)\, dy,$$

$$(K^2 f)(x) = [K(Kf)](x) = \int_G K(x, y') \left[\int_G K(y', y)\, f(y)\, dy\right] dy'$$

$$= \int_G \underbrace{\left[\int_G K(x, y')K(y', y)dy'\right]}_{K^2(x,y)} f(y)dy$$

...

In general the iterated kernel $K^n(x, y)$ is

$$K^n(x, y) = \int_G K(x, y')\, K^{n-1}(y', y)\, dy' = \int_G K^{n-1}(y', y)\, K(x, y')\, dy'.$$

The <u>resolvent</u> is a function defined by the infinite series

$$R(x, y, \lambda) = \sum_{k=0}^{\infty} \lambda^k K^{k+1}(x, y).$$

Hence, the solution to the integral equation $u = \lambda K u + f$ with continuous kernel $K(x, y)$ for $|\lambda| < \dfrac{1}{M(b-a)}$ is

$$\boxed{u(x) = f(x) + \lambda \int_a^b R(x, y, \lambda)\, f(y)\, dy.}$$

Example 13.1.6. Find the solution to the integral equation

$$u(x) = \frac{23}{6}x + \frac{1}{8}\int_0^1 x\, y\, u(y)\, dy \quad \text{using the resolvent method.}$$

Solution:

By inspection, $K(x, y) = xy$, $f(x) = \dfrac{23}{6}x$, $b - a = 1$, $\lambda = \dfrac{1}{8}$, and $M = 1$, where the

convergence criterion is satisfied: $|\lambda| < \dfrac{1}{M(b-a)} = \dfrac{1}{8} < \dfrac{1}{1 \cdot 1}$.

The iterated kernels are

$$K^1(x, y) = xy,$$

$$K^2(x, y) = \int_0^1 K^1(x, y')K(y', y)\,dy' = \int_0^1 xy'y'y\,dy' = xy\left[\frac{y'^3}{3}\right]_0^1 = \frac{xy}{3},$$

$$K^3(x, y) = \int_0^1 K^2(x, y')K(y', y)\,dy' = \int_0^1 \frac{xy'}{3}y'y\,dy' = \frac{xy}{3}\left[\frac{y'^3}{3}\right]_0^1 = \frac{xy}{3^2},$$

$$\cdots$$

$$K^n(x, y) = \frac{xy}{3^{n-1}}.$$

The resolvent is evaluated as

$$R(x, y, \lambda) = \sum_{k=0}^{\infty} \lambda^k K^{k+1}(x, y)$$

$$= xy + \frac{1}{8}\frac{xy}{3} + \frac{1}{8^2}\frac{xy}{3^2} + \frac{1}{8^3}\frac{xy}{3^3} + \cdots + \frac{1}{8^n}\frac{xy}{3^n} + \cdots$$

$$= xy\left[1 + \frac{1}{8}\cdot\frac{1}{3} + \frac{1}{8^2}\cdot\frac{1}{3^2} + \frac{1}{8^3}\cdot\frac{1}{3^3} + \cdots + \frac{1}{8^n}\cdot\frac{1}{3^n} + \cdots\right]$$

$$= xy\left[\frac{1}{1 - \dfrac{1}{24}}\right] = \frac{24}{23}xy.$$

Hence, the final solution is

$$u(x) = f(x) + \lambda \int_a^b R(x, y, \lambda)\,f(y)\,dy = \frac{23}{6}x + \frac{1}{8}\int_0^1 \frac{24}{23}xy\frac{23}{6}y\,dy$$

$$= \frac{23}{6}x + \frac{1}{2}x\int_0^1 y^2\,dy$$

$$= \frac{23}{6}x + \frac{1}{2}x\left[\frac{y^3}{3}\right]_0^1\,dy = 4x. \qquad \text{[answer]}$$

Successive approximation (Volterra equation of the second kind)

The Volterra integral equation of the second kind,

$$u(x) = \lambda \int_0^x K(x, y)\,u(y)\,dy + f(x),$$

can be solved using successive approximations, such that

$$u_0(x) = f(x),$$

$$u_n(x) = \sum_{k=0}^{n} \lambda^k K^k f = \lambda K u_{n-1} + f.$$

This methodology yields the solution

$$u(x) = \sum_{k=0}^{\infty} \lambda^k \left(K^k f \right)(x).$$

Example 13.1.7. Find the solution to the integral equation

$$u(x) = 1 + \int_0^x u(y)\, dy$$

using the method of successive approximations.

Solution:
By inspection, $K(x, y) = 1$, $f(x) = 1$, $\lambda = 1$, and $M = 1$.

The iterated kernels are

$$K^0 f = f(x) = 1,$$

$$K^1 f = \int_0^x K(x, y) \left(K^0 f \right)(y)\, dy = \int_0^x 1 \cdot 1\, dy = [y]_0^x = x,$$

$$K^2 f = \int_0^x K(x, y) \left(K^1 f \right)(y)\, dy = \int_0^x 1 \cdot y\, dy = \left[\frac{y^2}{2} \right]_0^x = \frac{x^2}{2},$$

$$K^3 f = \int_0^x K(x, y) \left(K^2 f \right)(y)\, dy = \int_0^x 1 \cdot \frac{y^2}{2}\, dy = \frac{1}{2} \left[\frac{y^3}{3} \right]_0^x = \frac{x^3}{2 \cdot 3},$$

$$\cdots$$

$$K^n f = \frac{x^n}{n!}.$$

Hence the solution is $u(x) = \sum_{k=0}^{\infty} \lambda^k \left(K^k f \right)(x) = \sum_{k=0}^{\infty} \frac{x^k}{k!} = e^x.$ [answer]

Laplace transform method

The convolution theorem for Laplace transforms in Section 3.1.4 can be used to specifically solve an integral equation of the form

$$u(x) = f(x) + \lambda \int_0^x K(x - t) u(t)\, dt,$$

where $f(x)$ and $K(x)$ are known functions. Applying a Laplace transform to this equation gives

$$\mathcal{L}\{u\} = \mathcal{L}\{f\} + \lambda\mathcal{L}\{K\}\mathcal{L}\{u\}.$$

Solving for $\mathcal{L}\{u\}$, $\mathcal{L}\{u\} = \dfrac{\mathcal{L}\{f\}}{1 - \lambda\mathcal{L}\{K\}}$, and taking the inverse transform, the solution for $u(x)$ follows:

$$u(x) = \mathcal{L}^{-1}\left\{\frac{\mathcal{L}\{f\}}{1 - \lambda\mathcal{L}\{K\}}\right\}.$$

Example 13.1.8. Solve the equation $u(x) = x - \int_0^x (x - t)u(t)dt$.

Solution:

Here $f(x) = x$, $K(x) = x$, and $\lambda = -1$. From Table 3.1, $\mathcal{L}\{x\} = \dfrac{1}{s^2}$.

Thus, making use of the inverse transform from Table 3.1, the solution is $u(x) =$
$\mathcal{L}^{-1}\left\{\dfrac{\mathcal{L}\{f\}}{1 - \lambda\mathcal{L}\{K\}}\right\} = \mathcal{L}^{-1}\left\{\dfrac{1}{s^2 + 1}\right\} = \sin x.$ [answer]

13.1.2 Integral/differential equation transformations

A differential equation can be recast as an integral equation or, vice versa, an integral equation can be converted into a differential equation. As mentioned, integral equations provide an important and powerful method of solution in place of differential equations since the boundary or initial conditions are built into the integral equation rather than imposed as the last step of the solution method. This section demonstrates the connection of integral equations with boundary and initial value problems.

Reduction of an initial value problem to a Volterra equation

Example 13.1.9. Reduce the following initial value problem to a Volterra integral equation: $u' - 3x^2 u = 0$, with initial condition $u(0) = 1$.

Solution.
Integrating the differential equation from 0 to x yields

$$\int_0^x \left[(u'(y)) - 3(y^2 u(y))\right] dy = 0,$$

$$\int_0^x (u') \, dy - 3\int_0^x (y^2 u) \, dy = 0.$$

Recognizing the antiderivative from the fundamental theorem of calculus for the first integral gives

$$u(x) - u(0) - 3\int_0^x (y^2 u) \, dy = 0.$$

Imposing the initial condition $u(0) = 1$ gives the final result for the Volterra equation:

$$u(x) = 1 + 3 \int_0^x y^2 u(y) \, dy. \qquad \text{[answer]}$$

Here the kernel to the differential equation is $K(x, y) = y^2$.

Reduction of the Volterra equation to an initial value problem

Reduction of the Volterra equation to an initial value problem involves consecutive differentiation of the integral equation with respect to the variable x and substitution of $x = 0$ for the setting of the initial conditions.

Example 13.1.10. Reduce the Volterra integral equation $u(x) = x^3 + \int_0^x (x - y)^2 u(y) \, dy$ to an initial value problem.

Solution.
Substitute $x = 0$ in the integral equation to obtain the initial condition

$$u(0) = 0^3 + \underbrace{\int_0^0 (x - y)^2 u(y) \, dy}_{0} = 0.$$

Differentiating the integral equation, using the Leibnitz rule for differentiation of an integral,

$$\frac{d}{dx} \int_{a(x)}^{b(x)} g(x, y) dy = \int_{a(x)}^{b(x)} \frac{\partial g(x, y)}{\partial x} dy + g[x, b(x)] \frac{db(x)}{dx} - g[x, a(x)] \frac{da(x)}{dx},$$

gives $u'(x) = 3x^2 + \int_0^x 2(x - y) u(y) \, dy$, which implies $u'(0) = 3 \cdot 0^2 + \int_0^0 2(x - y) u(y) \, dy = 0$.

Differentiating again gives

$$u''(x) = 6x + 2 \int_0^x u(y) \, dy, \quad \text{which implies } u''(0) = 6 \cdot 0 + 2 \int_0^0 u(y) \, dy = 0.$$

Differentiating one more time, using the simplified Leibnitz rule $\frac{d}{dx} \int_0^x g(y) dy = g(x)$, gives the final result with a third-order ordinary differential equation,

$$u''' - 2u = 6, \text{ with boundary conditions } u(0) = 0, \; u'(0) = 0, \text{ and } u''(0) = 0.$$
[answer]

Reduction of a boundary value problem to a Fredholm equation

Example 13.1.11. Reduce the boundary value problem $y''(x) + y(x) = x$ to the Fredholm integral equation. Here x belongs to the interval $(0, \pi)$ with boundary conditions (i) $y(0) = 1$ and (ii) $y(\pi) = \pi - 1$.

Solution.
Letting $y''(x) = u(x)$ and integrating gives

$$\int_0^x y''(t)dt = \int_0^x u(t)dt$$

so that

$$y'(x) - y'(0) = \int_0^x u(t)dt.$$

Integrating again yields

$$\int_0^x \left[y'(t_2) - y'(0) \right] dt_2 = \int_0^x \left[\int_0^{t_2} u(t_1)dt_1 \right] dt_2.$$

Carrying out the integration for the term on the left-hand side of the equation,

$$y(x) - y(0) - y'(0) x = \int_0^x \left[\int_0^{t_2} u(t_1)dt_1 \right] dt_2.$$

Using the repeated integration formula

$$\int_0^x \int_0^{t_n} \cdots \int_0^{t_3} \int_0^{t_2} f(t_1) \, dt_1 dt_2 \ldots dt_{n-1} dt_n = \frac{1}{(n-1)!} \int_0^x (x-t)^{n-1} f(t) \, dt$$

for the right-hand side of the equation gives

$$y(x) - y(0) - y'(0) x = \int_0^x (x-t) u(t) \, dt.$$

Employing the first boundary condition in the above equation produces

$$y(x) - 1 - y'(0) x = \int_0^x (x-t) u(t) \, dt.$$

However, $y'(0)$ is not known. This quantity can be determined by applying the second boundary condition with $x = \pi$ in the previous equation and then solving for $y'(0)$:

$$y(\pi) - 1 - y'(0) \pi = \int_0^\pi (\pi - t) u(t) \, dt$$

so that

$$y(\pi) = \pi - 1 = 1 + y'(0) \pi + \int_0^\pi (\pi - t) u(t) \, dt.$$

Therefore,

$$y'(0) = 1 - \frac{2}{\pi} - \frac{1}{\pi} \int_0^\pi (\pi - t) u(t) \, dt.$$

Substituting $y'(0)$ into the expression for $y(x)$ provides the result:

$$y(x) = 1 + x - \frac{2}{\pi} x - \frac{x}{\pi} \int_0^\pi (\pi - t) u(t) \, dt + \int_0^x (x - t) u(t) \, dt.$$

Thus, substituting in this expression for $y(x)$ and $y''(x) = u(x)$ into the original differential equation yields

$$u + \left[1 + x - \frac{2}{\pi} x - \frac{x}{\pi} \int_0^\pi (\pi - t) u(t) \, dt + \int_0^x (x - t) u(t) \, dt \right] = x$$

or

$$
\begin{aligned}
u &= \frac{2}{\pi} x - 1 + \frac{x}{\pi} \int_0^\pi (\pi - t) u(t) \, dt - \int_0^x (x - t) u(t) \, dt \\
&= \frac{2}{\pi} x - 1 + \int_0^x \left[\frac{x}{\pi} (\pi - t) - (x - t) \right] u(t) \, dt + \frac{x}{\pi} \int_x^\pi (\pi - t) u(t) \, dt \\
&= \frac{2}{\pi} x - 1 + \int_0^x \frac{t(\pi - x)}{\pi} u(t) \, dt + \int_x^\pi \frac{x(\pi - t)}{\pi} u(t) \, dt.
\end{aligned}
$$

Hence, the Fredholm equation is $u(x) = \frac{2}{\pi} x - 1 + \int_0^\pi K(x, t) u(t) \, dt$, with the kernel

$$
K(x, t) = \begin{cases} \dfrac{t(\pi - x)}{\pi} & \text{if} \quad 0 \le t \le x, \\[2mm] \dfrac{x(\pi - t)}{\pi} & \text{if} \quad x \le t \le \pi. \end{cases}
$$

Picard's iteration method

As discussed in Section 2.1.2, the first-order ordinary differential equation $y' = f(x, y)$, with the condition $y(x_o) = y_o$, can be recast on integration as an integral expression: $y' = y_o + \int_{x_o}^x f(t, y(t)) \, dt$. As demonstrated in Section 2.1.2, this integral equation can be solved using Picard's iteration method.

13.2 Green's function

As mentioned in Section 13.1, integral equations provide a means to incorporate the boundary conditions into a kernel. In fact, the nonhomogeneous Sturm–Liouville equation of Section 2.5.4 can be written as

$$\mathscr{L} y(x) + f(x) = 0, \tag{13.3}$$

where $\mathscr{L}$ is a self-adjoint differential operator acting on $y(x)$, where

$$\mathscr{L} = \frac{d}{dx}\left(r(x)\frac{d}{dx}\right) + q(x). \tag{13.4}$$

As a differential equation, Eq. (13.3) is solved for $y(x)$ at specific boundary conditions at the endpoints in the interval $[a, b]$. However, as shown by Arfken et al. (2013), Eq. (13.3) can be recast into an integral equation where the solution $y(x)$ involves a Green's function $G(x, t)$:

$$y(x) = \int_a^b G(x, t)f(t)dt. \tag{13.5}$$

The Green's function in Eq. (13.5) is the kernel of the Fredholm equation of the first kind given in Section 13.1. It has the important symmetry property through a reciprocity principle that $G(x, t) = G(t, x)$ (Arfken et al., 2013). The boundary conditions have been specifically built into the Green's function $G(x, t)$ via

$$G(x, t) = \begin{cases} -\dfrac{1}{A}u(x)v(t), & a \leq x < t, \\[2mm] -\dfrac{1}{A}u(t)v(x), & t < x \leq b, \end{cases} \tag{13.6}$$

where $u(x)$ and $v(x)$ are linearly independent and each satisfies the homogeneous Sturm–Liouville equation (that is, with $f(x) = 0$) as well as the given boundary conditions at $x = a$ and $x = b$, respectively. Moreover, the Wronskian of these two functions (see Eq. (2.56)) requires $W \neq 0$ for linear independence, thereby satisfying "Abel's formula"

$$u(x)v'(x) - v(x)u'(x) = \frac{A}{r(x)} \tag{13.7}$$

for this Sturm–Liouville problem, with A being a constant.

The problem in Eq. (13.3) and Eq. (13.4) can in fact be generalized further to the Sturm–Liouville eigenvalue problem of Eq. (2.178):

$$\mathscr{L}y(x) + \lambda p(x)y(x) = 0. \tag{13.8}$$

This latter equation is a homogeneous Fredholm equation of the second kind. Since there are no specific restrictions on $f(x)$ in Eq. (13.3), one can assume that $f(x) = \lambda p(x)y(x)$ so that the solution of Eq. (13.8) simply follows from Eq. (13.5) as

$$y(x) = \lambda \int_a^b G(x, t)p(t)f(t)dt. \tag{13.9}$$

Example 13.2.1. Consider the linear oscillator problem

$$y''(x) + \lambda y(x) = 0 \tag{13.10}$$

with the boundary conditions $y(0) = y(1) = 0$.

Constructing the Green's function, one needs to find solutions of the homogeneous Sturm–Liouville equation $\mathscr{L}y(x) = 0$ or $y''(x) = 0$. As mentioned, to satisfy boundary conditions, one solution for the Green's function must vanish at $x = 0$ and the other at $x = 1$. Thus, the functions u and v have a linear form $ax + b$ in order to satisfy $y''(x) = 0$. Imposing the required boundary condition on each solution yields the (unnormalized) functions

$$\begin{aligned} u(x) &= x, \\ v(x) &= 1 - x. \end{aligned} \tag{13.11}$$

Using Eq. (13.7), with $r(x) = 1$ for Eq. (13.10), we have $uv' - vu' = -1$ so that $A = -1$. The Green's function follows from Eq. (13.6) (see Fig. 13.1):

$$G(x, t) = \begin{cases} x(1 - t), & 0 \le x < t, \\ t(1 - x), & t < x \le 1. \end{cases} \tag{13.12}$$

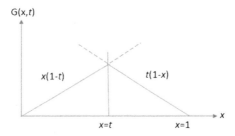

FIGURE 13.1

Green's function for a linear oscillator.

Thus, with $p(x) = 1$ and $f(x) = \lambda y(x)$ from Eq. (13.10), the solution using Eq. (13.9) is

$$y(x) = \lambda \int_0^1 G(x, t) y(t) dt. \tag{13.13}$$

The known solution of the differential equation in Eq. (13.10) is obtained from the result for the separated-variable function $X(x)$ in Example 5.2.3 (with $L = 1$):

$$y(x) = \sin \sqrt{\lambda} x \quad (\lambda = n^2 \pi^2, \text{ with } n = 1, 2, \ldots). \tag{13.14}$$

The exact solution in Eq. (13.14) does in fact satisfy Eq. (13.13). This result can be seen by noting the symmetry of $G(x, t)$ from the reciprocity principle (that is, exchanging x and t in the Green's function of Eq. (13.12) given that the dummy integration variable is t in Eq. (13.13)). Thus, inserting this symmetrical function $G(t, x)$ along with the exact solution of $y(t) = \sin n\pi t$ into Eq. (13.13) and integrating yields the solution function $y(x)$ in Eq. (13.14). This result therefore shows that the integral solution of Eq. (13.13) is in fact a solution of the Sturm–Liouville problem in Eq. (13.10).

Application of the Green's function for nonhomogeneous problems

Consider the Helmholtz equation in three dimensions with general spatial coordinates $\vec{r}$:

$$\nabla^2 u(\vec{r}) + k^2 u(\vec{r}) = 0, \tag{13.15}$$

where k is a constant. This equation also reduces to the Laplace equation in the limit that k approaches zero (see Chapter 5). The equation has to be solved in a volume V subject to boundary conditions specified on the bounding surface.

If the right-hand side of the Helmholtz or Laplace equation (see, for example, Eq. (13.15)) is zero, then the equation is said to be homogeneous. On the other hand, if the right-hand side of the equation is a function of position coordinates, it is nonhomogeneous. Such is the case for the Poisson equation as described in Chapter 5. The term on the right-hand side of the equation is sometimes called a driving term.

Consider the nonhomogeneous Helmholtz equation with a general driving term $f(\vec{r})$:

$$\nabla^2 u(\vec{r}) + k^2 u(\vec{r}) = f(\vec{r}). \tag{13.16}$$

The method of solution for Eq. (13.16) involves the use of the Green's function and is solved in the following steps (Towner, 2020; Boas, 2006):

(i) Solve the equation

$$\nabla^2 G(\vec{r}, \vec{r}') + k^2 G(\vec{r}, \vec{r}') = \delta(\vec{r}, \vec{r}'), \tag{13.17}$$

subject to the same boundary conditions as for Eq. (13.16). The function $\delta(\vec{r}, \vec{r}')$ is the so-called Dirac delta function.

(ii) Then, the general solution of Eq. (13.16) is

$$u(\vec{r}) = \iiint_V G(\vec{r}, \vec{r}') f(\vec{r}') d^3\vec{r}', \tag{13.18}$$

where the integration is over the volume V under consideration.

Here the Dirac delta function in Eq. (13.17) of step (i) has the following properties:

(a) $\delta(x) = 0$ for $x \neq 0$,
(b) $f(a) = \int_{-\infty}^{\infty} f(x)\delta(x - a)\,dx,$

where $f(x)$ is a well-behaved function, which is integrated over the point $x = a$. A special case of the property in item (b) is when $f(x) = 1$ and the function is integrated over the origin (that is, $a = 0$) so that $\int_{-\infty}^{\infty} \delta(x)\, dx = 1$. This function can also be integrated over a finite domain such that

$$\int_b^c f(x)\delta(x - a)\, dx = \begin{cases} f(a), & b \le a \le c, \\ 0, & \text{otherwise.} \end{cases}$$

The proof that Eq. (13.18) is a solution of Eq. (13.16) is given as Problem 13.4.

Most of the effort to find the solution in Eq. (13.18) is determining the Green's function in Eq. (13.17) for the given geometry and boundary conditions of interest. However, once this function is found, it can be used for many driving functions $f(\vec{r})$ provided that the geometry and boundary conditions are not changed.

Example 13.2.2. Solve the second-order ordinary differential equation $\dfrac{d^2 y}{dx^2} + y = f(x)$, subject to the boundary conditions $y = 0$ at $x = 0$ and $x = \pi/2$, using the Green's function method.

Solution.
One needs to determine the Green's function from Eq. (13.17) such that

$$\frac{d^2}{dx^2} G(x, x') + G(x, x') = \delta(x - x'). \tag{13.19}$$

When $x = x'$, the right-hand side of Eq. (13.19) is equal to zero. In this case the differential equation reduces to the homogeneous equation $\dfrac{d^2 y}{dx^2} + y = 0$, which from Section 2.2 has the solution as a linear combination of $\cos x$ and $\sin x$. If we drop the function $\cos x$ at $x = 0$ and $\sin x$ at $x = \pi/2$, the boundary conditions are automatically satisfied. Hence, one expects the Green's function to take the following form:

$$G(x, x') = \begin{cases} A(x')\sin x, & x < x', \\ B(x')\cos x, & x > x'. \end{cases}$$

The functions $A(x')$ and $B(x')$ need to be determined. At $x = x'$, the Green's function needs to (i) be continuous and (ii) have a finite discontinuity in its slope at $x = x'$, which can be determined from Eq. (13.19). Condition (i) gives

$$A(x')\sin x = B(x')\cos x. \tag{13.20}$$

The second condition follows on integrating Eq. (13.19) from $x = x' - \epsilon$ and $x = x' + \epsilon$, and then letting ϵ approach zero:

$$\int_{x'-\epsilon}^{x'+\epsilon} \frac{d^2 G(x, x')}{dx^2}\, dx + \int_{x'-\epsilon}^{x'+\epsilon} G(x, x')\, dx = \int_{x'-\epsilon}^{x'+\epsilon} \delta(x - x')\, dx.$$

Thus, integrating, noting that the second term on the left-hand side of the equation $\int_{x'-\epsilon}^{x'+\epsilon} G(x, x')dx$ tends to zero as $\epsilon \to 0$ since G is continuous, and using the property of the Dirac delta function for the term on the right-hand side of the equation yields

$$\frac{dG}{dx}\Big|_{x'-\epsilon}^{x'+\epsilon} = 1.$$

Substituting $G(x, x')$ into this latter expression gives

$$\left[\frac{d}{dx}B(x')\cos x\right]_{x'+\epsilon} - \left[\frac{d}{dx}A(x')\sin x\right]_{x'+\epsilon} = 1$$

or

$$- B(x')\sin x' - A(x')\cos x' = 1. \tag{13.21}$$

Solving Eq. (13.20) and Eq. (13.21) gives $A(x') = -\cos x'$ and $B(x') = -\sin x'$, so that the Green's function is determined as

$$G(x, x') = \begin{cases} -\cos x' \sin x, & x < x', \\ -\sin x' \cos x, & x > x'. \end{cases}$$

For the second step, using Eq. (13.18) and the Green's function, the solution for $y(x)$ that satisfies the boundary conditions is

$$\begin{aligned}
y(x) &= \int_0^{\pi/2} G(x, x')\, f(x')\, dx' \\
&= \int_0^x (-\cos x \sin x')\, f(x')\, dx' + \int_x^{\pi/2} (-\sin x \cos x')\, f(x')\, dx' \\
&= -\cos x \int_0^x \sin x'\, f(x')\, dx' - \sin x \int_x^{\pi/2} \cos x'\, f(x')\, dx'. \quad \text{[answer]}
\end{aligned}$$

The methodology of Example 13.2.2 can be generalized for the following second-order ordinary differential equation:

$$y'' + p(x)y' + q(x)y = f(x)$$

in the range $a \le x \le b$ and with boundary conditions $y(a) = 0$ and $y(b) = 0$. The first step is to find the Green's function, which must satisfy

$$\frac{d^2}{dx^2}G(x, x') + p(x)\frac{d}{dx}G(x, x') + q(x)G(x, x') = \delta(x - x'), \tag{13.22}$$

and the two boundary conditions so that $G(a, x')$ and $G(b, x') = 0$ when $x \ne x'$. Let $y_1(x)$ and $y_2(x)$ be the two independent solutions of the homogeneous equation $y'' + p(x)y' + q(x)y = 0$.

Similarly, one can postulate a solution of the Green's function of the form

$$G(x, x') = \begin{cases} A(x')y_1(x), & a \leq x < x', \\ B(x')y_2(x), & x' < x \leq b, \end{cases}$$

which satisfies the boundary conditions. Again the functions $A(x')$ and $B(x')$ at $x = x'$ need to ensure that the Green's function (i) is continuous and (ii) has a finite discontinuity in its slope at $x = x'$ that can be determined from Eq. (13.22). Similarly, the first condition yields

$$A(x') y_1(x') - B(x') y_2(x') = 0. \tag{13.23}$$

Again, the second condition follows on integrating Eq. (13.22) from $x = x' - \epsilon$ and $x = x' + \epsilon$, and then letting ϵ approach zero:

$$\underbrace{\int_{x'-\epsilon}^{x'+\epsilon} \frac{d^2G(x, x')}{dx^2} \, dx}_{I} + \underbrace{\int_{x'-\epsilon}^{x'+\epsilon} p(x) \frac{dG(x, x')}{dx} \, dx}_{II} + \underbrace{\int_{x'-\epsilon}^{x'+\epsilon} q(x)G(x, x') \, dx}_{III}$$

$$= \underbrace{\int_{x'-\epsilon}^{x'+\epsilon} \delta(x - x') \, dx}_{IV} \,.$$

$$\text{Integral } I = \frac{dG}{dx}\bigg|_{x'-\epsilon}^{x'+\epsilon} = \left[\frac{d}{dx} B(x')y_2(x)\right]_{x'+\epsilon} - \left[\frac{d}{dx} A(x')y_1(x)\right]_{x'-\epsilon}$$

$$= B(x') \frac{dy_2(x)}{dx}\bigg|_{x'+\epsilon} - A(x') \frac{dy_1(x)}{dx}\bigg|_{x'-\epsilon}$$

$$= B(x')y_2'(x' + \epsilon) - A(x')y_1'(x' - \epsilon)$$

$$= B(x')y_2'(x') - A(x')y_1'(x') \quad (\epsilon \to 0).$$

$$\text{Integral } II = \left[p(x) \frac{dG}{dx}\right]_{x'-\epsilon}^{x'+\epsilon} - \int_{x'-\epsilon}^{x'+\epsilon} \frac{dp(x)}{dx} G \, dx \to 0$$

$$\text{(since } p, G, \text{ and } \frac{dp}{dx} \text{ are continuous at } x = x').$$

$$\text{Integral } III = \int_{x'-\epsilon}^{x'+\epsilon} q(x) G \, dx \to 0 \quad \text{(since } q \text{ and } G \text{ are continuous at } x = x').$$

$$\text{Integral } IV = \int_{x'-\epsilon}^{x'+\epsilon} \delta(x - x') \, dx = 1.$$

Integral *III* is integrated by parts. Thus, the second condition leads to

$$B(x')y_2'(x') - A(x')y_1'(x') = 1. \tag{13.24}$$

Solving Eq. (13.23) and Eq. (13.24) gives $A(x') = y_2/W$ and $B(x') = y_1/W$, where $W(x')$ is the Wronskian as defined in Eq. (2.56):

$$W = y_1(x')y_2'(x') - y_2(x')y_1'(x').$$

Hence, the Green's function is determined as

$$G(x, x') = \begin{cases} y_2(x')y_1(x)/W(x'), & a \leq x < x', \\ y_1(x')y_2(x)/W(x'), & x' < x \leq b. \end{cases}$$

Given the solution for $G(x, x')$, the solution for $y(x)$ that satisfies the boundary conditions is

$$\begin{aligned} y(x) &= \int_a^b G(x, x') f(x') dx' \\ &= \int_a^x \frac{y_1(x')y_2(x)}{W(x')} f(x') dx' + \int_x^b \frac{y_2(x')y_1(x)}{W(x')} f(x') dx' \\ &= y_2(x) \int_a^x \frac{y_1(x')f(x')}{W(x')} dx' - y_1(x) \int_b^x \frac{y_2(x')f(x')}{W(x')} dx' \\ &= c_1 y_1(x) + c_2 y_2(x) + y_p(x). \end{aligned}$$

In the formulation of the second equality for the first integral, the limits for x' range from $a \leq x' \leq x$, hence $G(x, x') = G_2(x, x')$, while similarly, for the second integral, $G(x, x') = G_1$ since x' ranges from $x \leq x' \leq b$. Here c_1 and c_2 are the values of the integral taken at the limits b and a, respectively. The function $y_p(x)$ is determined from the integrals evaluated at the limit x:

$$y_p(x) = y_2(x) \int \frac{y_1(x)f(x)}{W(x)} dx - y_1(x) \int \frac{y_2(x)f(x)}{W(x)} dx, \qquad (13.25)$$

which are now written as indefinite integrals.

This result follows directly from the analysis of second-order ordinary differential equations in Section 2.2.5, where y_p in Eq. (13.25) is identical to Eq. (2.64). The general solution to the ordinary differential equation can be written as a general solution of the homogeneous equation (see Eq. (2.57)) called the <u>complementary function</u>,

$$y_h(x) = y_1(x) + y_2(x),$$

and a <u>particular solution</u> $y_p(x)$ involving the driving term. The Green's function therefore offers another method to find a particular solution.

The result in Example 13.2.2 easily follows from this general solution. The two linearly independent solutions to the homogeneous problem $y'' - y = 0$ are $y_1(x) = \sin x$ and $y_2(x) = \cos x$. Note that y_1 satisfies the boundary condition at $x = 0$ and y_2

satisfies the boundary condition at $\pi/2$. The general solution to the nonhomogeneous problem is

$$y(x) = y_2(x) \int_0^x \frac{y_1(x')f(x')}{W(x')} dx' + y_1(x) \int_x^{\pi/2} \frac{y_2(x')f(x')}{W(x')} dx',$$

where

$$W(x') = y_1(x')y_2'(x') - y_2(x')y_1'(x')$$
$$= \sin x'(-\sin x') - \cos x'(\cos x') = -\left(\sin^2 x' + \cos^2 x'\right) = -1.$$

Thus, the same result is obtained as given for the linear oscillator in Example 13.2.2:

$$y(x) = -\cos x \int_0^x \sin x' \, f(x') dx' - \sin x \int_x^{\pi/2} \cos x' \, f(x') dx'.$$

Problems

13.1   Solve the following integral equation using a Laplace transform method:

$$y(t) = \sin t + \int_0^t y(\tau)\sin(t-\tau) d\tau.$$

13.2   Consider the Sturm–Liouville problem in Example 13.2.1. This problem can be recast as the second-order ordinary differential equation $y''(x) = f(x)$ with boundary conditions $y(0) = y(1) = 0$ on letting $\lambda = -1$, $p(x) = 1$, and $f(x) = y(x)$ in Eq. (13.10). Thus, the solution follows from Eq. (13.13) as

$$y(x) = \int_0^1 G(x,s) \, f(s) ds.$$

Here the Green's function is symmetric (that is, $G(x,s) = G(s,x)$) and is defined as the negative quantity of Eq. (13.12) (which satisfies the boundary conditions):

$$G(x,s) = \begin{cases} s(x-1), & 0 \le s < x, \\ x(s-1), & x < s \le 1. \end{cases}$$

(a)   Show that for $f(x) = x^2$, the solution using the Green's function formalism is $y(x) = \dfrac{x}{12}\left(x^3 - 1\right)$.

(b)   Show that

$$\frac{\partial G(x,s)}{\partial s} = \begin{cases} x-1, & 0 \le s < x, \\ x, & x < s \le 1, \end{cases}$$

so that $(\partial G(x,s)/\partial s) = x - 1 + U(s-x)$, where $U(s-x)$ is the unitary step function (or Heaviside function) (see Section 3.1.3):

$$U(s-x) = \begin{cases} 0 & \text{if} \quad s < x, \\ 1 & \text{if} \quad s > x. \end{cases}$$

(c) Given that the derivative of the Heaviside function is the Dirac delta function, show that from part (b) $\dfrac{\partial^2 G(x,s)}{\partial s^2} = \delta(s-x) = \delta(x-s)$. Hence, using the property of the Dirac delta function $\displaystyle\int_{x-\epsilon}^{x+\epsilon} \delta(s-x) f(s)\, ds = f(x)$, show that one obtains the original ordinary differential equation $\dfrac{d^2 y}{dx^2} = \displaystyle\int_0^1 \dfrac{\partial^2 G}{\partial x^2}(x,s) f(s)\, ds = f(x)$. Thus, $y(x) = \displaystyle\int_0^1 G(x,s) f(s)\, ds$ is a solution of the nonhomogeneous boundary value problem. Moreover, the Green's function can be thought of as a response function to a unit impulse at $s = x$.

13.3 Consider Example 5.2.11 for heat conduction in a semiinfinite solid but where the temperature is initially zero in the solid, and at $t > 0$ a constant temperature of u_o is applied and maintained at the face $x = 0$. The initial and boundary conditions for this problem become $u(x,0) = 0$, $u(0,t) = u_o$, and $|u(x,t)| < M$.

(a) Using a Laplace transform method with $\mathscr{L}\{u(x,t)\} = U(x,s)$ and the transformed boundary condition $U(0,s) = \mathscr{L}\{u(0,t)\} = u_0/s$, show that the transformed solution is $U(x,s) = \left(\dfrac{u_o}{s}\right) e^{-\sqrt{s/k}\,x}$. Thus, using the inverse transform from Problem 10.3, the final solution is $u(x,t) = \mathscr{L}^{-1}\{U(x,s)\} = u_o\, \mathrm{erfc}[x/(2\sqrt{kt})]$.

(b) If the boundary condition $u(0,t) = u_o$ is replaced by the more general condition $u(0,t) = g(t)$, show that the transformed solution is $U(x,s) = G(s)\, e^{-\sqrt{s/k}\,x}$.

(c) From Problem 10.3, $\mathscr{L}^{-1}\{e^{-\sqrt{s/k}\,x}\} = \dfrac{x}{2\sqrt{\pi\kappa}}\, t^{-3/2} e^{-x^2/(4\kappa t)}$. Hence, show that using the convolution theorem (see Section 3.1.4) for the inverse transform of part (b), the solution for $u(x,t)$ is given by the integral equation

$$u(x,t) = \int_0^t \frac{x}{2\sqrt{\pi\kappa}} u^{-3/2} e^{-x^2/(4\kappa u)}\, g(t-u)\, du.$$

13.4 Prove that Eq. (13.18) is a general solution of the Helmholtz equation in Eq. (13.16).

13.5 Classify each of the following integral equations as a Fredholm- or Volterra-type integral equation, as linear or nonlinear, and as homogeneous or nonhomogeneous, and identify the parameter λ and the kernel $K(x,y)$:

(a) $u(x) = x + \int_0^1 xy\, u(y)\, dy$,

(b) $u(x) = 1 + x^2 + \int_0^x (x-y)\, u(y)\, dy$,

(c) $u(x) = e^x + \int_0^x y u^2(y) dy$,

(d) $u(x) = \int_0^1 (x - y)^2 u(y) dy$,

(e) $u(x) = 1 + \dfrac{x}{4} \int_0^1 \dfrac{1}{(x + y)} \dfrac{1}{u(y)} dy$.

13.6 Convert the following Volterra integral equation into an initial value problem: $u(x) = x + \int_0^x (y - x) u(y) dy$.

13.7 Convert the following initial value problem into a Volterra integral equation: $u''(x) + u(x) = \cos x$, subject to the initial conditions $u(0) = 0$ and $u'(0) = 1$.

13.8 Convert the following boundary value problem into a Fredholm integral equation: $y''(x) + y(x) = x$, for $x \in (0, 1)$, subject to the boundary conditions $y(0) = 1$ and $y(1) = 0$.

13.9 Given the Fredholm integral equation $u(x) = x + \lambda \int_0^1 (xy) u(y) dy$, solve this equation using (i) the successive approximation method, (ii) a Neumann series, and (iii) the resolvent method.

13.10 Given the Volterra integral equation $u(x) = 1 + \int_0^x (x - y) u(y) dy$, solve this equation using the successive approximation method.

Calculus of variations

Variational calculus employs an integral representation to find an optimum quantity to be minimized (or maximized). It complements ordinary differential calculus as presented in previous chapters. It has applications for problems in physics to find the motion of macroscopic objects (classical mechanics) or to determine an optimal path. It can also be used in engineering for assessment of an optimal shape of an object (for example, an aircraft wing in order to minimize wind resistance), or to find a "utility function" in economics for measurement of the usefulness of specific consumer goods.

As mentioned, this technique can be used to find the shortest route or distance on a surface (a geodesic). In this process, one searches for trajectories that minimize a so-called "action," which corresponds to classic equations of motion for a given system. Specifically, one minimizes the value of an integral using a "principle of stationary action." The action is an integral over time $S = \int_{t_1}^{t_2} L dt$ of a "Lagrangian" L. Thus, the action S represents an integral of the Lagrangian for an input evaluation over time for the development of a system from an initial time to a final time:

$$S = \int_{t_1}^{t_2} L[q(t), \dot{q}, t] dt. \tag{14.1}$$

Here q and $\dot{q}$ are generalized coordinates and the dot pertains to a derivative with respect to the independent variable time t. The endpoints of the evolution are fixed as $q_1 = q(t_1)$ and $q_2 = q(t_2)$. According to "Hamilton's principle," the *true* evolution of a system is one where the action is stationary (that is, $\delta S = 0$, thereby involving a minimum, maximum, or saddle point of S). This methodology has been historically termed "Lagrangian mechanics" and results in the equations of motion.

14.1 Euler–Lagrange equation

A fundamental differential equation that is derived from the calculus of variations is the Euler–Lagrange equation. This equation is derived from Eq. (14.1) by applying a stationary action:

$$\delta S = \delta \int_{t_1}^{t_2} L[q, \dot{q}, t] dt = \int_{t_1}^{t_2} \left(\frac{\partial L}{\partial q} \delta q + \frac{\partial L}{\partial \dot{q}} \delta \dot{q} \right) dt \tag{14.2a}$$

Advanced Mathematics for Engineering Students. https://doi.org/10.1016/B978-0-12-823681-9.00022-8

$$= \int_{t_1}^{t_2} \left(\frac{\partial L}{\partial q} \delta q + \frac{\partial L}{\partial \dot{q}} \frac{d(\delta q)}{dt} \right) dt = 0. \qquad (14.2b)$$

Integrating the last term of Eq. (14.2b) by parts on letting

$$u = \frac{\partial L}{\partial \dot{q}} \quad \text{and} \quad dv = d(\delta q),$$

$$du = \frac{d}{dt} \left(\frac{\partial L}{\partial \dot{q}} \right) dt \quad \text{and} \quad v = \delta q \qquad (14.3)$$

yields

$$\int_{t_1}^{t_2} \left(\frac{\partial L}{\partial \dot{q}} \frac{d(\delta q)}{dt} \right) dt = \int_{t_1}^{t_2} \frac{\partial L}{\partial \dot{q}} d(\delta q) = \left[\frac{\partial L}{\partial \dot{q}} \delta q \Big|_{t_1}^{t_2} - \int_{t_1}^{t_2} \left(\frac{d}{dt} \frac{\partial L}{\partial \dot{q}} dt \right) \delta q \right].$$
$$(14.4)$$

Inserting Eq. (14.4) into Eq. (14.2b) gives

$$\delta S = \frac{\partial L}{\partial \dot{q}} \delta q \Big|_{t_1}^{t_2} + \int_{t_1}^{t_2} \left(\frac{\partial L}{\partial q} - \frac{d}{dt} \frac{\partial L}{\partial \dot{q}} \right) \delta q \, dt = 0. \qquad (14.5)$$

Since only the path is varying and not the endpoints, we have $\delta q(t_1) = \delta q(t_2) = 0$ and Eq. (14.5) becomes

$$\delta S = \int_{t_1}^{t_2} \left(\frac{\partial L}{\partial q} - \frac{d}{dt} \frac{\partial L}{\partial \dot{q}} \right) \delta q \, dt = 0. \qquad (14.6)$$

Hence, for an arbitrary small change δq for a stationary action the integrand itself must equal zero, so that

$$\boxed{\frac{\partial L}{\partial q} - \frac{d}{dt} \frac{\partial L}{\partial \dot{q}} = 0,} \qquad (14.7)$$

which is the Euler–Lagrange equation.

Example 14.1.1 (Second law of motion). Use the Euler–Lagrange equation to derive Newton's second law of motion.

Solution.
Let the generalized coordinate be position for the dependent variable so that $q(t) \to x(t)$ and $q(\dot{t}) \to \dot{x}(t)$, where $\dot{x} = dx/dt$ is the velocity. The Euler–Lagrange equation in Eq. (14.7) can therefore be written as

$$\frac{d}{dt} \frac{\partial L}{\partial \dot{x}} - \frac{\partial L}{\partial x} = 0. \qquad (14.8)$$

In this case, the Lagrangian can be defined as the difference between the kinetic energy T and potential energy V of a system:

$$L = T - V. \qquad (14.9)$$

The kinetic energy of a particle of mass m is

$$T = \frac{1}{2}m\dot{x}^2. \tag{14.10}$$

The negative gradient of the potential energy $V(x)$ is defined as the force F:

$$F(x) = -\frac{dV(x)}{dx}. \tag{14.11}$$

Thus, substituting Eq. (14.9) into Eq. (14.8) and using Eq. (14.10) and Eq. (14.11) for the kinetic and potential energies, respectively, yields

$$\frac{d}{dt}(m\dot{x}) - \frac{\partial(-V)}{dx} = m\ddot{x} - F(x) = 0 \Rightarrow F = ma, \tag{14.12}$$

which is Newton's second of law of motion, where a is the acceleration such that $a = \ddot{x}$.

Example 14.1.2 (Straight line between two points). Using the Euler–Lagrange equation, show that a straight line is the shortest distance between two points on a flat xy plane.

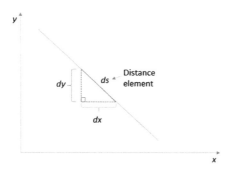

FIGURE 14.1

Schematic of a distance element in flat space.

Solution.
Using the Pythagorean theorem (see Fig. 14.1), the small distance element on a flat surface is

$$ds = [(dx)^2 + (dy)^2]^{1/2} = [1 + y_x^2]^{1/2}dx. \tag{14.13}$$

Hence, the distance may be written as

$$S = \int_{x_1,y_1}^{x_2,y_2} ds = \int_{x_1,y_1}^{x_2,y_2} [(dx)^2 + (dy)^2]^{1/2} = \int_{x_1}^{x_2} [1 + y_x^2]^{1/2}dx. \tag{14.14}$$

Letting the generalized coordinates in Eq. (14.1) and Eq. (14.7) become $q \rightarrow y$, $\dot{q} \rightarrow \frac{dy}{dx} = y_x$, and $t \rightarrow x$,

$$S = \int_{x_1}^{x_2} L[y, y_x, x] dx \qquad (14.15)$$

and

$$\frac{\partial L}{\partial y} - \frac{d}{dx}\frac{\partial L}{\partial y_x} = 0. \qquad (14.16)$$

Comparing Eq. (14.14) with Eq. (14.15), the Lagrangian L for this problem is

$$L = [1 + y_x^2]^{1/2}. \qquad (14.17)$$

Therefore, substituting Eq. (14.17) into Eq. (14.16) yields

$$-\frac{d}{dx}\left[\frac{y_x}{(1 + y_x^2)^{1/2}}\right] = 0 \Rightarrow \frac{y_x}{(1 + y_x^2)^{1/2}} = C \quad \text{(a constant)}. \qquad (14.18)$$

The solution of Eq. (14.18) is

$$y_x = a \quad \text{(a constant)} \Rightarrow y = ax + b. \qquad (14.19)$$

The second relation in Eq. (14.19) is the familiar equation of a straight line where a and b are constants so that the line passes through (x_1, y_1) and (x_2, y_2). Thus, the Euler–Lagrange equation predicts that the shortest distance between two fixed points is a straight line on a flat plane.   [answer]

This result is further generalized in Example 14.1.3, where the shortest distance in four-dimensional spacetime is a geodesic, which is a basic construct of general relativity.

Example 14.1.3 (Geodesic equation).   Derive the shortest distance on a curved surface using the Euler–Lagrange equation.

Solution.
The methodology in Example 14.1.2 can be generalized using a metric $g_{\mu\nu}$ for four-dimensional spacetime to account for any type of surface and for any general coordinate system:

$$ds^2 = g_{\mu\nu}dx_\mu dx_\nu. \qquad (14.20)$$

For instance, to represent Eq. (14.13) with two dimensions (i.e., with indices $\mu = 1, 2$ and $\nu = 1, 2$), Eq. (14.20) can be written in the following component form using the Einstein convention of summing on repeated indices (Lieber and Lieber, 1966):

$$ds^2 = g_{11}dx_1 \cdot dx_1 + g_{12}dx_1 \cdot dx_2 \\ + g_{21}dx_2 \cdot dx_1 + g_{22}dx_2 \cdot dx_2, \qquad (14.21)$$

where g, considering the previous example of a flat (Euclidean) surface, is the 2×2 matrix $\begin{bmatrix} g_{11} & g_{12} \\ g_{21} & g_{22} \end{bmatrix} = \begin{bmatrix} 1 & 0 \\ 0 & 1 \end{bmatrix}$.

Following the methodology of Example 14.1.2, Eq. (14.14) can be similarly written using the generalized Lagrangian function

$$L = \sqrt{g_{\alpha\beta} \dot{x}_\alpha \dot{x}_\beta}, \tag{14.22a}$$

where the action is

$$S = \int_{\lambda_0}^{\lambda} L(x_\alpha, \dot{x}_\alpha, \lambda) \, d\lambda. \tag{14.22b}$$

Here the path is parametrized by λ, in which $\dot{x}_\alpha(\lambda) = \dfrac{dx_\alpha(\lambda)}{d\lambda}$ (see Fig. 14.2).

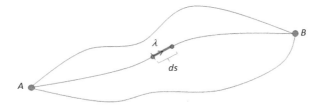

FIGURE 14.2

Various paths between two fixed points A and B.

Thus, in summary, the action for this problem is

$$S = \int_{\lambda_0}^{\lambda} ds = \int_{\lambda_0}^{\lambda} \sqrt{g_{\alpha\beta} \dot{x}_\alpha \dot{x}_\beta} \, d\lambda. \tag{14.23}$$

The Euler–Lagrange equation with generalized coordinates in Eq. (14.7) can be written to minimize the path length as

$$\frac{\partial L}{\partial x_\mu} - \frac{d}{d\lambda} \frac{\partial L}{\partial \dot{x}_\mu} = 0. \tag{14.24}$$

As follows from the derivation in appendix B using the Lagrangian formulation, the equation of a geodesic is

$$\boxed{\ddot{x}_\nu + \dot{x}_\alpha \dot{x}_\delta \Gamma^\nu_{\delta\alpha} = 0,} \tag{14.25}$$

where the Christoffel symbol $\Gamma^\nu_{\delta\alpha} = \frac{1}{2} g^{\mu\nu} [\partial_\delta g_{\mu\alpha} + \partial_\alpha g_{\delta\mu} - \partial_\mu g_{\alpha\delta}]$ and $\partial_k g_{ij} \equiv \dfrac{\partial g_{ij}}{\partial x_k}$. For instance, for ordinary three-dimensional (Euclidean) flat space where $\nu = 1, 2, 3$

with coordinates x_1, x_2, and x_3 and choosing Cartesian coordinates (that is, $g_{11} = g_{22} = g_{33} = 1$ and $g_{\mu\nu} = 0$ for $\mu \neq \nu$), $\Gamma^{\nu}_{\delta\alpha} = 0$ since the derivatives of these constant quantities of the metric in Eq. (B.9) are zero. Thus, Eq. (14.25) reduces to $\ddot{x}_\nu = \frac{d^2 x_\nu}{d\lambda^2} = 0$, or in component form, $\frac{d^2 x_1}{d\lambda^2} = 0$, $\frac{d^2 x_2}{d\lambda^2} = 0$, $\frac{d^2 x_3}{d\lambda^2} = 0$. The solution of each component equation is a straight line for the geodesic, analogous to Eq. (14.19) in Example 14.1.2. In particular, the solution of these equations (that is, for $x \equiv x_1$, $y \equiv x_2$, $z \equiv x_3$) yields: $x = x_0 + \lambda a$, $y = y_0 + \lambda b$, $z = z_0 + \lambda c$ where all other quantities are integration constants. Thus, solving for λ yields the Cartesian equation of a line: $\frac{x - x_0}{a} = \frac{y - y_0}{b} = \frac{z - z_0}{c}$.

Example 14.1.4 (Equivalent form of the Euler–Lagrange equation). The Euler–Lagrange equation in Eq. (14.8) can be written as

$$\frac{\partial f}{\partial y} - \frac{d}{dx}\frac{\partial f}{\partial y'} = 0,$$

where f is a function of $f(x, y, y')$. Show that this formulation is equivalent to

$$\frac{d}{dx}\left\{ f - y'\frac{\partial f}{\partial y'} \right\} - \frac{\partial f}{\partial x} = 0.$$

This latter form is useful if f is not explicitly a function of x, as shown in Problem 14.1 and Example 14.1.6.

Solution:
Using the chain rule,

$$\begin{aligned}
\frac{df}{dx} &= \frac{\partial f}{\partial x}\frac{dx}{dx} + \frac{\partial f}{\partial y}\frac{dy}{dx} + \frac{\partial f}{\partial y'}\frac{dy'}{dx} \\
&= \frac{\partial f}{\partial x} + \frac{\partial f}{\partial y}y' + \frac{\partial f}{\partial y'}y''.
\end{aligned} \tag{14.26}$$

In order to eliminate the last term in Eq. (14.26), one can use the chain rule on the following relation:

$$\frac{d}{dx}\left(y'\frac{\partial f}{\partial y'} \right) = y'\frac{d}{dx}\left(\frac{\partial f}{\partial y'} \right) + \frac{\partial f}{\partial y'}y''. \tag{14.27}$$

Thus, subtracting Eq. (14.27) from Eq. (14.26) yields

$$\frac{d}{dx}\left\{ f - y'\frac{\partial f}{\partial y'} \right\} - \frac{\partial f}{\partial x} = y'\underbrace{\left\{ \frac{\partial f}{\partial y} - \frac{d}{dx}\frac{\partial f}{\partial y'} \right\}}_{=0}. \tag{14.28}$$

Since the term on the right-hand side of Eq. (14.28) in curly brackets is the Euler–Lagrange equation that equals zero, the result follows:

$$\boxed{\frac{d}{dx}\left\{ f - y'\frac{\partial f}{\partial y'}\right\} - \frac{\partial f}{\partial x} = 0.}$$

(14.29)

Example 14.1.5. The solution of the brachistochrone problem is an important historical result in the 17th century solved by Bernoulli for finding the optimal path from which an object will fall from one point to another (see Table 1.1). Consider a bead of mass m that is able to slide down a wire by gravity without friction to the point P_2 in Fig. 14.3. Find the shape of the wire so that the bead falls from the origin 0 to point P_2 in the least amount of time.

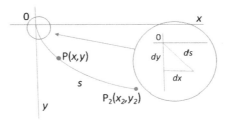

FIGURE 14.3

Schematic of a bead sliding down a wire without friction for a brachistochrone problem.

Solution.

The equivalent form of the Euler–Lagrange equation in Eq. (14.29) can be used to solve this problem. Using a conservation of kinetic and potential energy at the point 0 and P_2, $0 + mgy_1 = \frac{1}{2}m\left(\frac{ds}{dt}\right)^2 + mg(y_1 - y) \Rightarrow \frac{ds}{dt} = \sqrt{2gy}$. Using this result, a separation of variables yields the total time T: $T = \int_0^T dt = \frac{1}{\sqrt{2g}}\int_0^{x_2}\frac{\sqrt{1 + y'^2}}{\sqrt{y}}dx$.

Here (from the insert picture), the relation $ds = \left(1 + \frac{dy}{dx}\right)^{1/2} dx$ has been used to convert from ds to dx in the second integral. The function $f = \sqrt{1 + y'^2}/\sqrt{y}$ does not depend on x, so that Eq. (14.29) reduces to $f - y'\frac{\partial f}{\partial y'} = c$, where c is a constant. Thus, inserting f into the reduced Euler–Lagrange equation and solving for y' yields $y' = \frac{dy}{dx} = \sqrt{\frac{a - y}{y}}$, where $\sqrt{a} = 1/c$. This relation can be separated and integrated on both sides, yielding $\int dx = \int\sqrt{\frac{y}{a - y}}dy$. Using a change of variables for y, where $y = a\sin^2\theta$, this integral can be evaluated as $x = 2a\int\sin^2\theta\, d\theta =$

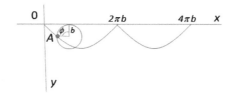

FIGURE 14.4

Curve represented by a cycloid with a fixed point A on a circle of radius b as it rolls along the x axis.

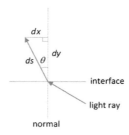

FIGURE 14.5

Derivation of Snell's law.

$a \int (1 - \cos 2\theta) \, d\theta = \dfrac{a}{2} (2\theta - \sin 2\theta) + k$. The integration constant $k = 0$ since the curve passes through the origin. Letting $b = a/2$ and $\phi = 2\theta$, the curve in Fig. 14.4 is given by $x = b(\phi - \sin \phi)$ and $y = b(1 - \cos \phi)$, which represents a *cycloid*.   [answer]

Example 14.1.6 (Snell's law in optics). Consider the propagation of light through a medium. The <u>index of refraction</u> is defined as $n = c/v$, where c is the speed of light in a vacuum and v is the phase velocity of light in the medium (e.g., $n = 1.3$ for water) (see Fig. 14.5). The refractive index determines how light is bent from the normal to the surface when entering a material. Derive <u>Snell's law</u> using a variational technique since the light in a medium takes a minimum travel time so that $\delta S = \displaystyle\int_{x_1}^{x_2} \dfrac{ds}{v} = 0$.

Solution.

From Example 14.1.2, $ds = \sqrt{1 + (y'^2)} \, dx$, so that $\delta S = \displaystyle\int_{x_1}^{x_2} \dfrac{\sqrt{1 + y'^2}}{v(y)} \, dx = \displaystyle\int_{x_1}^{x_2} f(y, y') \, dx$. Since $f(y, y')$ does not depend on x, the equivalent form of the Euler–Lagrange equation in Eq. (14.29) can be used, yielding $\dfrac{d}{dx}(f - y' f_{y'}) = 0$. On integrating, $f - y' f_{y'} = \text{constant}$ so that $\dfrac{\sqrt{1 + y'^2}}{v} - \dfrac{y'^2}{v(\sqrt{1 + y'^2})} = \dfrac{1}{v\sqrt{1 + y'^2}} = \text{constant}$. Letting θ in Fig. 14.5 be the angle between the tangent to the surface and the light path, $\sin\theta = \dfrac{dx}{ds} = \dfrac{1}{\sqrt{1+y'^2}}$. Therefore, as follows, $\dfrac{1}{v}\sin\theta = \text{constant}$.

However, from the definition of the refractive index, $n \propto 1/v$, so that Snell's law follows:

$$\boxed{n \sin\theta = \text{constant.}} \qquad \text{[answer]}$$

Generalizations of the Euler–Lagrange equation

Multidimensional generalizations of the Euler–Lagrange equation for determining an extremum with several functions and/or variables are given as follows:

(i) For several functions $(f_1, f_2, ..., f_m)$ of a single variable x and derivative, the extremum of the functional

$$I[f_1, f_2, ..., f_m] = \int_{x_0}^{x_1} L\left\{x, f_1, f_2, ..., f_m, f_1', f_2', ..., f_m'\right\} dx; \ f_i' \equiv \frac{df_i}{dx}$$

satisfies the Euler–Lagrange equations

$$\frac{\partial L}{\partial f_i} - \frac{d}{dx}\left(\frac{\partial L}{\partial f_i'}\right) = 0.$$

(ii) For a single function with n variables x with a single derivative for a surface Ω, the extremum of the functional

$$I[f] = \int_\Omega L\left\{x_1, x_2, ..., x_n, f, f_1, f_2, ..., f_n\right\} d\mathbf{x}; \ f_j \equiv \frac{\partial f}{\partial x_j}$$

satisfies the Euler–Lagrange equation

$$\frac{\partial L}{\partial f} - \sum_{j=1}^{n} \frac{\partial}{\partial x_j}\left(\frac{\partial L}{\partial f_j}\right) = 0.$$

(iii) For several functions with several variables with a single derivative, the extremum of the functional

$$I[f_1, f_2, ..., f_m]$$
$$= \int_\Omega L\left\{x_1, ..., x_n, f_1, ..., f_m, f_{1,1}, ..., f_{1,n}, ..., f_{m,1}, ..., f_{m,n}\right\} d\mathbf{x},$$
$$f_{i,j} \equiv \frac{\partial f_i}{\partial x_j}$$

satisfies the system of Euler–Lagrange equations

$$\frac{\partial L}{\partial f_1} - \sum_{j=1}^{n} \frac{\partial}{\partial x_j}\left(\frac{\partial L}{\partial f_{1,j}}\right) = 0_1,$$

$$\frac{\partial L}{\partial f_2} - \sum_{j=1}^{n} \frac{\partial}{\partial x_j} \left(\frac{\partial L}{\partial f_{2,j}} \right) = 0_2,$$

$$\cdots$$

$$\frac{\partial L}{\partial f_m} - \sum_{j=1}^{n} \frac{\partial}{\partial x_j} \left(\frac{\partial L}{\partial f_{m,j}} \right) = 0_m.$$

For example, from case (ii), if the function $f = f(t, x, y, z, u, u_t, u_x, u_y, u_z)$, where the dependent variable $u = u(t, x, y, z)$ and the independent variables are t, x, y, z, the Euler–Lagrange equation can be generalized as

$$\boxed{\frac{\partial f}{\partial u} = \frac{\partial}{\partial t} \left(f_{u_t} \right) + \frac{\partial}{\partial x} \left(f_{u_x} \right) + \frac{\partial}{\partial y} \left(f_{u_y} \right) + \frac{\partial}{\partial z} \left(f_{u_z} \right).} \qquad (14.30)$$

Example 14.1.7 (Vibrating string equation). Derive the equation of motion from the Euler–Lagrange equation for the vibration of an elastic string of length ℓ in Fig. 5.2 with a displacement $u(x, t)$.

Solution.
Let an element of an unstretched string at equilibrium be dx and let the corresponding length for the stretched string be ds. Hence, using Eq. (14.13) along with a binomial expansion, the lengthening of the string is $ds - dx = \sqrt{dx^2 + du^2} - dx \simeq$
$\left[1 + \frac{1}{2} \left(\frac{\partial u}{\partial x} \right)^2 + \ldots \right] dx - dx \simeq \frac{1}{2} \left(\frac{\partial u}{\partial x} \right)^2 dx$. The potential energy density is related to the restoring force: $F = \dfrac{dV}{(ds - dx)}$. Given that the restoring force is the string tension T, $dV = T(ds - dx) = \frac{1}{2} \left(\frac{\partial u}{\partial x} \right)^2$, on integrating, the potential energy is $V = \frac{1}{2} T \int_0^\ell u_x^2 dx$. The kinetic energy density of the string with a constant mass per unit length ρ is $dK = \frac{1}{2} \rho \left(\frac{\partial u}{\partial t} \right)^2$. On integrating, $K = \frac{1}{2} \rho \int_0^\ell \left(\frac{\partial u}{\partial t} \right)^2 dx$.

In accordance with Hamilton's principle over the interval from t_1 to t_2, the path is minimized such that

$$\delta S = \delta \int_{t_1}^{t_2} (K - V) dt = 0.$$

Thus, in this example,

$$\delta S = \delta \int_{t_1}^{t_2} (K - V)\,dt$$

$$= \delta \int_{t_1}^{t_2} \int_0^\ell \frac{1}{2}(\rho u_t^2 - T u_x^2)\,dx\,dt \qquad (14.31)$$

$$= 0,$$

which has the form $\delta \int_{t_1}^{t_2} \int_0^\ell L(u_t, u_x)\,dx\,dt$, where the Lagrangian $L = \frac{1}{2}(\rho u_t^2 - T u_x^2)$. Hence, as follows from Eq. (14.30),

$$\frac{\partial L}{\partial u} - \frac{\partial}{\partial t}\left(\frac{\partial L}{\partial u_t}\right) - \frac{\partial}{\partial x}\left(\frac{\partial L}{\partial u_x}\right) = 0. \qquad (14.32)$$

Using the Lagrangian L, Eq. (14.32) yields

$$\rho \frac{\partial}{\partial t}(u_t) = T\frac{\partial}{\partial x}(u_x)$$

or

$$\frac{\partial^2 u}{\partial t^2} = c^2 \frac{\partial^2 u}{\partial x^2}, \qquad \text{[answer]}$$

where $c = \sqrt{T/\rho}$ is the speed of the string wave. This equation is the one-dimensional wave equation for a vibrating string in Eq. (5.10).

14.2 Lagrange multipliers

Lagrange multipliers can be used for minimization problems when there is a constraint. For instance, given a function with three independent variables $f(x, y, z)$, the minimization (or extremum) of the function requires $df = 0$. Given the constraint

$$\varphi(x, y, z) = 0, \qquad (14.33)$$

the function $f(x, y, z)$ can be minimized with the use of a <u>Lagrange multiplier</u> such that (Arfken et al., 2013)

$$\boxed{\begin{aligned} \frac{\partial f}{\partial x} + \lambda \frac{\partial \varphi}{\partial x} &= 0, \\ \frac{\partial f}{\partial y} + \lambda \frac{\partial \varphi}{\partial y} &= 0. \end{aligned}} \qquad (14.34)$$

The Lagrange multiplier can also be incorporated into the calculus of variations. By Hamilton's principle, one seeks a path that is stationary:

$$\delta \int \left[L(q_i, \dot{q}_i, t) + \sum_k \lambda_k(t)\varphi_k(q_i, t) \right] dt = 0. \qquad (14.35)$$

The Lagrangian equations of motion now have an added constraint that includes the Lagrangian multiplier λ_k:

$$\frac{d}{dt}\frac{\partial L}{\partial \dot{q}} - \frac{\partial L}{\partial q} = \sum_k a_{ik}\lambda_k, \tag{14.36}$$

where

$$a_{ik} = \frac{\partial \varphi_k}{\partial q_i}. \tag{14.37}$$

In particular, if q_i is a length, then $a_{ik}\lambda_k$ represents the force of the kth constraint in the q_i direction analogous to Eq. (14.11).

Example 14.2.1 (Minimum critical volume for a cylindrical reactor). Find the minimum critical volume for a finite cylindrical nuclear reactor of height H and radius R. There is a constraint from neutron diffusion theory that for the geometric buckling of the reactor B_g

$$\varphi(R, H) = \left(\frac{2.4048}{R}\right)^2 + \left(\frac{\pi}{H}\right)^2 = B_g^2 = \text{constant}. \tag{14.38}$$

Solution:
Using Lagrange multipliers in Eq. (14.34), the function f for the volume of a cylinder is (Spiegel, 1973)

$$f(R, H) = V = \pi R^2 H. \tag{14.39}$$

Hence Eq. (14.34) yields

$$\frac{\partial f}{\partial R} + \lambda \frac{\partial \varphi}{\partial R} = 2\pi R H - 2\lambda \frac{(2.4048)^2}{R^3} = 0,$$
$$\frac{\partial f}{\partial H} + \lambda \frac{\partial \varphi}{\partial H} = \pi R^2 - 2\lambda \frac{\pi^2}{H^3} = 0. \tag{14.40}$$

Multiplying the first equation by $R/2$ and the second one by H,

$$\pi R^2 H = \chi \frac{(2.4048)^2}{R^2} = \chi \frac{2\pi^2}{H^2}, \tag{14.41}$$

yields

$$H = \frac{\sqrt{2}\pi R}{2.4048} = 1.847R. \quad \text{[answer]} \tag{14.42}$$

Moreover, using Eq. (14.38), Eq. (14.39), and Eq. (14.42), the critical volume in terms of the buckling B_g is $V = \dfrac{148.3}{B_g^3}$. These results are given in Table 2.14 of Lewis et al. (2017).

Problems

14.1 Consider a bead on a rotating wheel, as shown in Fig. 14.6. Here $F = \dfrac{mv^2}{r}$

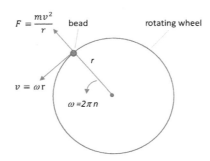

FIGURE 14.6

Schematic of a bead on a rotating wheel.

is the centrifugal force, m is the mass of the bead, v is the tangential velocity at the radius r of the wheel, and n is the number of rotations per second such that $\omega = 2\pi n$.

  (a) Show that the kinetic energy in the problem is $T = \dfrac{1}{2}m\dot{r}^2$. Given that $v = \omega r$ and using Eq. (14.11) such that $F = -\dfrac{dV}{dr}$, show that $V = -\dfrac{m\omega^2 r^2}{2}$.

  (b) Using the Euler–Lagrange equation, show that the equation of motion is $\ddot{r} = \omega^2 r$, which has the solution $r(t) = Ae^{-\omega t} + Be^{\omega t}$.

14.2 Consider a pendulum as shown in Fig. 14.7 of mass m and length l. The angle of the pendulum from the vertical is ϕ with a velocity of $v = l\,d\phi/dt$. Using the

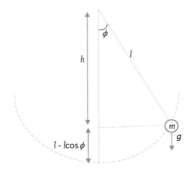

FIGURE 14.7

Schematic of the pendulum for an Euler–Lagrange analysis.

Euler–Lagrange equation in Eq. (14.8), with the Lagrangian L defined by Eq. (14.9) as the difference between the kinetic energy $(T = \frac{1}{2} m v^2)$ and the potential energy $(V = m g h)$, show that the equation of motion for the pendulum is given by

$$\frac{d^2\phi}{dt^2} + \frac{g \sin \phi}{l} = 0.$$

Hint: The state variable for this problem is the angle ϕ that the pendulum makes from the vertical.

14.3 Consider a small block of mass m sliding a distance s down a moving wedge of mass M, as shown in Fig. 14.8. The wedge is also sliding a distance x along the floor. Neglecting frictional effects, calculate the following:

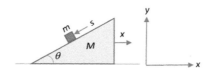

FIGURE 14.8

Schematic of a sliding block on a moving wedge.

(a) Given that the velocity of the wedge is simply $\dot{x}$, show that the x- and y-components of the velocity of the small block are $v_x = \dot{x} - \dot{s} \cos \theta$ and $v_y = -\dot{s} \sin \theta$, respectively.

(b) Show that the total kinetic energy of the combined system is $T = \frac{1}{2} M \dot{x}^2 + \frac{1}{2} m (\dot{x}^2 + \dot{s}^2 - 2\dot{x}\dot{s} \cos \theta)$. Also show that the potential energy for the small block is $V = -m g s \sin \theta$, where g is the acceleration due to gravity.

(c) Determine the Euler–Lagrange equations for this problem for the two variables s and x and their derivatives.

(d) Solving the Euler–Lagrange equations in part (c), show that the two accelerations $\ddot{s}$ and $\ddot{x}$ are given by

$$\ddot{s} = \frac{(M + m)}{(M + m \sin^2 \theta)} g \sin \theta = c_1 \quad \text{and} \quad \ddot{x} = \frac{M}{(M + m \sin^2 \theta)} g \sin \theta \cos \theta = c_2,$$

where c_1 and c_2 are simply constants.

14.4 Show that the shortest curve that has an area A below it is a circular arc,

$$(\lambda x - c)^2 + (\lambda y - d)^2 = 1,$$

as shown in Fig. 14.9. Here λ is a Lagrange multiplier constant and c and d are constants of integration. For this constrained problem, $y(0) = a$, $y(1) = b$, and $A = \int_0^1 y(x) \, dx$.

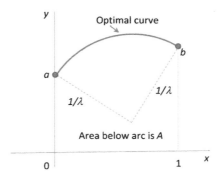

FIGURE 14.9

Schematic of a circular arc for the optimal shape of a curve with an area under it.

Hint: Use the Euler–Lagrange equation in Eq. (14.36) with the constraint $\varphi = A$. From Example 14.1.2, the Lagrangian under consideration is given by Eq. (14.17).

Maple software package

This appendix provides a list of key commands for the commercial software package Maple (Maplesoft, 2015). Maple is an analytical and numerical solver tool that can be used for general mathematical analysis. This tool can be used in the book for rapid calculation of various mathematical constructs.

It reviews basic Maple commands and reserved symbols, as well lists some common functions that arise in engineering. Various operations are described involving differentiation, integration, and the solution of a single ordinary differential equation or systems of differential equations. Problems in linear algebra are also briefly considered involving basic matrix operations such as addition, subtraction, and multiplication of matrices. Also provided are techniques for evaluating the transpose, determinant, and inverse of a matrix, as well as the solution of a linear system.

For the syntax of the Maple commands listed below, a "Text" input button is chosen (this input style also encompasses earlier versions of the software). The "Text" button option is found in the legend showing separate buttons for: "Text," "Math," "Drawing," "Plot," and "Animation." It is located just above the worksheet and formatting legend on the graphical user interface.

A.1 Maple commands

Arithmetic operations

Operation	Symbol	Example
Addition	$+$	$2 + 2$
Subtraction	$-$	$10 - x$
Multiplication	$*$	$3 * y * z$
Division	$/$	x/z
Exponent	$**$ or $\wedge$	$x \wedge z$
Factorial	$!$	$10!$

Reserved symbols: $\pi \to \text{Pi}$ $\sqrt{-1} \to \text{I}$
$\quad\quad\quad\quad\quad\quad\quad$ $e \to \exp(1)$ $\infty \to \text{infinity}$

Functions
exp(x), ln(x), log10(x), sqrt(x), abs(x)
sin(x), cos(x), tan(x), cot(x), sec(x), csc(x)
arcsin(x), arccos(x), arctan(x), arccot(x),arcsec(x), arccsc(x)
sinh(x), cosh(x), tanh(x), coth(x), sech(x), csch(x)
arcsinh(x), arccosh(x), arctanh(x), arccoth(x), arcsech(x), arccsch(x)
GAMMA(x), BesselJ(v, x), BesselY(v, x), BesselI(v, x), BesselK(v, x)

Note: > is the Maple prompt after which is what you type:

> $expr := (x \wedge 2) + (y \wedge 3)$; (one hits enter after the semicolon)

Differentiation (first way)

$$\boxed{\text{diff(expression, variables)}}$$

> diff(expr1, x);
> diff(expr1, x, y);
> diff(expr1, x, x);

Differentiation (second way)
> $f := x- > (x \wedge 5) + (x \wedge 4)$; (−> is the minus symbol followed by an inequality
symbol)
> D(f)(x)   (i.e., $f'(x)$);
> (D@@3)(f)(x)   (i.e., $f'''(x)$);

Integration
> $expr1 := exp(a * x) * cos(b * x)$;
> int(expr1, x);
> diff(%, x);   (% refers to the last result)

Summation of series
> sum(f, i = k..n);

Commands
> simplify();   (simplifies an expression)
> evalf();   (numerically evaluates functions and parameters)
> expand();   (expand an expression)
> factor();   (factor an expression)

Differential equations

$$\boxed{\text{dsolve(equation, dependent variable(independent variable))}}$$

For the differential equation: $(x^2 y' = y^2 + 5xy + 4x^2)$ this can be written as

```
> eq1 := (x ∧ 2) * diff(y(x), x) = (y ∧ 2) + (5 * x * y) + 4 * (x ∧ 2);
> dsolve(eq1, y(x));
```

```
> eq2 := (D@ @4)(y)(x) + 4 * y(x) = 0;
> dsolve(eq2, y(x));
```

For an initial value problem (note: it is important to use only integers with no floating points! (>? topic;)):
```
> eq3 := (D@ @3)(y)(x) − D(y)(x) = 0;
> ({eq3, y(0) = 6, D(y)(0) = −4, (D@ @2)(y)(0) = 2} , y(x));
```

Systems of differential equations
```
> sys {D(y)(x) = 4 * y2(x), D(y2)(x) = 4 * y1(x) + 2 − 16 * x ∧ 2};
> fcns := {y1(x), y2(x)} ;
> dsolve(sys, fcns);
```

```
> sys{D(y1)(x) = y1(x) + 2 * y2(x), D(y2)(x) = −8 * y1(x) + 11 * y2(x),
   y1(0) = 1, y2(0) = 7};
> fcns := {y1(x), y2(x)} ;
> dsolve(sys, fcns);
```

Linear algebra

One must load the "linalg" package first
```
>with(linalg):
```
Given the matrix: $A = \begin{bmatrix} -1 & -2 & 4 \\ 2 & -5 & 2 \\ 3 & -4 & -6 \end{bmatrix}$

```
> A:=matrix(3,3,[−1,  −2,     4,
                  2,  −5,     2,
                  3,  −4,   −6]);
```

The returned result is

$$A = \begin{bmatrix} -1 & -2 & 4 \\ 2 & -5 & 2 \\ 3 & -4 & -6 \end{bmatrix}$$

Addition and subtraction of matrices
```
> 5 * A − 3 * B;
```

Multiplication of matrices
```
> A.B;
```

Transpose of matrix
> transpose(A);

Inverse of matrix (see Example 4.2.1 and Example 4.3.4)
> inverse(A);

$$\begin{bmatrix} \dfrac{-19}{23} & \dfrac{14}{23} & \dfrac{-8}{23} \\[2mm] \dfrac{-9}{23} & \dfrac{3}{23} & \dfrac{-5}{23} \\[2mm] \dfrac{-7}{46} & \dfrac{5}{23} & \dfrac{-9}{46} \end{bmatrix}$$

Determinant of matrix
> det(A);

$$-46$$

Solution of a linear system $Ax = b$
For example (see Example 4.3.1):

$$\begin{bmatrix} -1 & -2 & 4 \\ 2 & -5 & 2 \\ 3 & -4 & -6 \end{bmatrix} \begin{bmatrix} x_1 \\ x_2 \\ x_3 \end{bmatrix} = \begin{bmatrix} -3 \\ 7 \\ 5 \end{bmatrix}$$

> b := vector([$-3, 7, 5$]);

$$b := [-3, 7, 5]$$

> x := linsolve(A, b);

$$x := [5, 1, 1]$$

An alternative method to solve the system of equations is with a <u>Gauss elimination</u> technique (see Example 4.3.3):
> A1 := augment(A, b);   (augmented matrix of the system $Ax = b$)

$$A1 = \begin{bmatrix} -1 & -2 & 4 & -3 \\ 2 & -5 & 2 & 7 \\ 3 & -4 & -6 & 5 \end{bmatrix}$$

> b1 := gausselim(A1);   (A1 triangularized by row operation)

$$b1 = \begin{bmatrix} -1 & -2 & 4 & -3 \\ 0 & -9 & 10 & 1 \\ 0 & 0 & \dfrac{-46}{9} & \dfrac{-46}{9} \end{bmatrix}$$

```
> x1 := backsub(b1);
```

$$x1 := [5, 1, 1]$$

Solution of a linear system in Example 6.2.2:

$$\begin{bmatrix} -4 & 1 & 1 & 0 \\ 1 & -4 & 0 & 1 \\ 2 & 0 & -4 & 1 \\ 0 & 2 & 1 & -4 \end{bmatrix} \begin{bmatrix} u_{11} \\ u_{21} \\ u_{12} \\ u_{22} \end{bmatrix} = \begin{bmatrix} 1 \\ -3 \\ -1 \\ -8 \end{bmatrix}$$

```
> A:=Matrix([[ -4,   1,   1    0],
             [ 1,  -4,   0    1 ],
             [ 2,   0,  -4    1 ],
             [ 0,   2,   1  -4 ]]);
```
Note that a slightly different format is employed for inputting the matrix (where the size of the matrix is not included). Here, the matrix is input using the automated "Insert Matrix" command in Maple. The returned result is

$$A = \begin{bmatrix} -4 & 1 & 1 & 0 \\ 1 & -4 & 0 & 1 \\ 2 & 0 & -4 & 1 \\ 0 & 2 & 1 & -4 \end{bmatrix}$$

```
> b := vector([1, -3, -1, -8]);
```

$$b := [1, -3, -1, -8]$$

```
> u := linsolve(A, b);
```

$$u := \left[\frac{78}{161}, \frac{267}{161}, \frac{206}{161}, \frac{507}{161} \right]$$

```
> evalf(%);
```

$$[0.4844720497 \quad 1.658385093 \quad 1.279503106 \quad 3.149068323]$$

Fourier series analysis (see Example 2.5.15)
```
> cn := 1/(2 * Pi) * int(exp(-Pi * x) * exp(-I * n * x), x = -Pi..Pi);
```

$$c_n = \frac{1}{2} \frac{\left(e^{2\pi^2 + 2In\pi} - 1\right) e^{-\pi^2 - In\pi}}{\pi(\pi + In)}$$

```
> simplify(%);
```

$$\frac{1}{2} \frac{e^{\pi(\pi + In)} - e^{-\pi(\pi + In)}}{\pi(\pi + In)}$$

Geodesic formulation

The following appendices provide the supporting analysis for derivation of the geodesic, in terms of a brief definition of tensors in Section B.1 and the derivation of the geodesic equation using the formalism of the Lagrangian in Section B.2.

B.1 Tensors

The metric $g_{\mu\nu}$ is in fact a <u>covariant tensor</u> of rank two (since there are two indices). A scalar is a tensor of rank zero (having only a magnitude), while a vector is a tensor of rank one (having both a magnitude and direction) (see Section 4.4.1). For a change in coordinate system from an unprimed to a primed one, the corresponding transformation laws for a covariant tensor (lower indices) and <u>contravariant tensor</u> (upper indices) of rank two are given as (Lieber and Lieber, 1966):

- covariant tensor: $A'_{\alpha\beta} = \dfrac{\partial x_\gamma}{\partial x'_\alpha} \dfrac{\partial x_\delta}{\partial x'_\beta} A_{\gamma\delta}$;

- contravariant tensor: $A'^{\alpha\beta} = \dfrac{\partial x'_\alpha}{\partial x_\mu} \dfrac{\partial x'_\beta}{\partial x_\nu} A^{\mu\nu}$.

One can also multiply a covariant tensor by a contravariant one to yield a <u>mixed tensor</u> with the transformation law $C'^\mu_\lambda = \dfrac{\partial x_\alpha}{\partial x'_\lambda} \dfrac{\partial x'_\mu}{\partial x_\beta} C^\beta_\alpha$.

The physical difference for a covariant and contravariant tensor is depicted in Fig. B.1. Here a vector $\vec{V}$ is represented in a nonorthogonal plane Cartesian system of coordinates in the Euclidean plane, with (normed) basis vectors $\hat{e}_1$ and $\hat{e}_2$. As seen in the figure, the covariant components (x_1, x_2) of $\vec{V}$ are the perpendicular projections (dot product of each of the basis vectors) on the coordinate axis, in contrast to the contravariant components (x^1, x^2) of $\vec{V}$, which are parallel projections.

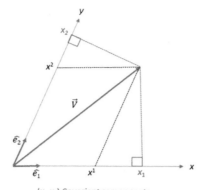

(x_1, x_2) Covariant components

(x^1, x^2) Contravariant components

FIGURE B.1

Geometric representation for covariant and contravariant components of a vector.

B.2 Lagrangian and action

It is easier to deal with L^2 than L in Eq. (14.24) because of the square-root sign in Eq. (14.23) (d'Inverno, 1992). Multiplying Eq. (14.24) through by $2L$ gives

$$\frac{\partial(L^2)}{\partial x_\mu} - 2L\frac{d}{d\lambda}\frac{\partial L}{\partial \dot{x}_\mu} = 0. \tag{B.1}$$

Using the chain rule successive times for the expression $\dfrac{d}{d\lambda}\left[\dfrac{\partial(L^2)}{\partial \dot{x}_\mu}\right]$ gives

$$\frac{d}{d\lambda}\left[\frac{\partial(L^2)}{\partial \dot{x}_\mu}\right] = \frac{d}{d\lambda}\left[2L\frac{\partial L}{\partial \dot{x}_\mu}\right] = 2\frac{dL}{d\lambda}\frac{\partial L}{\partial \dot{x}_\mu} + 2L\frac{d}{d\lambda}\left[\frac{\partial L}{\partial \dot{x}_\mu}\right] \tag{B.2a}$$

$$\Rightarrow 2L\frac{d}{d\lambda}\left[\frac{\partial L}{\partial \dot{x}_\mu}\right] = \frac{d}{d\lambda}\left[\frac{\partial(L^2)}{\partial \dot{x}_\mu}\right] - 2\frac{dL}{d\lambda}\frac{\partial L}{\partial \dot{x}_\mu}. \tag{B.2b}$$

Substituting the expression in Eq. (B.2b) into Eq. (B.1) yields

$$\underbrace{\frac{\partial(L^2)}{\partial x_\mu}}_{(I)} - \underbrace{\frac{d}{d\lambda}\left[\frac{\partial(L^2)}{\partial \dot{x}_\mu}\right]}_{(II)} = \underbrace{-2\frac{dL}{d\lambda}\frac{\partial L}{\partial \dot{x}_\mu}}_{(III)}. \tag{B.3}$$

Inserting L from Eq. (14.22a) into each of the three terms in Eq. (B.3) gives

$$(I) = \frac{\partial}{\partial x_\mu}[g_{\alpha\beta}\dot{x}_\alpha \dot{x}_\beta] = [\partial_\mu g_{\alpha\beta}\dot{x}_\alpha \dot{x}_\beta],$$

$$(II) = \frac{d}{d\lambda}\frac{\partial}{\partial \dot{x}_\mu}[g_{\alpha\beta}\dot{x}_\alpha \dot{x}_\beta] = \frac{d}{d\lambda}[g_{\alpha\beta}\underbrace{\frac{\partial \dot{x}_\alpha}{\partial \dot{x}_\mu}}_{\delta_{\alpha\mu}}\dot{x}_\beta + g_{\alpha\beta}\underbrace{\frac{\partial \dot{x}_\beta}{\partial \dot{x}_\mu}}_{\delta_{\beta\mu}}\dot{x}_\alpha]$$

$$= \frac{d}{d\lambda}[\underbrace{g_{\mu\beta}\dot{x}_\beta}_{g_{\mu\alpha}\dot{x}_\alpha} + g_{\mu\alpha}\dot{x}_\alpha] = \frac{d}{d\lambda}[2g_{\mu\alpha}\dot{x}_\alpha] = 2g_{\mu\alpha}\ddot{x}_\alpha + 2\frac{dg_{\mu\alpha}}{d\lambda}\dot{x}_\alpha \qquad \text{(B.4)}$$

$$= 2g_{\mu\alpha}\ddot{x}_\alpha + 2\underbrace{\frac{\partial g_{\mu\alpha}}{\partial x_\delta}}_{\partial_\delta g_{\mu\alpha}}\underbrace{\frac{dx_\delta}{d\lambda}}_{\dot{x}_\delta}\dot{x}_\alpha,$$

$$(III) = 2\frac{dL}{d\lambda}\frac{\partial}{\partial \dot{x}_\mu}\left[\sqrt{g_{\alpha\beta}\dot{x}_\alpha \dot{x}_\beta}\right] = 2\frac{dL}{d\lambda}[\frac{1}{2}(g_{\alpha\beta}\dot{x}_\alpha \dot{x}_\beta)^{-1/2}\underbrace{\frac{\partial}{\partial \dot{x}_\mu}\{g_{\alpha\beta}\dot{x}_\alpha \dot{x}_\beta\}}_{2g_{\mu\eta}\dot{x}_\eta}].$$

The last term in (III) arises from the result of the intermediate expression in (II) with a change of indices from α to η. Moreover, for the derivation of Eq. (B.4), the Kronecker delta function $\delta_{ij} = \begin{cases} 1, & i = j, \\ 0, & i \neq j \end{cases}$ has been used (also $\delta_{ij} = \delta^i_j$). A change of index variable is also employed in some cases when the repeated index indicates a dummy variable for summing. In addition, the function $\partial_k g_{ij} \equiv \frac{\partial g_{ij}}{\partial x_k}$. There is also a symmetry that can be exploited for the metric and delta functions such that $g_{ij} = g_{ji}$ and $\delta_{ij} = \delta_{ji}$.

From Eq. (14.22b),

$$S = \int_{\lambda_0}^{\lambda} L(x_\alpha, \dot{x}_\alpha, \lambda)\, d\lambda$$

$$\Rightarrow \dot{S}(\lambda) \equiv \frac{dS(\lambda)}{d\lambda} = L(x_\alpha, \dot{x}_\alpha, \lambda)\Big|_{\lambda_0}^{\lambda} = L(x_\alpha, \dot{x}_\alpha, \lambda) - L(x_\alpha, \dot{x}_\alpha, \lambda_0).$$

Hence, $\ddot{S}(\lambda) = \frac{dL}{d\lambda}$ and the third term (III) in Eq. (B.4) can be written using Eq. (14.22b) as

$$(III) = 2\frac{\ddot{S}}{L}g_{\mu\eta}\dot{x}_\eta. \qquad \text{(B.5)}$$

Since the path length $S = \lambda$ for the parametrization of the curve in Fig. 14.2, $\ddot{S} = 0$ in Eq. (B.5). Thus, the third term (III) vanishes in these equations. This result also follows by recalling that for the variation of the action $\delta S = 0$ in Eq. (14.6) so that

$\dfrac{\delta S}{\delta \lambda} \approx \dfrac{dS}{d\lambda} = 0$. Using the final results of Eq. (B.4) with the term $(III) = 0$, Eq. (B.3) becomes

$$\partial_\mu g_{\alpha\beta} \dot{x}_\alpha \dot{x}_\beta - 2g_{\mu\alpha} \ddot{x}_\alpha - 2\partial_\delta g_{\mu\alpha} \dot{x}_\alpha \dot{x}_\delta = 0 \tag{B.6a}$$

$$\Rightarrow 2g_{\mu\alpha} \ddot{x}_\alpha + \dot{x}_\alpha \dot{x}_\delta [-\partial_\mu g_{\alpha\delta} + 2\partial_\delta g_{\mu\alpha}] = 0 \tag{B.6b}$$

$$\Rightarrow 2g_{\mu\alpha} \ddot{x}_\alpha + \dot{x}_\alpha \dot{x}_\delta [-\partial_\mu g_{\alpha\delta} + \partial_\delta g_{\mu\alpha} + \partial_\delta g_{\mu\alpha}] = 0. \tag{B.6c}$$

The expression in Eq. (B.6b) was obtained by multiplying through by -1, changing the dummy subscript of β to δ, and collecting terms. Moreover, since α and δ are repeat (dummy) indices, these indices can be interchanged in the last term in square brackets in Eq. (B.6c), yielding

$$2g_{\mu\alpha} \ddot{x}_\alpha + 2\dot{x}_\alpha \dot{x}_\delta [-\frac{1}{2}\partial_\mu g_{\alpha\delta} + \frac{1}{2}\partial_\delta g_{\mu\alpha} + \frac{1}{2}\partial_\alpha g_{\mu\delta}] = 0 \tag{B.7a}$$

$$\Rightarrow 2g_{\mu\alpha} \ddot{x}_\alpha + 2\dot{x}_\alpha \dot{x}_\delta [-\frac{1}{2}\partial_\mu g_{\alpha\delta} + \frac{1}{2}\partial_\delta g_{\mu\alpha} + \frac{1}{2}\partial_\alpha g_{\delta\mu}] = 0. \tag{B.7b}$$

In Eq. (B.7a), a factor of 2 was factored outside of the square brackets. For the last term of Eq. (B.7b), the μ and δ subscripts were interchanged because of the symmetry of the metric. Multiplying Eq. (B.7b) through by the contravariant tensor $g^{\mu\nu}$ yields

$$\cancel{2}g^{\mu\nu} g_{\mu\alpha} \ddot{x}_\alpha + \cancel{2}\dot{x}_\alpha \dot{x}_\delta \underbrace{\frac{1}{2} g^{\mu\nu} [\partial_\delta g_{\mu\alpha} + \partial_\alpha g_{\delta\mu} - \partial_\mu g_{\alpha\delta}]}_{\Gamma^\nu_{\delta\alpha}} = 0. \tag{B.8}$$

The term

$$\Gamma^\nu_{\delta\alpha} = \frac{1}{2} g^{\mu\nu} [\partial_\delta g_{\mu\alpha} + \partial_\alpha g_{\delta\mu} - \partial_\mu g_{\alpha\delta}] \tag{B.9}$$

is identified as the Christoffel symbol. There is a symmetry in the lower symbols of $\Gamma^\nu_{\delta\alpha}$. In addition, $g^{\mu\nu} g_{\mu\alpha} = \delta^\nu_\alpha$ so that Eq. (B.8) reduces to the final equation of a geodesic:

$$\boxed{\ddot{x}_\nu + \dot{x}_\alpha \dot{x}_\delta \Gamma^\nu_{\delta\alpha} = 0.} \tag{B.10}$$

Bibliography

Arfken, G.B., Weber, H.J., Harris, F.E., 2013. Mathematical Methods for Physicists, seventh edition. Academic Press, Boston.

Boas, M., 2006. Mathematical Methods in the Physical Sciences, third edition. Wiley.

BYU, 2020. www.et.byu.edu/~vps/ET502WWW/NOTES/CH7m.pdf. (Accessed 22 February 2020). Brigham Young University, Provo, UT.

COMSOL Inc., 1998–2018. Comsol Multiphysics®-Reference Manual, Version 5.4. Burlington, Massachusetts.

d'Inverno, R., 1992. Introducing Einstein's Relativity. Oxford University Press.

Dixon, W.J., 1951. Ratios involving extreme values. The Annals of Mathematical Statistics 22 (1), 68–78.

Drazin, P., 1985. Solitons. Cambridge University Press.

Foltz, B., 2020. One-way ANOVA: understanding the calculation. https://www.youtube.com/watch?v=0Vj2V2qRU10&t=4s. (Accessed 14 June 2020). Video on YouTube.

Hutton, D., 2004. Fundamentals of Finite Element Analysis. McGraw-Hill Series in Mechanical Engineering. McGraw-Hill.

Jackson, J., 1975. Classical Electrodynamics, second edition. Wiley.

Koelink, E., 2006. Scattering Theory: wi4211 Advanced Topics in Analysis.

Kreyszig, E., 1993. Advanced Engineering Mathematics. Wiley International Editions: Mathematics. Wiley.

Lewis, B.J., Onder, E.N., Prudil, A.A., 2017. Fundamentals of Nuclear Engineering. Wiley International Editions: Nuclear Engineering. Wiley.

Lieber, L.R., Lieber, H.G., 1966. The Einstein Theory of Relativity. Holt, Rinehart and Winston.

Maplesoft, 2011–2015. Maple Programming Guide. Toronto, Canada.

Onder, E., 2004. Characterization of the Slug Flow Formation in Vertical-to-Horizontal Channels with Obstructions. PhD thesis. Ecole Polytechnique of Montreal.

Oxford, 2020. http://www.robots.ox.ac.uk/~sjrob/Teaching/SP/l7.pdf. (Accessed 25 May 2020). Oxford University, Oxford, UK.

Peregrine, D., 1966. Calculations of the development of an undular bore. Journal of Fluid Mechanics 25, 321–330.

Press, W., Flannery, B., Teukolsky, S., Vetterling, W.T., 1986. Numerical Recipes. Cambridge University Press.

Rahman, M., 2007. Integral Equations and Their Applications. WIT Press, Southhampton, United Kingdom.

Sasser, J., 2018. History of ordinary differential equations the first hundred years. http://citeseerx.ist.psu.edu/viewdoc/download. (Accessed 3 July 2018).

Savitzky, A., Golay, M., 1964. Smoothing and differentiation of data by simplified least squares procedures. Analytical Chemistry 36 (8), 1627–1639.

Skoog, D., West, D., 1982. Fundamentals of Analytical Chemistry. Saunders Golden Sunburst Series. Saunders College Pub.

Smith, S., 1997. The Scientist and Engineer's Guide to Digital Signal Processing. California Technical Publishing.

Spiegel, M., 1964. Complex Variables. Schaum's Outline Series. McGraw-Hill.

Spiegel, M., 1971. Advanced Mathematics for Engineers and Scientists. Schaum's Outline Series. McGraw-Hill.

Spiegel, M., 1973. Mathematical Handbook of Formulas and Tables. Schaum's Outline Series. McGraw-Hill.

Systat Software, Inc., 2018. https://systatsoftware.com/. (Accessed 3 January 2019).

Taha, T., Ablowitz, M., 1984. Analytical and numerical aspects of certain nonlinear evolution equations. III. Numerical, Kortweg-de Vries equation. Journal of Computational Physics 55, 231–253.

Towner, I., 2020. PHYS 312: Mathematical Methods in Physics. Queen's University, Kingston, Ontario.

Whitham, G., 1974. Linear and Nonlinear Waves. Wiley-Interscience.

Zabusky, N., Kruskal, M., 1965. Interaction of solitons in a collisionless plasma and the recurrence of initial states. Physical Review Letters 15, 240–243.

Zienkiewicz, O., Taylor, R., Zhu, J., 2013. The Finite Element Method: Its Basis and Fundamentals, seventh edition. Butterworth-Heinemann, Oxford.

Index